The Hidden Pattern

A Patternist Philosophy of Mind

Ben Goertzel

BrownWalker Press
Boca Raton • 2006

The Hidden Pattern: A Patternist Philosophy of Mind

BrownWalker Press
Boca Raton, Florida
USA • 2006

ISBN: 1-58112-989-0 (paperback)
ISBN: 1-59942-404-5 (ebook)

BrownWalker.com

Contents

Preface

In this book I present some interim results from a quest I've been on for many years – a quest for a coherent conceptual framework making it possible to understand the various aspects of mind and intelligence in a unified way. The underlying goals of this quest have been twofold: a general desire to understand myself and the universe, and a specific interest in understanding how to create "artificial" minds (in the form of computer software or else novel engineered physical systems). So, this is a work of philosophy-of-mind, yet haunted consistently throughout by the spectre of AI. Much of what's interesting in these pages probably results from the crosspollination between these two aspects, as well as from the crosspollination between two different approaches to understanding: the informal and introspective approach, versus the nitty-gritty analytical/scientific/engineering approach, both of which are well-represented here.

The key themes of this book have been following me around at least since 1982, when I was 16 years old and first made a significant effort to synthesize everything I knew toward the goal of achieving a coherent understanding of human and digital mind. The task was well beyond my knowledge and capabilities at that point, but yet the basic concepts that occurred to me then were essentially the same ones that I present in these pages now, as I approach my (not too enthusiastically anticipated) 40th birthday.

I recall in particular a big moment about 5 months into my 16th year, when I spent a Spring Break visit back home from college writing down my thoughts on intelligence on little scraps of notebook paper, hoping to conceive something practical in the area of AI design. I failed at my attempt to make a practical AI design that Spring Break, but it was a conceptually productive time nonetheless. Various ideas from the back of my mind crystallized into a holistic understanding, and I came to the realization to which the title of this book refers – which is simply that *the mind and world are themselves nothing but pattern – patterns among patterns, patterns within patterns...* Where I got stuck was in trying to invent a general yet computationally efficient and humanly-plausible-to-program algorithm for pattern recognition (a hard problem that I'll talk a bit about in Chapter 15 below: after more than two decades of hard thinking I think I've finally made significant progress!).

When I first came to this grandiose "patternist" conclusion I had years before read Douglas Hofstadter's "Godel, Escher Bach," (1979) which had put the concept of "pattern" firmly into my mind. I hadn't yet read Charles Peirce (1935) (who modeled the universe in terms of "habits", a close-synonym for "pattern"), nor Nietzsche (1968, 1997, 2001), who spoke of "the world as will to power and morphology" (or in other words, the universe as a system of patterns struggling for power over each other). Nor had I yet read Gregory Bateson (1979), who articulated what he called "The MetaPattern: that it is pattern which connects." Nor Benjamin Lee Whorf (1964), who interpreted the universe as a web of linguistic patterns, but interpreted the notion of language so generally that he ultimately was proposing a universal pattern-theory. Each of these thinkers, when I encountered their work during the next couple years after my early Eureka experience, gave me a feeling of great familiarity, and also a stab of frustration. Each of them, within a small degree of error, seemed to be trying to get across the same basic concept as my "Hidden Pattern" insight. But none of them had developed the idea nearly as deeply or extensively as seemed necessary to me. Now, a couple decades later, I find that I haven't developed the idea nearly deeply or extensively as I would like either – but I believe I've moved it forward a fair bit, and hopefully created some tools that will accelerate future progress.

Patternist philosophy isn't something with a fixed number of axioms and conclusions. It's a fluid and shifting set of interlocking ideas – most of all, it's a *way of thinking* about the mind. Transmitting a deep, broad and idiosyncratic "way of thinking" to others isn't an easy task (it's easier just to use one's peculiar, personal "way of thinking" for oneself, to draw interesting conclusions, and then present others with the conclusions), but I've been driven to attempt it.

The word "philosophy" is carefully chosen here – though many of the ideas presented here developed in the context of my scientific work in various disciplines, I don't consider "patternist philosophy" a scientific theory per se; it is more general and abstract than that. Patternist philosophy may be used to inspire or refine scientific theories; but it may also be used for other purposes, such as non-scientific introspective self-understanding. However, as will be clear when I discuss the philosophy of science in depth in these pages, I do think the difference between science and non-science is subtler and fuzzier than generally recognized; and that, in a sense, an abstract construction like patternist philosophy may be validated or refuted in the same way as scientific research programs.

I mentioned above my twin goals of general self- and world-understanding and AI design. Another, more fine-grained goal that has motivated my ongoing work on pattern philosophy has been the desire to create a deep, abstract framework capable of unifying the multiple visions of mind I've encountered. Each of these visions seems to embody some valid insights: the introspective view I have of my own mind; the views of mind that the community of scientists has arrived at via their investigations in brain science, mathematical logic, cognitive psychology and other disciplines; and the views of mind that Eastern philosophers and "mystics" around the world have derived via their collective experiments in introspection and mind-clarification. Patternist philosophy spans all these perspectives, and I think this is one of its strengths.

The human-psychology aspects of patternism were my focus in the mid-1990s, when I was spending much of my time doing research in theoretical cognitive science. That was the stage when my study of Eastern metaphysics was also at its peak — largely as part of an (ultimately failed) attempt to build a connection between my own world-view and that of my wife at the time, who during that period was becoming a Zen Buddhist priest. From 1997 till the present, on the other hand, I've been spending much of my time designing, building and analyzing AI software systems (first a system called Webmind, and now a system called Novamente); and the recent development of patternist philosophy has thus had a strong AI bias.

Reflecting my intellectual focus in recent years, this book contains a lot of discussion of the implications of the patternist philosophy for AI, including numerous mentions of my own current AI project, the Novamente AI Engine. However, the technical details of the Novamente design aren't entered into here – many of those are described in two other books that I've been working on concurrently with this one: *Engineering General Intelligence* (coauthored with Cassio Pennachin) and *Probabilistic Logic Networks* (coauthored with Matt Ikle', Izabela Freire Goertzel and Ari Heljakka).

My schedule these last few years has been incredibly busy, far busier than I'm comfortable with – I much prefer to have more "open time" for wandering and musing. My time has been full with Novamente AI research, with several other book projects, and with a great deal of business work associated with the AI-based businesses my colleagues and I have started in the last few years…and my personal life has also been busy, what with the demands of raising three children and repeatedly moving house — not to mention getting divorced and remarried…In short, modern human life in all its variety and chaos. Given all this, it's been rather difficult for me to find time to work on this book. Every hour I've worked on this book, I've been intensely aware that I could have been spending time generating business for Biomind, or working on the details of the Novamente system, or giving my wife or kids more attention than the amount they manage to squeeze out of my over-packed schedule…But I have taken the time to write the book anyway, because I believe the ideas I present here are very important ones.

I don't think the "problem of the mind" or the "problem of the fundamental nature of the universe" are problems that can ever be completely and finally solved, and I certainly don't claim to have done so in the train of thought reported here. However, I do believe I've arrived at a number of important insights regarding the mind and world – insights which connect together in a network, and which taken together provide powerful answers for some questions that have preoccupied people for a long time.

As well as stimulating interest and further thinking and research, it is my hope that these ideas will play a part in the process of going beyond "people" altogether. Like Nietzsche I believe that "Man is something that must be overcome," both by the creation of improved humans and the creation of superhuman intelligences whose nature we can only dimly conceive. Along the path toward this self-overcoming, a deep understanding of mind and pattern is bound to be critical. I have certainly found it so in my own scientific work (both in AI and in other areas like bioinformatics), as well as in my personal quest for spiritual and mental development.

Next, I'll make a few boring comments about this prose you are reading. In writing about these ideas, I have chosen a style that is natural to me but may seem eccentric to some others – a kind of mix between informal conversational prose and more technical academic-ish prose. Whatever you think of them, my choices regarding style and level of exposition have been quite deliberate. While I admire the elegance of philosophical stylists such as Nietzsche, Goethe and Pascal (to toss out the names of a few of my favorites), and I'm sometimes tempted to try to emulate them (though I'm not sure how well I'd succeed), in this book I've opted for clarity over stylistic beauty nearly every time. Abstract philosophy ideas are hard enough to communicate when one strives for clarity; and though I love reading Nietzsche, I have less appreciation than he did for the subtle aesthetics of being misunderstood. I hope I've managed to communicate the ins and outs of my topics reasonably adequately.

As a corollary to the stylistic decision mentioned in the previous paragraph, I have chosen not to adopt a fully thorough style of referencing here. I have given a decent number of references here and there in the text, but by and large I've only referenced those sources that seemed extremely important for the subject under discussion – ones containing ideas that I directly and significantly use in the text, or else that I really think the reader should bother to look up if they want to fully understand the ideas I'm presenting. You shouldn't assume from this that I'm unaware of the vastness of the extant literature discussing the themes I consider. I've read many hundreds, probably thousands of books and papers on these topics; and by eschewing more exhaustive referencing, I'm not trying to give a false impression that everything I say and don't explicitly reference is completely original and unprecedented.

Finally, a note about my own current research direction. Part of the reason I've finally decided to write these "patternist" ideas down systematically and publish them is that they've become somewhat old and tiresome to me. I think patternist philosophy is extremely important, but it's also not something I think about that much anymore, because it's become second nature to me. I'm working on a lot of practical scientific and engineering projects these days, but in the philosophical portion of my life I'm devoting my time to a somewhat different set of ideas than the ones I discuss here: the creation of a general, qualitative theory of the development of complex pattern-systems over time. This pursuit relates closely to my current work on Novamente, which has to do with figuring out how to get a "baby Novamente" to evolve the right mind-structures through interaction with a simulated world; and also relates to other themes touched in this book, such as the possibility of deriving physical law from considerations related to subjective reality, and the possibility of general laws of emergent complex systems dynamics. So, I've been feeling I should write up these "patternist philosophy" ideas before I get so bored with them that I can't do a good job of writing them up anymore. The fact that my latest conceptual obsession (pattern-based development-theory) uses the patternist philosophy as a launching-pad is encouraging to me, and provides some additional validation within my own subjective universe for the hypothesis that the patternist philosophy is useful ("progressive" in the Lakatosian sense discussed in Chapter 13) as well as sensible and comprehensive.

Acknowledgements

Two people — my wife Izabela and my former wife Gwen — have indulged me in more lengthy, productive and creative discussions on these issues than anyone else. My primary research and business partner Cassio Pennachin has been very helpful in collaboratively working out the principles of the Novamente AI design, and many of the insights I've achieved via working with him on Novamente have found their way into this book. Other friends and colleagues over the years have also been wonderful thought-partners, a list including but by no means limited to (in no particular order): Lisa Pazer, Meg Heath, Stephan Bugaj, Mike Kalish, John Dunn, Cate Hartley, Allan Combs, Eliezer Yudkowsky, Moshe Looks, Debbie Duong, Ari Heljakka, Mike Ross and Gui Lamacie.

Ed Misch, my university philosophy professor, first got me seriously interested in Western philosophy of mind way back when I was 16 (roughly 6 months after I started obsessing on AI … Ed kindly and tactfully clued me in to how many other thinkers had already trod some of the paths I was exploring). My interest in Eastern philosophy of mind, on the other hand, was kindled by my mother's study of Oriental philosophy in my early youth, and then by my first wife Gwen's adventures in Buddhism. My interest in AI and other nonhuman types of mind grew largely out of a host of science fiction novels and stories that I read in my youth – in that sense I feel highly grateful to the genre of sci-fi for existing and urging me to think even more obsessively about "what if?" than I would have done on my own.

My mother Carol, father Ted and grandfather Leo are due copious thanks for decades ago, getting me started on the path of independent intellectual inquiry – and for encouraging me as this path has led me so many strange places over the years.

Finally, my wife Izabela and children Zarathustra, Zebulon and Scheherazade are due much gratitude for their tolerance of my seemingly endless busy-ness these days, as I've worked on this book along with a dozen or so other projects.[1]

---

[1] How frustrating the human body can be as an interface for the human mind, which has already, at this pre-Singularity phase, evolved so far beyond the limitations of its container – if I fail to survive into the transhuman era, only a small percentage of the interesting and potentially useful ideas in my mind are ever going to get communicated, due purely to the slowness of the physical mechanics of communication, and of the low-level cognitive processes of verbalizing and otherwise linearizing already-formed concepts and relationships.

Chapter 1
Meta-Philosophy

The main focus of this book is on the philosophy of mind – a topic that, obviously, is extremely broad with a huge variety of aspects. However, before addressing philosophy of mind, I will – in this brief chapter — give a little attention to philosophy in a more general sense, with a view toward conceptually positioning the philosophy of mind to follow.[2]

Pragmatic Relativism and the Value of Philosophy

As a not-very-dramatic prelude, I'll begin by explaining my personal meta-philosophy – my philosophy of philosophy. This is quite simply a philosophy of *pragmatic relativism*.

Relativism means I don't believe there is any one correct way of looking at the universe. I don't believe any philosophy or any science is going to find the ultimate essence of the universe, break down the universe into its essential components, give an ultimate explanation for why things are the way they are, etc. I don't believe there is any one big meaning underlying this universe we see all around us. This doesn't make me a nihilist – it's quite possible to avoid believing anything is absolutely true or objective, while still avoiding the cognitive and emotional excesses of nihilism.[3] I also note that this isn't an absolute or dogmatic belief. If someone finds a single ultimate meaning for it all, that's great, and I'm open to that possibility! But as a working assumption, I've chosen to assume that, in all probability, no such thing exists.

On the other hand, I do believe that the quest for ultimate meanings and foundational analyses and reductions of the universe is extremely valuable – even though it's ultimately bound to fail because the universe has no ultimate meaning and foundation.

The value system I've chosen for myself consists of three primary values: freedom, joy and growth. I will elaborate on this value system in the final chapter of the book, interpreting each of these concepts in a pattern theory context, and elucidating how they fit together. Of course, every one of these three things is difficult to define in a rigorous way, but nevertheless, these are the values I have chosen, fully realizing that they are defined only relative to the human

---

[2] In later chapters I will also have something to say about *philosophy of science* and *ethical philosophy*, but according to the patternist perspective, those topic are best addressed after the essential concepts of pattern theory and philosophy of mind have already been presented.

[3] This is one of those "existential" points that you don't *really* understand until you've experienced the opposite – i.e., until you've lived, for a while, the "cognitive and emotional excesses of nihilism." In my late teens I was highly confused on these issues, and liked to go around giving people detailed logical proofs that they didn't exist. Reading Nietzsche and taking psychedelic drugs, among other experiences such as simply growing up, nudged my mind in the direction of what I came to call "transnihilism" – or what Philip K. Dick (1991) called the recognition of "semi-reality": the attitude that the world may not be objectively real in the way some people naively think, but is incontrovertibly *there* nonetheless.

cultural and cognitive patterns existing in my human mind/brain.

And I believe that the quest for ultimate foundations of the universe is an important contributor to the values of freedom, joy and growth. As a highly relevant example of this, I've found that developing my own speculative metaphysics has been absolutely essential to developing my own philosophy of mind – and developing my own philosophy of mind has been absolutely essential to developing my Novamente AI design … which, if it works even vaguely as planned, will be a very useful thing, and even if not (egads!) will surely lead to a lot of valuable research insights and useful applications (in fact it already has done so, to an extent).

The Value of Metaphysics

For those who believe in the absolute reality of the physical world and the illusoriness of everything else, metaphysics is basically a trivial pursuit. But I'm a dyed-in-the-wool relativist[4], and I've never been able to accept *anything* as possessing absolute reality, nor to condemn subjectively apparent realities as "illusory" in any strong sense. I'm aesthetically and intuitively attracted to developing ideas with maximum explanatory power based on the minimum possible assumptions — where minimality is concerned, an absolutely real physical world seems not to fit the bill; whereas, rejecting subjective reality as totally illusory seems to fail on the grounds of weak explanatory power.

I'm strongly drawn to the "phenomenological" perspective in which physical reality is something the mind constructs, based on patterns it recognizes among its perceptions – and then based on information it gathers via communication; but communication is only recognized as a reliable information source via observation of many instances where communicated information agrees with perceived information. And yet, of course, as a would-be mind engineer, I also recognize the value of the alternate view, in which physical structures and dynamics give rise to the mental realm as an emergent phenomenon. I think both of these views – "mind creates reality" and "reality creates mind" – are important though limited.

And so I'm attracted to a more metaphysical/metamental perspective, in which one identifies some simple concepts common to both mind and reality, and develops these concepts as clearly and simply as possible, without reference to theories of either psychology or physics. One might protest that there is no basis on which to judge theories of this type – but this is just a (half-useful) half-truth. In a later chapter, I'll discuss the philosophy of science, and will argue that the quality of a scientific theory must be judged by the quantity of interestingness and surprisingness among the conclusions to which it leads. Similarly, one may judge one metaphysical theory against another by assessing the usefulness of the theories for generating more concrete, not-just-metaphysical ideas. By this standard, for example, the speculative metaphysics of Leibniz's *Monadology* (1991) has proved itself fairly useless – subsequent thinkers and doers have found little use for Leibniz's idea that the universe consists of a swarm of monads that continue forever to enact their initial programs set in place by God. And by this same standard, I believe, the patternist metaphysics given here has already begun to prove its worth, via its power at leading to interesting conclusions in domains such as psychology, artificial intelligence, and evolutionary biology.

---

[4] As Ben Franklin wrote, in my favorite of his maxims, "Moderation in all things – including moderation."

Patternist Metaphysics

Some metaphysical theories are concerned with identifying the most basic stuff of the universe and defining everything else in terms of this stuff.[5] One then gets involved in debates about whether entities or processes are more primal, whether time emerges from space or space emerges from time, etc. I've never found debates like this to be very productive, and I prefer to think about "identifying a foundational network of concepts for describing the universe", without worrying too much about which concepts within the network are more foundational. In a network of interrelated abstract concepts, it's often possible to create many possible "spanning subtrees" of the network (to use some terminology from graph theory), thus tracing many alternative pathways from "foundational" to "derived" concepts.

But one must begin somewhere, and – echoing Faust's "In the beginning was the Act" – in articulating my metaphysics I choose to begin with *events*. An event is simply an occurrence. Speaking phenomenologically, events may be effectively instantaneous, or events may have directionality. The "directionality" in an individual event may be thought of as a local time axis, but need not be considered as a global time axis – i.e. at the very general level of foundational metaphysics we don't need to assume all the local directionalities are aligned with each other, or even that there is any fundamental sense in which "aligned with each other" is meaningful. This kind of generality is useful when one is talking about topics like quantum gravity or quantum gravity based computation, where the very origin of the substance of "time" is a topic of analysis.

If an event is directional, it has parts, and the parts have a directional ordering to them, a before and after. These parts may be considered as events themselves. We may consider the beginning part of an event as the "input" of the event and the ending part of the event as the "output" part of the event. The middle parts of the event may be considered as "processing."

Next, suppose that events are divided into categories. This division may be done in a lot of different ways, of course. Here we may introduce the distinction between objective and subjective metaphysics. As a relativist, I don't believe in true objectivity of mind or reality or anything else, but I do think it's meaningful to talk about "quasi-universal" metaphysics, in the sense of metaphysics that holds for every subjective reality of any significant complexity. In the case of the division of events into categories, one can introduce subjectivity in a fully explicit way, by stating that the event-categories in question are defined by some specific mind's categorization of its world. Or, one can be quasi-universalist, and define "event-categories" relative to subjective reality, so that events E1 and E2 are put in the same event-category if they are judged highly similar by a particular judging mind. This is quasi-universalist because, intuitively, any significantly intelligent mind is going to have an explicit or implicit notion of similarity, and hence one may define event-categories as "similarity clusters" in any interesting subjective reality.

Now we may define a *process* as a collection of events whose inputs are all in the same category, and whose outputs are all in the same category. In logic terms, events are instances and processes are classes of these instances.[6]

---

[5] A friend who was involved in the creation of the SUMO formal ontology (a DARPA funded project at Teknowledge Inc.; see Niles and Pease, 2001) says that the scientists involved with SUMO spent three weeks debating whether to put "thing" or "stuff" at the top of the ontological hierarchy. (Of course, he was kidding — but only partially)

[6] As an aside, I'm aware the word "process" has a deep meaning in Whitehead's philosophy – which I've read and enjoyed but never studied carefully – and my use of the word here is definitely not intended to be

Processes set the stage for the introduction of patterns, the key player in patternist metaphysics. To obtain patterns, one needs processes, and one also needs the notion of *simplicity*. By a "simplicity measure," I mean a process whose inputs are processes, and whose outputs are entities belonging to some ordered domain. As in mathematics, by an ordered domain I mean any set of entities that is endowed with some ordering operation that is reflexive, antisymmetric, transitive and total.[7]   The classic examples of ordered domains are integers and real numbers. Most of the simplicity measures used in mathematical pattern theory so far map into either the positive integers, the positive real numbers, or the interval [0,1].

Here, again, we touch the issue of subjectivism. If we are studying the subjective metaphysics of some particular mind, then we may assess simplicity as "simplicity measured relative to that mind." But this notion requires some unraveling – essentially what I mean is "X is simpler than Y relative to mind M if, when all else is equal, M prefers X as an explanation than Y." That is, I define simplicity relative to a mind as "that which, if we define it as simplicity, makes that mind work as closely as possible according to Occam's Razor."

If we are speaking in general, outside of any particular mind, then in order to define simplicity we need to adopt some kind of basic representational framework. For instance, if we define events as strings of zeros and ones, then processes become functions on bit strings, which may be represented e.g. as bit strings representing programs for some particular computer. Since we need to introduce a "reference computer" here, we have not really escaped from subjectivity, we've merely introduced a mathematical way to represent our assumption of subjectivity. Computation theory tells us that, as the bit strings we're dealing with get longer and longer, the choice of computer matters less and less – but it still matters.

Finally, with a simplicity measure in hand, we can interpret a space of processes as a space of patterns. A pattern in some X may be defined as a process P whose output is X, and so that P and its input, taken together, are simpler than X. We may then envision the space of processes as a network of patterns – patterns being processes that transform processes into processes, thus making the process network more complex by introducing a network of simplifications.

We may also define relative patterns, i.e. patterns that are to be considered relative to some particular "system" (i.e. some set of patterns). A pattern in X relative to M is a process P whose output is X, and whose input consists of some subset of M and some auxiliary input – and so that P and its auxiliary input, taken together, are simpler than X. For instance, there is an obvious pattern in the series 1, 2, 4, 8... relative to my mind (which contains a lot of mathematical background knowledge), but this pattern may not exist in the series relative to a young child's mind.

Given the notion of pattern, we can then develop the full apparatus of patternist philosophy, as will be enlarged upon in subsequent chapters. We can make definitions like:

- *Complexity*: the amount of pattern in an entity,

---

strictly Whiteheadian in nature. In Hegelian terms, roughly speaking, an entity is a Being and events and processs are Becomings.
[7] If we denote the ordering relation by $\leq$, then the axioms that make this relation an ordering are: $a \leq a$ (reflexivity) , if $a \leq b$ and $b \leq a$ then $a = b$ (antisymmetry), if $a \leq b$ and $b \leq c$ then $a \leq c$ (transitivity) , $a \leq b$ or $b \leq a$ (totalness) . The natural numbers are the smallest possible totally ordered domain with no upper bound.

- *Intelligence*: the ability to achieve complex goals in complex environments,
- *Mind*: the set of patterns associated with an intelligent system (i.e. an intelligent set of processes),
- *Emergence*: the existence of patterns in a set of processes that are not patterns in any of the individual processes in the set,
- *Relative complexity*: the complexity of an entity, where the patterns in the entity are defined relative to the knowledge in the system,
- *Simplicity and relative simplicity*: the inverses of complexity and relative complexity.

These definitions allow us to go full circle and return to an earlier point in our metaphysical development. We may define a simplicity measure that's dependent on a particular mind, where a mind is defined as the set of patterns associated with an intelligent system. Similarly, going back further, we may define processes in terms of categories of events, where the categorization is defined either by the categories explicitly in some mind, or by similarity as measured by the possession of similar sets of patterns relative to that mind.

Physical law, from this perspective, consists of a collection of processes that are extremely intense, powerful patterns in the event-space that is the universe. These patterns are highly "intense" in that they provide massive simplification of the universe. Quantum particles, gravitational fields and so forth are then understood as patterns defined relative to the pattern-set that is physical law. That is, when the notion "quantum particle" is used to simplify the observation of a path in a bubble chamber, this simplification is conditional on the body of knowledge that we call "physical law." Without this body of knowledge as implicit background information, the path in the bubble chamber may display patterns such as "the left half is shaped the same way as the right half" but not patterns such as "this is an intermediate vector boson."

Everyday physical reality – baseballs, galaxies, Ministers of Finance, chinchilla toenails and so forth – consists of patterns that embodied minds use to organize their sense perceptions. It's well-known that these patterns correspond only complexly and loosely with the patterns that exist conditional on physical law – for instance, humanly perceived colors are explicable in terms of electromagnetic phenomena only after a great deal of calculation and hand-wringing.

In short, patternist metaphysics portrays ultimate and everyday physical reality as well as mind in terms of sets of patterns. Patterns themselves are processes identified as patterns via the imposition of some simplicity measure; and processes are defined in terms of categories of primal events, where categories may be defined as clusters in the context of a similarity measure on sets of events. In addition to the basic concepts of directionality and similarity, the concept of simplicity is key here. Identifying similarities turns events into processes; and then, by the act of defining a way of measuring simplicity, one turns a universe of scattered and disorganized processes into a network of patterns, in which one may detect structures and processes identifiable as "mental" or "physical."

I have chosen to take pattern as the foundational concept. One could try to derive patternist philosophy from yet more basic concepts, but my feeling is that, conceptually, one obtains diminishing returns by reducing the fundamental concept-set beyond what I've done here. For instance, one can use a formalism like G. Spencer-Brown's "Laws of Form" (1994) to define events, directionality, simplicity and similarity. This is an interesting exercise but, at least in the ways I've tried so far, it doesn't really add anything to the philosophy, and has more of the ring of mathematical, computational or physics modeling.

I have here discussed the notion of pattern in a sort of semi-formal, intuitive way. One may also seek to formalize patternist metaphysics mathematically, and this is potentially very valuable, but I feel that assigning this kind of formalization a position at the heart of patternist philosophy would be a mistake. The point of a metaphysical theory is to give conceptual interpretations not precise models. In Appendices to this book, I present formalizations of many of the ideas of patternist metaphysics, in the specific context of computation theory and probability theory. Developing patternist thinking in this mathematical manner is a fascinating pursuit and I hope it will lead to deep conclusions one day, but I don't think it makes sense to put mathematics at the basis of a philosophical theory. Rather, one needs a philosophical theory to tell one what mathematics to construct. It is true that mathematics can then sometimes turn around and feed information back into philosophical theory (Godel's Theorem being the paradigm case), but this feedback occurs within the foundational philosophical theory within which the mathematics is being interpreted. For instance, to get to Godel's Theorem, one must start with a logicist philosophy, which leads one to construct formal-logic mathematics, which leads to Godel's Theorem, which then enriches logicist philosophy. Pattern-theoretic mathematics has not yet led to any major revisions or advances in patternist philosophy but this may potentially happen in future as the mathematics is further developed – this is my hope!

And so I've reached the end of my core metaphysics after just a few pages! We will revisit these basic metaphysical concepts many, many times in the rest of the book – for instance, the question "what is mind?" becomes the question "what kind of pattern network constitutes a mind?" The question "what is consciousness?" becomes "how is consciousness associated with pattern networks?" The question "how to create a thinking machine" becomes "what physical patterns cause the emergence of pattern-networks of the type characteristic of minds." And so on. I certainly don't claim that this is the only useful perspective on mind and reality and AI and so forth — but it seems to me to be significantly more useful than the other frameworks I've studied. (Time will tell if this impression is correct or not.)

Chapter 2
Kinds of Minds[8]

Now I will proceed in the direction of philosophy of mind, beginning, in this chapter, with the relatively simple step of defining the various sorts of mind that may exist. It seems to me that many (though nowhere near all) of the confusions that one finds in the literature on cognitive science, AI and philosophy of mind have to do with the failure to correctly draw distinctions between different kinds of mind. Very few theorists make the effort to distinguish properties of minds in general from properties of human minds in particular.

One may argue that this kind of conflation is reasonable — perhaps, since human minds are the only clear examples of highly generally intelligent systems that we have, we have no basis for distinguishing properties of the human mind from properties of minds in general. But this excuse doesn't hold up to scrutiny. In some cases, indeed, it's hard to tell whether some property of the human mind is a necessary aspect of mind-ness or merely a peculiarity of human cognitive nature. But even so, the distinction can be made much more often than anyone bothers to do.

Before proceeding further, I'll make a few comments on how the notion of "mind" in general may be defined in terms of patternist philosophy. Intelligence, I have defined in my prior works (Goertzel, 1993) as the achievement of complex goals in complex environments. In Chapter 4 I will enlarge upon this definition extensively, and also introduce a related notion of "efficiency-scaled intelligence", which measures intelligence per unit of processing power. A mind, then, I define as *the set of patterns associated with an intelligent system*. This means patterns in the intelligent system, and also patterns emergent between the system and its environment (including patterns emergent between the system and its tools, other intelligent systems, etc.).

Note that both intelligence and mind are fuzzy concepts[9] — intelligence comes in degrees, and a pattern belongs to a mind to a certain degree. These definitions make no commitments about the internal structure or dynamics of mind — raising the question of whether there are any

---

[8] This chapter was inspired largely by conversations with Meg Heath during her visit to my home in late summer 2004. In listening to her theories of distributed cognition, I became repeatedly frustrated with what seemed like confusion between properties of mind-in-general and properties of human-mind or humanlike-mind. I then realized that I had succumbed to this type of confusion a few times in my own thinking and writing, and decided it was necessary to clarify the matter explicitly by making a typology of minds, even though the effort to do so seemed a bit pedantic at first. I should add however that Meg has her own theory and typology of minds and surely doesn't agree with everything I say in this chapter. Hopefully by the time you read this she will have put her own very interesting perspective in writing, but at the moment she has not done so, so I have no pertinent references to give.

[9] "Fuzzy" is meant here not in the sense of "ill-understood" or "vague", but rather in the sense of fuzzy set theory, where a fuzzy set is defined as a set to which membership is naturally considered as having various degrees (e.g., tall, hot, etc.).

universal principles regarding what goes on inside minds, and what governs their interactions? I think that such principles do exist, but that they are few in number, and that one can give a larger number of more specific principles if one moves from analyzing "mind in general" to analyzing particular kinds of mind — still keeping things at a high level of abstraction.

But how can we meaningfully ontologize the space of possible mind-types? What are the kinds of minds? In this chapter, I will address this question, and then position with this space both the human mind and the Novamente AI system that my colleagues and I are currently developing.[10]

Huge versus Modest Resources

At the highest level, I propose, we should distinguish minds operating under severely finite resources ("modest-resources minds"), from minds operating under essentially infinite resources ("huge-resources minds").

Human minds fall into the former category, as do the minds of other animals on Earth, and the minds of any AI software programs humans will create in the foreseeable future. Most of human psychology is structured by the resource limitations of the human brain and body, and the same will be true of the psychology of our AI programs.

On the other hand, whether the latter category of minds will ever exist in reality is unclear. Marcus Hutter (2004), following up earlier work by Solomonoff (1964, 1964a) and others, has explored the mathematical nature of infinite-computational-resources-based intelligence. He has proved theorems stating, in essence, that an appropriately designed software system, if allocated arbitrarily much memory and processing power, can achieve an arbitrarily high level of intelligence.[11] He describes a particular software design called AIXI that achieves maximal intelligence, if given infinite computational resources. And then he describes an approximation to AIXI called AIXTtl that achieves close-to-maximal intelligence, if given an extremely huge — but finite — amount of computational resources.

The laws of physics, as we currently understand them, would seem to preclude the actual construction of systems like the infinite AIXI or even the very large finite AIXItl, due to the bound that special relativity places on the rate of information transmission. Since information can't spread faster than the speed of light, there is a limit to how much information processing any physical system occupying a fixed finite amount of space can do per unit time. This limit is very

---

[10] Note that, here and in later chapters, I will mostly discuss Novamente as it is intended to be when it's finished and fully operational, rather than in its current highly incomplete – though useful for many practical purposes — form. Whether a fully complete and operational Novamente system ever comes about depends on a lot of boring practical issues regarding funding and staffing of the project, but the design itself is interesting as a theoretical object no matter how the practical project comes out.

[11] While this may seem an obvious conclusion, to prove it rigorously using the language of mathematics is no easy trick. Hutter wrote a chapter for a book that I co-edited, and his original title was something like "A Gentle Introduction to AIXI and AIXItl" – but it was by far the most difficult and mathematical chapter in the book, so he changed the title. But his original title had a point, because his chapter was far less technical than his previous expositions of his ideas. The fact that it's so hard to mathematically formalize such a conceptually simple point is an indication that contemporary mathematics is badly suited for the analysis of intelligent systems. This is not surprising since it evolved mainly for the description of simple physical systems, and then more recently for the description of simple computational systems. Attempts to create mathematics better suited for complex living and intelligent systems – such as Robert Rosen's (2002) work – haven't yet succeeded well.

high compared to the amount of information processing done by human brains or contemporary computers, but nevertheless it's trivially small compared to what would be required to run a program like AIXI or AIXItl. Of course, our current understanding of the laws of physics is far from perfect, and it may happen that when we understand the universe better, we'll see that there are in fact ways to achieve superluminal information transmission or something with a similar impact on a computer.

One may define efficiency-scaled intelligence as the the ratio of intelligence of behavior displayed to computational resources utilized. Note that AIXI, because it uses infinite resources, doesn't have a well-defined "efficiency scaled intelligence." Whether AIXItl has a high efficiency-scaled intelligence is a tough mathematical question, but intuitively it seems clear that the answer is no. But these are basically irrelevant questions since both of the designs are most likely impossible to create in practice.

Juergen Schmidhuber, Hutter's former PhD advisor, is trying to get around AIXItl's impracticality problem with an AI design called OOPS, that embodies a vaguely AIXItl-like design, but is capable of solving some simple problems in a reasonable period of time using modest computational resources (Schmidhuber, 1997). In my view, however, OOPS has very little chance of scaling up beyond very simple problems. It seems to me that the conceptual gap between huge-resources mind-design and modest-resources mind-design is too big to be bridged by clever variations like OOPS. Huge-resources and modest-resources mind design and analysis appear to be almost entirely different problems.

The essence of a modest-resources mind consists of a set of strategies for avoiding combinatorial explosions. This is an insight achieved by AI theory during the second half of the 20th century, and it's one that's clearly applicable to human minds as well as software intelligence. A mind with limited space resources — or with space resources that are implicitly limited due to time constraints combined with information propagation limitations as described by special relativity — requires mechanisms for deciding what to retain in memory and what to forget. A mind with limited time resources requires methods to determine which possibilities to explore when deciding which action to take in a certain situation. There is then interplay between the knowledge-paring and possibility-paring processes.

Knowledge-paring and possibility-paring are difficult problems, and modest-resources minds must learn how to do them. This learning process in itself consumes resources. It is often easier to do this learning in a manner that's specialized to some particular domain of experience, rather than in a completely general way. Thus modest-resources minds will tend to contain specialized subsystems, each of which deals with a certain type of knowledge (e.g. visual knowledge, acoustic knowledge, social knowledge, self-knowledge). They must then confront problems of unification — of making their various specialized subsystems communicate with each other effectively. Furthermore, not every needed instance of knowledge and possibility paring can be handled by a specialized subsystem, so a more generalized paring subsystem is also required. A modest-resources mind hence requires some kind of "Mind OS" (OS = operating system) that is able to connect a general-purpose paring subsystem with a family of specialized paring subsystems, which may sometimes overlap with each other in functionality. Of course the form this "Mind OS" takes will depend hugely on the particular physical substrate of the mind in question; we will revisit this issue in the context of the Novamente AI design in later chapters.

These arguments show that the assumption of modest resources gives one a lot of information about a mind — so that the structures and dynamics of a modest-resources mind will necessarily be very different from that of a huge-resources mind.

Varieties of Embodiment

Within the domain of modest-resources minds, many additional abstract distinctions can be usefully drawn, including:

- Embodied versus non, and within the "embodied" category: singly versus multiply versus flexibly embodied, and tool-dependent versus non,
- Mindplexes versus fully unified minds,
- Socially-dependent versus non.

By a "singly embodied mind," I mean a mind that is particularly associated with a certain "physical" system (the "body" of the mind). Note that this "physical" system need not be physical in the narrowest sense — it could for instance be an aspect of a computer-simulation world. The important thing is that it displays the general properties of physicality — such as, involving a rich influx of "sensations" that present themselves to the mind as not being analyzable except in regard to their interrelations with one another, and that are in large part not within the mind's direct and immediate control. An embodied mind uses its body to get sensations about some portion of the physical universe within which its body is in contact; and it carries out all, or nearly all, of its direct actions on the physical world via its body. A lot of the patterns constituting the mind should be patterns in this body-system.

A multiply embodied mind is a mind that, in a similar sense, is particularly associated with a certain set of more than one physical system, which are disconnected (or nearly disconnected) from each other. "Disconnected" here means that the bandwidth of information transmission between the "disconnected" systems is vastly less than the bandwidth of transmission between the parts within the individual systems. Human and animal minds are singly embodied, but a single computer program simultaneously controlling a dozen robots might be multiply embodied.

Finally, a flexibly-embodied mind is a one that may have one or more bodies at any given point in time, but isn't specifically attached to any one of the bodies – it can flexibly switch embodiments based on whatever criteria suit it at the moment. A good example would be a human hooked into a video game environment via powerful virtual reality style sensors and actuators. The human mind would be there invariantly regardless of which character, or group of characters, was being played.

Not all minds need to be embodied at all – for instance, one could have a mind entirely devoted to proving mathematical theorems, and able to communicate only in mathematics. Such a mind would have no need for a body; its sensations and actions would consist of statements in a formal language.

One may also delineate a type of mind that is not necessarily embodied in the above sense, but possesses what I call "body-centeredness." A body-centered mind is one that is particularly associated with a certain physical system, but, most of the mind consists not of patterns in the physical system, but rather patterns emergent between the physical system and the environment. So in this case the body is not the substrate of the bulk of the mind-patterns, but only the partial substrate. Of course, a system may be singly or multiply body-centered; and the boundary between embodiment and body-centeredness is fuzzy. Humans have some embodiment and some body-centeredness to them. The development of language, it would seem, moved us further from strict embodiment toward the direction of body-centeredness. Future developments like Internet-connected neural implants might move us even further in that direction — a topic that leads us on to the next kind of mind in our ontology, mindplexes.

A mindplex is a mind that is composed (at least in part) of a set of units that are, themselves, minds. A mindplex may be a singly or multiply embodied mind, or it might not be embodied at all. The notion of a mindplex begs the question of how one distinguishes an autonomous, coherent "mind" from a part of a mind. For instance, if we conceive a mind as the set of patterns associated with intelligent system, then why isn't the set of patterns in human society considered a mind? The solution to this dilemma lies in the recognition that the notion of "mindness" is fuzzy. A human society is a mind, to an extent — and how much mindness it has relative to an individual human mind, is subject to debate. It seems clear to me, intuitively, that human society has less "efficiency-adjusted intelligence" than a typical individual human mind — but does it have less unadjusted, raw intelligence? Perhaps human society is a mindplex. But clearly there would be much more mindplexness involved if, for instance, a massive AI system were connected to the Internet and instructed to participate in online dialogues of various forms in such a way as to encourage the achievement of the essential goals of humanity. In this situation, we'd have a hybrid human/AI mindplex of the type I've called elsewhere a "global brain mindplex."

What about the population of the USA? Is this set of people a mindplex? Here mindplexness interacts with embodiment in an interesting way. Both the USA and global human society as a whole are multiply embodied minds — and mindplexes — but global human society has a much stronger degree of embodiment than the population of the USA, because the USA is not that isolated, so that very many of the patterns constituting the mind of the USA population are emergent patterns between the brains/bodies of Americans and the brains/bodies of other people. The population of the USA is more multiply body-centered, whereas the population of the Earth is more multiply embodied.

Next, the notion of body-centeredness may be decomposed. There are two cases of mind-patterns extending beyond the body centering a mind: mind-patterns spanning the body and other intelligences, and mind-patterns spanning the body and inanimate, mostly unintelligent objects. Of course this is a fuzzy distinction since intelligence is in itself a fuzzy notion. But like the other fuzzy distinctions I'm making in this ontology of minds, I think it's one worth making. A body-centered mind many of whose defining patterns span its body-center and a set of inanimate objects may be thought of as a "tool-dependent" mind, whereas a body-centered mind many of whose defining patterns span its body-center and a set of other significantly mind-bearing systems, may be thought of as a "socially-dependent" mind. Theorists of the human mind haven't reached any consensus as to the extent to which human minds are body-centered versus embodied, and the extent to which they're tool-dependent and/or socially-dependent. It seems clear that the tool-dependence and social-dependence of human minds has been drastically underestimated by the bulk of recent cognitive science theory; research during the next couple decades will likely go a long way toward telling us how much. Our tool-dependence is obvious from our dependence on various physical implements to shape our world in accordance with the concepts in our minds; as well as by our use of mnemonic and communicative tools like diagrams and writing. Our social-dependence is obvious above all from our dependence upon language, which gains much of its meaning only emergently in the interaction between individuals.

One might wonder how — in abstract, general terms – it is possible to distinguish a tool used by a mind from a part of the body about which the mind is centered. Of course, this is clear in a human context, but it might be less clear in other contexts. And it may become less clear in the human context as body-augmenting technologies become more advanced. A conventional prosthetic limb is somewhere between tool and a part of the body; but a prosthetic limb with sufficiently advanced sensors and actuators becomes definitively a part of the body. Qualities that fuzzily distinguish body parts, in an abstract sense, seem to be: one always keeps them rather than using them only occasionally and then discarding them; they grow out of other body parts rather than being found fully formed or being built by a rational process; and one can both sense and act through them. Not all human body parts meet all of these criteria though; and it's not clear that the distinction between tools and body parts will hold up in the post-human era.

Of course, varieties of embodiment lead to varieties of cognitive habits. An embodied mind will tend to use physical-world and body related metaphors for a lot of its internal thinking, and a socially dependent mind will tend to hold inner dialogues with others, even when they're not present. Modest-resources minds depend on heuristics for paring down combinatorial possibilities; embodiment and sociality push minds toward certain sorts of heuristics, based on heuristics that are useful for solving physical or social problems. Physical embodiment pushes minds toward geometric and energy-based heuristics, for example; whereas sociality pushes minds towards heuristics that are based on notions of agents and roles.

Varieties of Social Interaction

Sociality is obviously a critical aspect of human intelligence, and would seem to evolutionarily precede the development of language, since primate intelligence is also heavily focused on social interaction. Fairly strong arguments have been made that our powerful cognitive abilities arose largely out of our early ancestors' need to model one another (cf Calvin and Bickerton, 2000). It is not clear that AI's will need to be social to this extent; however, from the point of view of pragmatically teaching AI's to think, as well as from the point of view of teaching them ethical behavior, building sociality into AI's makes a lot of sense. Having other similar beings around to study is a powerful heuristic for learning self-modeling; and having other similar beings to provide feedback on one's changes through time can be a valuable guide through the process of self-modification.

An obvious question, in an AI context, is whether, given a fixed amount of computational resources, it's better to put them all into one mind, or to split them among a number of socially interacting minds. I don't think we know the answer to this – we lack a sufficiently precise theoretical cognitive science to derive an answer in advance of extensive practical experimentation. My own guess is that the optimal thing may be a community of minds that have a certain degree of mutual separation, but also have more unity than exists between human minds. Having a community of minds is good in that it allows experimentation with different ways of structuring minds – allowing "evolutionary learning" of what makes a good mind. Of course, it may be that the future will reveal learning techniques that work so well they obsolete the whole notion of evolutionary learning, but I'm guessing that explicit diversity-generation is going to have a role in learning forevermore. However, it also seems to me that, in a "society of individual minds communicating through language and physical coexistence" situation like human society, nearly everything that's learned is wasted. It would be much better, I think, if:

- We had a language that involved simply putting our ideas in a common conceptual vocabulary, without requiring a projection of thoughts into a linear sequence of symbols (or diagrams or other such strongly physical-world-limited media)

- As a sometime alternative to language, we could directly exchange thoughts with each other (understanding that in many cases the exchanged thoughts would not be comprehensible to the recipient)
- There were data-mining processes able to scan vast numbers of minds at one time and find patterns correlated with various beneficial properties of minds (in computer science terms, this would correlate to a movement from a pure evolutionary development of minds, to a mind-evolution scenario based on Estimation of Distribution Algorithms (EDA's)[12], wherein one solves a problem via maintaining an evolving population of candidate solutions, and then running a data-mining process that studies the whole population and figures out which patterns characterize good solutions and creates new candidate solutions embodying these patterns).

These kinds of communication will be easy to build into AI systems one day, and they are also in principle achievable in humans via various sorts of brain implants. A collection of telepathically-enabled mind endowed with a EDA-style central datamining process becomes a special sort of mindplex. One possible future for the human race is that it fuses with computer technology to become a "tele-EDA-mindplex" or "global brain mindplex" of this sort.

With these futuristic ideas in view, it seems important to distinguish between socially-dependent minds that live in telepathy-enabled communities, those that live in linguistics-enabled communities, and those that lack both language and telepathy. Obviously the lack of either language or telepathy greatly restricts the kinds of collective cognition that can occur (and by "language" I mean any system for abstract symbolic representation, not including the protolanguage of chimps in the wild, but including, say, systems for communication using pictures or smell, not necessarily just sounds and written words.) The need to communicate through gesture and language rather than through various forms of thought-exchange, has a huge impact on the structure of a mind, because it means that a large part of cognition has to be devoted to the careful projection of thoughts into highly simplified, relatively unambiguous form. The projection of thoughts into highly simplified, relatively unambiguous form is a useful heuristic for shaping ideas, but it is not the only possible sort of heuristic, and minds not centrally concerned with linguistic interaction would probably make far less use of this particular cognitive move. Of course, linguistically-focused minds don't need to formulate all their ideas in language – but the fact is that, in a community of minds, there is a lot of benefit obtained by getting feedback on one's ideas, so there is a strong motivation to formulate one's ideas in language whenever possible, a fact that very strongly biases the thoughts that occur.

Self-Modification

Change and self-improvement are essential to any intelligent system. However, different minds may vary in the extent to which they can modify themselves with a modest amount of effort.

In this vein, one may distinguish radically self-modifying minds from conservatively-structured minds. A radically self-modifying mind is one that has the capability to undertake arbitrarily large modifications of itself with relatively little effort – its main constraint in

---

[12] Examples of EDA's are the Bayesian Optimization Algorithm (BOA) developed by Martin Pelikan (2002) and to the MOSES algorithm (Looks, 2006) used in Novamente,

modifying itself is its willingness to do so, which ties in with its ability to figure out subjectively-good self-modifications. On the other hand, a conservatively-structured mind is one that cannot make large modifications in itself without undertaking huge effort – and maybe not even then, without destroying itself.

Clearly, human brains are conservatively-structured, whereas digital AI programs have the potential to be radically self-modifying. Radical self-modification is both powerful and dangerous – it brings the potential for both rapid growth and improvement, and rapid unpredictable cognitive and ethical degeneration.

Quantum versus Classical Minds

Finally, there is one more kind of distinction that is worth making, though it's basically orthogonal to the ones made above. This has to do with the sort of physics underlying a mind. Quantum computing is still in a primitive phase but it's advancing rapidly, so that it no longer seems so pie-in-the-sky to be thinking about quantum versus classical minds.

The strange properties of the quantum world have been discussed in many other books and I won't try to do full justice to them here. However, I will enlarge on this topic a bit at the end of Chapter 8. What I'll argue there is that it's almost surely possible to build quantum cognitive systems with properties fundamentally different from ordinary cognitive systems – thus creating a whole new class of mind, different from anything anyone has ever thought about in detail.

Humans versus Conjectured Future AI's

Given all these ontological categories, we may now position the human mind as being: singly-embodied, singly-body-centered, tool and socially dependent and language-enabled, conservatively-structured, and either essentially classical or mostly-classical in nature (meaning: clearly very limited in its ability to do quantum-based reasoning).

The Novamente AI system – the would-be artificial mind my colleagues and I are currently attempting to create — is intended to be a somewhat different kind of mind: flexibly embodied, flexibly body-centered, tool and socially dependent, language and telepathy enabled, and radically self-modifying, and potentially fully quantum-enabled. Also, Novamentes are explicitly designed to be formed into a community structured according to the "telepathic EDA mindplex" arrangement.

It might seem wiser, to some, to constrain one's adventures in the AI domain by sticking more closely to the nature of human intelligence. But my view is that the limitations imposed by the nature of humans' physical embodiment pose significant impediments to the development of intelligence, as well as to the development of positive ethics. Single embodiment and the lack of any form of telepathy are profound shortcomings, and there seems no need to build these shortcomings into our AI systems. Rather, the path to creating highly intelligent software will be shorter and simpler if we make use of the capability digital technology presents for overcoming these limitations of human-style embodiment. And the minds created in this way will lack some of the self-centeredness and parochialism displayed by humans – much of which is rooted precisely in our single-bodiedness and our lack of telepathic interaction.

I would argue, tentatively, that flexible embodiment and telepathic enablement and EDA mindplexing are not only cognitively but ethically superior to the human arrangement. Being able to exchange bodies and minds with other beings will lead to a psychology of one-ness and sharing quite different from human psychology. And, being able to choose the best mind-forms from an evolving population and perpetuate them will avoid the manifold idiocies of the purely evolutionary process, in which minds are stuck with habits and structures that were appropriate only for their ancient evolutionary predecessors, but are still around because evolution is slow at getting rid of things.

Of course, there could be hidden problems with the flexibly-embodied, telepathic EDA-mindplex approach. However, we lack a mathematical framework adequate to uncover such problems in advance of doing extensive experimentation. So our plan with the Novamente project is to do the experiments. If these experiments lead to the conclusion that singly-embodied, no-telepathy minds are cognitively and/or ethically superior, then we may end up building something more humanlike. But this seems an unlikely outcome.

Chapter 3
Universal Mind[13]

Although this is a philosophy book, the perspective taken is largely a scientific one. I've approached the philosophy of mind via introspection of my own mind and via logical analysis, but also through a careful survey of the brain science and cognitive psychology literature, and through years of theoretical research and practical experimentation in the AI field.

However, I'm well aware that science is not the only domain of human endeavor to say interesting things about the mind; and my thinking about the mind has also drawn on various spiritual traditions in which I've taken an interest at various points in my life. In this chapter I'll briefly explore the relationship between the spiritual and scientific understandings of the mind.

In the previous chapter, I argued that human minds, as diverse as they are, occupy only a very tiny corner of the total space of possible minds. And I argued that mind may be considered as a very general phenomenon, related to abstract notions such as pattern and process. Now I will dig further into the abstract nature of mind, exploring the way "mind" interacts with "universe" – not merely in the sense of the physical universe, but in a more general and "cosmic" way.

It's worth noting that I am a non-religious person; in fact, at various points in my life I've been actively — sometimes rabidly — anti-religious. However, there is no doubt in my mind that "spiritual insights" have led various humans to profound knowledge of both personal and universal nature. Yes, the creation of social institutions ("religions") based on these insights has fairly often had destructive consequences; and the creation of belief systems based on these insights has often resulted in the widespread adoption of mind-stultifying beliefs contradicting the essence of the original spiritual insights. Yet these unpleasant consequences that have sometimes occurred upon the collision of spiritual insights with the rest of human culture do not eliminate the positive value and meaning of spiritual insights in themselves.

I will portray embodied minds like human minds as part of an overall universal mind-network, culminating in what may fairly be thought of as a "universal mind." I will then explore the similarities and differences between the universal mind and our embodied minds – and also pay some attention to the apparent limitations of the universal mind. (Indeed, this is cosmic stuff, but I've never been one to shy away from the "big questions.")

In this chapter I will draw eclectically from various spiritual traditions, just as in other chapters I draw eclectically from various branches of science. Hopefully it is clear to you that I'm

---

[13] This chapter draws heavily on text from Chapter 1 of my rou manuscript "The Unification of Science and Spirit," which is available at http://www.goertzel.org, but which will probably never be published on paper or in any finalized electronic version, because I'm not very happy with it. It seems to me now, in the "wisdom" of my old age, terribly immature and silly in several ways. However, I doubt that I'll ever find time to revise it in accordance with my current tastes and opinions. I've left it online in a spirit of openness and communication, even though it contains plenty of statements that I don't agree with now (and some that I don't think I even fully agreed with when I provisionally and experimentally wrote them down!).

doing more than just creating a pastiche of other peoples' ideas: I'm presenting my own coherent perspective, and could easily do so without mentioning anybody else's work all. However, my hope is that highlighting connections with prior thinkers and bodies of knowledge, where these exist, will help my own ideas to go down more easily.

As with the other early chapters in this book, the ideas here are presented in a somewhat sketchy way, because the full scope of the patternist philosophy has not yet been presented, so there's not yet much context. The most logically correct way to write the book would have been to create some sort of explicitly nonlinear "hyperlinked" book structure, in which each chapter would depend on all the others, taking care that the various interdependencies form a system of equations with a fairly tightly delimited semantic space of solutions. However, such a mode of writing is obviously very difficult from the reader's perspective, and so I've opted for a more traditional approach, in which the early chapters go over their topics more lightly, in a way that minimizes dependencies on ideas from the later chapters, and then the later chapters then delve deeper into their subjects, making use of the earlier chapters as foundation. The topics of the earlier chapters (like this one) are then revisited and reviewed from time to time in the later chapters, when particular points arise that shed particularly large amounts of light on early-chapter topics. But this structure means that it may definitely be of value to reread the earlier chapters again after the contents of the later chapters have sunk in. This chapter is a particularly good example of that. The "theological" notions given here are useful for "priming" the more psychological concepts in following chapters – but after the psychological ramifications of pattern theory have been understood, these theological explorations will hopefully be perceived with greater richness. For instance, the last two chapters of the book deal with immortality and ethics, themes that tie in deeply with nearly every theological tradition – but I can't talk about those topics in this chapter because my way of approaching them relies too heavily on the patternist theories of reality and cognition, to be elucidated in the chapters between here and the end of the book.

The Perennial Philosophy

I have long admired Aldous Huxley's crosscultural survey of spiritual thought and experience, *The Perennial Philosophy* (Huxley, 1990). More so than any other author, Huxley makes a powerful case for a unifying core underlying the diversity of spiritual thought and experience across the world's cultures.[14]

The "Perennial Philosophy," as Huxley calls it, is a collection of ideas that has emerged as the part of the "wisdom" of a huge variety of cultures and traditions.

It holds, first of all, that everything in the mind and in the universe is an aspect of the "divine." The divine is immanent in all things.

Next, there is a hierarchical structure to the universe. There are lower, middle and higher planes of being — the exact number varies from tradition to tradition.

---

[14] I encountered the book when I was 18 or so, after being led to Huxley's briefer book "The Doors of Perception" via reading somewhere that this was where the rock band "The Doors" had taken their name. Jim Morrison of the Doors is the one who also implanted the phrase "Universal Mind" (used in this chapter) in my vocabulary, via his lyrics in the song Freedom Man: "I was doing time/ in the Universal Mind / I was feeling fine ..."

There is also a pattern of interpenetration in the universe: each entity, when looked into deeply enough, can be seen to contain other entities — ultimately, all other entities. The part contains the whole, just as the whole contains the part.

A person must possess certain qualities in order to consistently live in contact with the higher levels of being: a certain humility and compassion, a respect and love for one's fellow beings.

And, finally, there is something standing in the way of our achieving these virtues and gaining a clear vision of the world. This is our thought-emotion complexes, our habitual mind.

The Perennial Philosophy teaches that the physical world is the lowest realm of being. It is unreal, a construct of the higher levels, which the refined mind can see right through. On the other hand, the thinking and feeling "mind" as we generally understand it is just an intermediate level in the universal hierarchy. The mind is bound up in "knots," self-preserving thought systems, which prevent it from seeing the true nature of the world. Freed from these knots, the consciousness ascends beyond thinking and feeling, on to a higher plane of pure intuition.

The Perennial Philosophy is not vacuous; it embodies a specific, and interesting, philosophy of mind. It teaches that the individual, embodied human mind is a very limited thing, a collection of habit-patterns that sees only a small part of the universe. It also, in Huxley's formulation, leaves open a question regarding the role of mind-in-general in the universe. Suppose there is a hierarchy of being, in which human minds lie at an intermediate level? Do the entities exist at higher levels properly understandable as *mind,* or must they be thought of differently? I will return to this question at the end of this chapter, after digging into the hierarchical aspect of the Perennial Philosophy in more detail. One of my main conclusions is that the universe as a whole – including the "higher levels of being" – can indeed meaningfully be considered as a mind. The complex goal that it pursues is specifically the goal of understanding itself.

My own take on the Perennial Philosophy is a bit less orthodox than Huxley's. I think the Perennial Philosophy is mostly correct but that there are some subtleties of interpretation that often lead people astray.

I agree with the Perennial Philosophy that there is a hierarchical structure to the universe, that in a sense each entity in the universe can be understood to contain all other entities (I'll explain what "contains" should be taken to mean in this context, a little later), that certain qualities related to openness and compassion are likely to help a mind to experience and live harmoniously with the greater universe, and that the mental habit-patterns of the individual mind are often obstacles to the mind recognizing its oneness with the cosmos. I prefer to recast these insights in my own language of patterns and processes – but of course Huxley's main point in his book was to abstract beyond various cultures' and individuals' preferential languages.

However, I do have some difficulty with the notion that there is a "divine principle" immanent in all things. The problem, I think, is that the word "divine" has been so badly overloaded. Any statement concerning "divinity" needs to be clarified with particular meticulousness.

The notion of "divinity" relates to a key issue that arises in many different wisdom traditions – though Huxley doesn't dwell on it as explicitly as I'd like — the "problem of evil." In the language introduced above, the problem is: Why is there a blockage that prevents each part of the world from realizing its interpenetration with the rest of the world? Why do "thought-emotion complexes" of a restricting and stultifying nature needs to exist? This is an abstract version of the old problem: If God is so good and so powerful, why does he let so many rotten things happen down here on Earth?

Of course, this is only a logical and conceptual problem if one assumes that there is some "divine" entity underlying everything in the universe, which is all-powerful. One then has the question: Why doesn't this all-powerful divine entity just get rid of the blockages and knots that keep the world from truly understanding and experiencing itself in all its aspects, and make everything nice and glorious and wonderful throughout? Why doesn't God just feed the innocent starving children?

The only clear way out of the paradox, obviously, is to give up the notion of the divine as all-powerful. My view is that the "divine" immanent in all things is a kind of raw, unanalyzable, essence of pure Being – what Charles S. Peirce called a First, as I'll discuss in Chapter 6 below. There is nothing in First that has to do with absolute power. Power is a kind of relationship, it's a Peircean Third; it's a different kind of animal than "divine essence."

In Chapter 20, after a series of chapters presenting my own theory of mind in detail, I will present my own twist on this age-old "inadequate God" solution to the problem of evil. As I see it, the reason "evil" exists in the universe – the reason that the universe contains isolated minds that don't fully understand themselves and don't fully realize their oneness with the rest of the universe – may be simply that the universe is finite. Of course, we can never know if the universe is finite or infinite, because our own minds have limits, and we can't perceive anything that goes beyond these limits. The universe could be infinite in ways that we can't currently understand. However, it's clear that the universe as we know it displays one of the key characteristics of finite systems – a lack of complete self-understanding. If we assume the universe is finite, then the fact that some parts of the universe lack a full understanding of themselves and their role in the whole universe is not at all surprising – because, except in the case of trivial systems, about which there is virtually nothing to understand; only an infinite system can have full self-understanding. (I'll give arguments later as to why this must be true, derived from the branch of mathematics called algorithmic information theory.) On the other hand, if we assume the universe is infinite, then we have the problem of explaining why it's so perversely structured that it doesn't make use of its infinite capability to make everything beautiful and wonderful for everyone. To me, the "finite universe" hypothesis has Occam's Razor on its side.

The Vedantic Hierarchy

Now I will delve a little further into one aspect of the Perennial Philosophy, the notion of a "cosmic hierarchy." There are many different ways of drawing such a hierarchy, but while the details may differ, the basic idea remains the same. The scheme that I will use here is loosely adapted from the hierarchy found in the Vedantic school of Indian philosophy.

Vedanta is one of the six major schools of classic Indian philosophy: Purva Mimamsa, Vedanta, Sankhya, Yoga, Nyaya and Vaisesika. The ultimate origin of all these schools is the Vedas, a corpus of hymns composed between 4000 and 6000 B.C. In true mythic fashion, the Vedic hymns are addressed to the deities ruling the various forces of nature. But there is also a more metaphysical side to their teachings, in that the gods are understood as expressions of a deeper, impersonal order of being. And this impersonal order of being is itself understood as an expression of a yet deeper realm of pure formlessness. This realm of formlessness is *Brahman* — or, in its individual aspect, *Atman*.

For the moment, I'll restrict discussion to a single, central part of Vedantic philosophy: the doctrine of the five sheaths or *koshas*. These sheaths are supposed to be understood as covers, obscuring ultimate being. Removing one level reveals the next higher level, and brings one closer to the center.

The great mystic Sri Aurobindo, in the early part of this century, explained and re-interpreted the Vedantic *koshas* in a way which is particularly relevant here (Aurobindo et al, 2001). Sri Aurobindo took each of the *koshas* and associated it with a certain type of *mental process*, a certain kind of inner experience.

The lowest level is *annamaya kosha*, the food sheath, which Sri Aurobindo associates with the physical mind, or sense-mind. This is the level of thought about physical circumstances, immediate surroundings.

Next comes *pranamaya kosha*, the energy sheath, which Sri Aurobindo calls the life-mind or vital mind. This level of being is associated with the breath, the *prana*, and the fundamental life-force. It is also associated with the feelings, the emotions.

Then comes *manomaya kosha*, the mental sheath. This represents inventive, creative thought: the making of novel connections, the combination of ideas. Some people may go through their lives and hardly ever encounter this level of being. However, all creative individuals spend much of their time here.

Vijnanamaya kosha, the intellect, represents higher *intuitive* thought. It is not experienced by all creative people, but rather represents a higher order of insight. When a person experiences a work of art or an idea popping into their mind full-blown, without explicit effort or fore-thought, as though it has come from "on high" — this is *visnanamaya kosha*, or true creative inspiration.

Finally, *anandamaya kosha*, the sheath of bliss, represents the "causal world." It is what the Greeks called the *Logos*; the source of abstract, mythical, archetypal forms. The forms in the *Logos*, the sheath of bliss, are too general and too nebulous to be fully captured as creative inspirations. They extend out inall directions, soaking through the universe, revealing the underlying interdependence of all things.

Beyond the sheath of bliss, finally, there is only the Self or *Atman*: pure, unadulterated being, which cannot be described. It is absolute simplicity.

I am not a Sanskrit scholar and will not attempt to do justice to the detailed meanings of the Vedantic terms. Aurobindo's slant on the terms is somewhat different from the traditional Vedantic view anyway. However, the Vedantic hierarchy serves here as a motivation for the hierarchy that I will consider, which is at least loosely Vedantic in nature and has seven levels:

1. Quanta (Quantum Reality)
2. World (Everyday Physical Reality)
3. Body
4. Mind
5. Intuition
6. Bliss (Transpersonal Reality)
7. Being (Ultimate Reality)

The level of everyday physical reality corresponds to *annamaya* — it is the everyday physical world, the "solid" world that we normally interact with. Body reality is *pranamaya,* the world of our feelings: pains, pleasures, hungers, desires, sex drives, and so forth. Mental reality, *manomaya*, is the domain of reason: it is what distinguishes humans from ordinary animals. Intuitive reality, *vignanamaya*, is beyond reason and represents the higher reaches of the individual mind. Transpersonal reality — *anandamaya*, the Realm of Bliss — is the spiritual realm in which boundaries between individuals are felt to dissolve.

The lowest level in my hierarchy, Quantum Reality, was not known to the ancient Indians or any pre-modern culture. Tongue-in-cheek, one might call it *quantum-maya*. It is a striking fact that quantum reality bears a fair amount of intuitive resemblance to transpersonal reality. Chapter 8 is devoted to an in-depth discussion of the philosophical implications of quantum physics so I won't digress too much on the subject here. But suffice it to say that: In the quantum world of submicroscopic particles, the boundaries between individual entities dissolve and distant entities are subtly interconnected. Modern theories of physics are focused on the emergence of particles and space-time out of nothingness, out of pure Being — in much the way that spiritual traditions envision transpersonal forms to emerge out of pure Being. This direct connection between the lowest part of the hierarchy and the highest part of the hierarchy indicates the fundamental incompleteness of the hierarchical map of the universe. Although hierarchy is a significant pattern in the universe, it does not capture *all* the structure in the universe — no one pattern does. Numerous heterarchical connections exist within each level of the hierarchy, and spanning the different levels. The nature of the heterarchical connections is to bind together forms and processes that are related to each other, and that mutually produce each other.

Science versus Spiritual Wisdom

I have found the hierarchy of being somewhat useful for understanding the relation between science and spirituality. In hierarchical terms, the key difference between wisdom and science has to do with the *direction of information flow* through the hierarchy of being. Science is dominated by bottom-up flow, emanating from the physical level, while wisdom is dominated by top-down flow, emanating directly from the higher Self.[15]

In wisdom traditions, the main source of knowledge is inner experience. Those with the most advanced inner experience are trusted as gurus or sages. Ideas about the external world — the lowest level of being — are formulated in accordance with the higher intuitions brought down by the spiritually advanced. The higher levels of being are given precedence, and the most important criterion for judging an idea is, not its empirical validity, but rather *how well it fits in with spiritual, intuitive insights.*

In science, on the other hand, the focus is generally on the lowest level of being, the physical level. Interference by the emotional level, associated with the level of *pranamaya* and the lower parts of *manomaya*, is strongly discouraged. The bottom line is always *physical, empirical data.* Higher levels of being are of course intimately involved in the conception of scientific theories. Every scientist works on the level of *manomaya* (Mind), and some scientists, including many of the greatest, have had regular experiences of higher Intuition or even transpersonal Bliss (*anandamaya*). However, the ultimate justification is always found on the *annamaya*, physical world level; the other levels are considered merely as sources of inspiration. The strength of science is precisely its focus on the decisive importance of physical data. Even in quantum theory, no matter how bizarre the nature of quantum reality is, everything "bottoms out" in the behavior of lab instruments that inhabit the ordinary, everyday world.

This point is worth enlarging on for a few paragraphs. According to Feynman (1997): "The first principle of science is not to fool yourself. And you're the easiest person to fool." This quote highlights the main function of empiricism in science: to overcome emotional thought-complexes in the minds of scientists. Scientists, because of their culture of empiricism, admit their

---

[15] The brief comments in this section fail to do justice to the subtlety of either scientific or spiritual inquiry. Chapter 13, later on, is devoted entirely to the philosophy of science, and there I enlarge on these remarks and expand them in entirely new directions.

errors far more often than individuals engaged in any other human endeavor. One does not often see a philosopher, politician or religious person, who has devoted years of their life to a particular point of view, suddenly change their mind and adopt an opposite perspective. However, this is not all that uncommon in science; in fact it is expected. Once the facts have shown a theory to be inadequate, the good scientist will put the theory aside, in spite of their emotional attachment to it.

This kind of transcendence of emotional attachment to scientific theories doesn't *always* happen of course – science is far from a truly objective enterprise. (And I wouldn't want it to be fully objective in all regards – but in some ways I do think it could use more objectivity; e.g. it could use more open and less biased mechanisms for the evaluation of new ideas.) There's a famous saying that old scientific theories don't get disproved, their proponents just die, and the new young scientists look at things in new ways. Scientists can be incredibly dogmatic and narrowminded – but even so, they do admit their erroneous ideas a lot more readily than philosophers or theologians. I've been extremely frustrated with the nature of science – time and time again I've had the experience of creating or advocating a better approach to some scientific problem, and having an extremely hard time "selling" it to scientists because of their attachment to their existing way of thinking. But yet, I am always confident that when I make a sufficiently convincing case for a few years on end, everyone but a few crusty old people and egomaniacal cretins will understand the point and change their thinking. For example, I've spent a lot of time recently trying to convince biologists to use some new techniques my colleagues and I developed for analyzing a particular kind of biological data called "microarray data." It's very, very hard to get most biologists to look at their data in the (to them) unusual way that this method requires. I'm frustrated that it's so hard to make them see the obvious value that these new techniques will add to their work. But yet I know that if I do a bunch of analysis using my new method myself, and publish the results in decent journals (after lots of annoying arguments with journal referees who are phobic regarding new ideas that they don't understand), then eventually (after years or in the worst case decades – I don't want to dismiss the frustration of the process!) enough young scientists will read the papers and get excited about the work that the new techniques will become commonplace. This process often takes much longer than it should, with the result that science advances far more slowly than it should – but the process does eventually function; science progresses and changes and improves its ideas based on a process of dialogic interaction among the minds of scientists and between scientists and the physical world.

One often hears, among scientists, sayings such as "You can't let a fact get in the way of a good theory." What this aphorism means is mainly that, in some cases, the higher levels are really more important than the physical level. A scientist who has received an idea from an experience of deep intuition will probably never be induced to give it up. He will keep on pushing and pushing, convinced (rightly or wrongly) that the facts must eventually yield themselves to his guiding vision. The same flexibility that lets dogmatic scientists pooh-pooh novel but reasonably-well-demonstrated ideas in favor of familiar ones, lets adventurous scientists speculatively set aside well-demonstrated and familiar ideas in favor of exciting but tentative, intuitively-bolstered possibilities.

The reticence of scientists to accept Einstein's special theory of relativity is a good example of the dogmatic side of science. Many older scientists never accepted relativity theory, despite the convincing nature of the data. It went against their belief systems. They were ignoring the information filtering up from World to Mind, in favor of emotional thought-complexes resident at the Mind and Body levels.

On the other hand, the overwhelming and rapid acceptance of quantum theory, at around the same time, speaks in favor of the self-correcting nature of science. Here the quantity and quality of data was so tremendous that no one could deny it. Conservative scientists consoled themselves with the thought that quantum physics, with all its bizarre aspects, would someday be supplanted by a theory that made more sense to them. That is, they did not accept quantum theory as *absolute truth*. But they accepted it as an outstanding scientific theory.

Even where the issues involved border on the philosophical, scientists have an impressive ability to change their minds based on new data. Consider, for instance, the question of the origin of the universe. In the middle of this century, there were two leading theories. There was the Big Bang theory, which proposed that the universe began at some point in the finite past, exploding out of an initial singularity. And there was the steady state theory, which argued that the universe had always existed, and would always exist; and that matter was continuously created out of empty space. Disputes between Big Bang and steady state theorists were lengthy, arcane and sometimes acrimonious. However, the discovery of the cosmic background radiation, as predicted by Big Bang theory, put an end to the matter. Even the staunchest steady state theorists were forced to admit defeat — to conclude that the idea to which they had devoted their lives was incorrect.

Steady state theorist Dennis Sciama expressed a common feeling when he said: "For me, the loss of the steady-state theory has been a source of great sadness. The steady-state theory has a sweep and beauty that for some unaccountable reason the architect of the universe appears to have overlooked. The universe is in fact a botched job, but I suppose we shall have to make the best of it." This quote from Sciama is beautifully demonstrative of the power of science. Scientists, under force of compelling data, are able to put aside belief systems that are deeply important to them, emotionally and intellectually. Science, in this sense, is relatively non-dogmatic. It is impressively self-correcting. Very, very few philosophers have ever concluded: "I am forced to admit that my philosophy was wrong. The universe is botched; it's not the way I wanted it to be." Ludwig Wittgenstein comes to mind as a partial example — but he is the exception that proves the rule. Science has a rigor that overpowers belief systems, even belief systems that science, itself, has set up.

So – getting back to our quasi-Vedantic perspective — hard science focuses on the bottom levels of the hierarchy of being; traditional spiritual wisdom focuses on the top. And what about the middle? A moment's thought indicates that the middle is occupied by the "softer" sciences: psychology, sociology and anthropology. In this book, my primary focus is on individual minds, and so I will spend most of my time at the middle levels of the hierarchy – looking at structures and dynamics that bridge the experiential and physical realms.

Psychology and the Hierarchy of Being

I just said that psychology focuses on the middle levels of the hierarchy of being – but while this is largely true, it is just an approximation; in fact the interplay between levels of being in the discipline of psychology is highly complex. On the one hand, there are areas of psychology such as behaviorism, and neuroscience, which are entirely focused on the physical world. Pursued narrowly, these parts of psychology can be revoltingly reductionist. For example, some behaviorists have taught that the mind does not exist: that the only reality humans have is on the *annamaya*, physical level. And some neuroscientists teach that mind is brain: that what we *are*, fundamentally, is a lump of neurons, seething with chemicals. These points of view are widely recognized as absurdly oversimplified by other psychologists.

There are other areas of psychology, such as cognitive and personality psychology, which are far more accepting of different orders of experience. These areas of psychology are based on constructing abstract *models* of thought and feeling. These models are evaluated based on empirical data, but also based on *inner intuition*, i.e. based on direct comparison with human experiences on the Body and Mind levels of being. This aspect of mental modeling is not often explicitly emphasized: it is said that a certain mental model is "elegant" or "natural," rather than that it matches up with personal experience. But there is no doubt that, in practice, comparison with inner experience is the major factor guiding the construction and judgement of abstract psychological models.

Finally, there is transpersonal psychology, the psychology of spiritual experience. Here one has abstract models that pertain to the higher levels of being, as well as just feeling and thinking. These models are judged based on a combination of empirical, physical evidence, and comparison with direct experience of the higher levels of being, Intuition and Bliss. At the present time, transpersonal psychology is not a very well developed area of research. But one can expect to hear much more from it in the future.

Transpersonal psychology intersects with the study of creative inspiration and intuition – the process by which fully-formed ideas can simply pop into the mind, whole and beautiful.... I will have more to say about this in later chapters. Nietzsche, in *Ecce Homo*, gave an amazing description of this process:

> Has anyone at the end of the eighteenth century a clear idea of what poets of strong ages have called **inspiration**? If not, I will describe it. — If one has the slightest residue of superstition left in one's system, one could hardly resist altogether the idea that one is merely incarnation, merely mouthpiece, merely a medium of overpowering forces. The concept of revelation — in the sense that suddenly, with indescribable certainty and subtlety, something becomes **visible**, audible, something that shakes one to the last depths and throws one down — that merely describes the facts. One hears, one does not seek; one accepts, one does not ask who gives; like lightning, a thought flashes up, with necessity, without hesitation regarding its form — I never had any choice.
>
> A rapture whose tremendous tension occasionally discharges itself in a flood of tears — now the pace quickens involuntarily, now it becomes slow; one is altogether beside onself, with the distinct consciousness of subtle shudders and one's skin creeping down to one's toes; a depth of happiness in which evern what is painful and gloomy does not seem something opposite but rather conditioned, provoked, a **necessary** color in such a superabundance of light...
>
> Everything happens involuntarily in the highest degree but as in a gale a feeling of freedom, of absoluteness, of power, of divinity. — The involuntariness of image and metaphor is strangest of all; one no longer has any notion of what is an image or metaphor: everything offers itself as the nearest, most obvious, simplest expression.

In a sense, pattern-dynamic phenomena like this lie outside the scope of ordinary human comprehension and consciousness. And yet, we may study experiences like this – even wild, overpowering ones like Nietzsche's — as patterns in ordinary mental processes, looking at the mechanisms that give rise to them. Some specific computational models underlying intuition will be reviewed in Chapter 18. Intuitive patterns come into the mind and present it with new structures which were not immediately predictable based on the structures already there – thus they appear to have "emerged from above," as if by divine inspiration. And yet, we know from the science of complex systems that when you place a large number of components together, sometimes collective dynamics and structures emerge that weren't transparently predictable from the nature of the components – any more than the properties of water are transparently, simply predictable from the molecular structure of bonded hydrogen and oxygen molecules.

My own work in theoretical psychology has involved the construction of a very general mathematical and conceptual model of the structure of mind, which is based on patternist ideas and which covers all the different levels of mind (which I'm here discussing using the Vedantic hierarchy as a communicative instrument). In some prior writings (Goertzel, 1997, 2001) I have called my patternist approach to mind the "*psynet model*"; here I will not use that terminology much, but in a number of chapters to follow I will review the key ideas underlying what I've previously called the "psynet model of mind". The psynet-model/patternist-perspective-on-mind portrays the mind as a network of interacting, intercreating pattern/processes, giving rise to emergent, large-scale structures, such as an hierarchical perception-control structure, and an heterarchical associative memory structure. Thoughts, feelings, perceptions, actions and intuitions, are viewed as "attractors" or self-organizing subsystems of this emergent, multiply-structured system. The Vedantic "realm of bliss" is viewed as a shifting pool of emergent patterns, sometimes crystallizing into coherent attractors, sometimes not, continually generating deep structural and dynamical novelty.

The patternist perspective on mind fits very neatly into the hierarchical view of the universe. It views the mind as a collection of *emergent patterns*, patterns emergent in the brain and related body systems. It speaks largely of *manomaya*, and the emergence of *manomaya* from *pranamaya*, but also speaks of the other levels as well, to an extent. Rather than reducing the one level to the other, it understands the nature of the hierarchy of emergence.

The patternist perspective accounts – in a general way — for a variety of neurobiological and psychological data; and it also fits in very naturally with the more general map of the world being presented here. The individual mind, which is the middle levels of the hierarchy of being, is in fact a sort of microscopic image of the entire hierarchy. And for this reason, the same concepts that have been useful for understanding the structure of the individual mind are also useful for understanding the structure of the universe as a whole.

A metaphor that largely captures the potential role of psychology in the study of the universe is the "Great Smoky Dragon." A Great Smoky Dragon is an animal whose middle section is entirely obscured by smoke. Its tail is sharp and clear, and its mouth vividly apparent, but no one can see what's in between. Through science, we can see one end of the universe — the "bottom end," where mind interfaces with brain function and easily replicable behaviors. Through spiritual practice, we can see the other end of the universe — the "top end," where mind loses its individuality and merges with the All. But the middle, the essence of ordinary human mind and experience, remains obscure. Whether we humans can pursue psychology well enough to reveal that "middle section" with any real thoroughness remains to be seen – at any rate, I've given it a shot.

Buddhist Psychology

An interesting twist on the relation between psychology and the Vedantic hierarchy is given by Buddhist psychology – a huge and deep field of inquiry that I'll only mention very loosely and briefly here. One of my favorite psychology/philosophy books is a strange, obscure two-volume set called "Buddhist Logic," a treatment by the Russian scholar Th. Stcherbatsky of the work of the medieval Buddhist logicians Dharmakirti and Dignaga (Stcherbatsky, 1958). Dharmakirti taught the desirability of clear, perfect perception, which he defined as perception without preconception and error. And he enlarged in great depth on the nature of preconception – by which he meant, not only the obvious conceptual and linguistic preconceptions, but also the stream of "false images" coming out of memory that are falsely taken to be real entities. Just as the shadow of a tree can be mistaken for a snake, so a person's perception of another person's face is an unholy combination of real percepts with memory-based inventions. Modern experimental psychology has amply validated Dharmakirti's insight that most perception is full of memory-based confabulation (Rosenfield, 1989).

Dharmakirti's notion of "perception" was not as narrow as one might think from construing this term in the context of modern psychology. In his view, perception through the five senses is only one kind of perception, and not the most important. There are also types of perception corresponding to the upper levels of the cosmic hierarchy. There is perception by the mind, encompassing the objects of dreams and contemplation. There is deep, inner self-consciousness. And there is the self-transcending perception of the accomplished meditator. Perception on all these levels may be clouded by errors and preconceptions: for Dharmakirti the goal of a Buddhist must be to have his mind clearly perceive on all the different levels.

In the accurate, perfect, error-free world alluded to by Dharmakirti, there is no continuous flow of time – there are only instants. Duration is created via the coarseness of the perceiving mind, which blurs instants together. There are no coherent objects or selves – these also are created via the coarse action of the perceiving mind, which carries out inferences generalizing from individual percepts to general classes of percepts, and then uses these generalizations to blur and bias its perceptions of the individual percepts themselves. When objects of perception are stripped of all mental accretions, they are *paramarthasat,* absolutely real.

When I first tried to interpret these ideas in the language of pattern-theory, it seemed to me that Dharmakirti was portraying patterns as instruments of delusion. In his theory, raw percepts from multiple levels of the cosmic hierarchy are viewed as the only true thing – and the inferring, remembering mind, which projects its patterns on these percepts, is the source of the illusions that distort our perceptions and keep us from enlightenment. But one can also take a different perspective. The patterns that make up the universe are real, and perceiving these patterns and their reality is part of the true perception. What is required is to recognize these patterns as what they are – as patterns. Seeing a woman's face talking as a woman's face talking is not wrong – even though this is a construct created by the mind based on inference and memory, even though the spatial and temporal continuity here is "illusory" and each expression is distorted in observation by prior knowledge of what she has looked like in prior contexts. What is wrong is not explicitly realizing the role that inference and memory are playing here. The correct thing is to see all the patterns that make up our hierarchical/heterarchical universe, and to explicitly see them as patterns, rather than misperceiving them as something else (objects, solid selves, realities, etc.).

This is basically the same point made by the classic Zen parable: "The novice perceives mountains as mountains and men as men; then after some spiritual advancement, he sees that men are mountains and mountains are men – but then after he becomes a master he see that mountains are mountains and men are men after all."[16] Here we may say: "The novice perceives mountains and men as mountains and men. Then, after some study, he realizes that these are just delusions, and he sees instead swarms of unorganized percepts coming from multiple levels of the cosmic

---

[16] A (possibly apocryphal) twist on the parable: After telling this, the Zen master Suzuki was asked "What is the difference between before & after?" His reply: "No difference — only the feet are a little bit off the ground."

hierarchy. And then once he achieves mastery, he sees mountains and men and at the same time understands that these are swarms of unorganized percepts coming from multiple levels of the cosmic hierarchy." Or, more compactly: "The novice perceives patterns as if they were realities; after more study, he sees that they are just subjective ways of organizing a fundamentally disorganized swarm of percepts; then once achieving mastery, he sees that these patterns have a basic reality at the same time as being defined as subjective ways of organizing a fundamentally disorganized swarm of percepts."

Of course, these more abstract formulations lack the poetry and beauty of the original parable about men and mountains – but so do the theoretic formulations of Dharmakirti and Dignaga, which are much more intricate, complex and obscure than the ideas I present in this book.[17] Zen Buddhism is focused on leading the individual to Enlightenment via the cessation of cognition and the primal experience of oneness/noneness. On the other hand Dharmakirti was working within a strain of Buddhism in which intellectual understanding of the nature of the universe is understood as a potentially key part of the path to Enlightenment. His formulations were aimed at leading the intellect to a correct understanding of the universe, with the intention that the rest of the multi-level mind can follow the intellect down the correct path. Relatedly, my formulations in this chapter are aimed at integrating the core insights of spiritual wisdom traditions into my overall patternist perspective – which is intellect-centered but goes beyond the restricted domain of the intellect.

The Universal Mind

I now return to the theme of explicitly unifying the scientific and spiritual perspectives on the world, within the patternist framework.

In the Perennial Philosophy perspective, the individual, embodied mind is just a limited segment of the universe – and so is the physical world. This is a very different perspective from the typical scientific empiricist view, in which the physical world is what's "real," and mind is to be defined based on relationships among physical entities. Summing up the ideas of the past few pages, I will now loosely sketch a view that unifies both of these perspectives, based on the notion that pattern is foundational.

From this patternist perspective, it's interesting to look at minds as patterns in physical systems. On the other hand, it's also interesting to look at the "physical universe" as a pattern collectively constructed by minds to help them understand their experience together. And the Vedantic hierarchy, as well, may be viewed from a patternist perspective. Each level in the Vedantic hierarchy may be viewed as a collection of patterns. Each level in the Vedantic hierarchy consists of patterns that may be viewed as emergent from the layer beneath – or, may be viewed as "realities unto themselves," experienced directly in the manner described by Buddhist psychology, and used as keys for interpreting the levels below.

First, quantum events collectively combine to form large-scale patterns that are interpretable as classical physical phenomena — which then have a fundamentally different character from their quantum underpinnings. On the other hand, classical measuring instruments are the way our human minds (we humans in our manomaya-Mind aspect) detect and define the

---

[17] Much of the beauty I find in Buddhist logic stems from the irony I see in constructing ornate, complex conceptual and literary edifices, improvised around the theme that "nothing is real." If nothing is real, then why bother to write long books? But then why bother to *not* write long books? These paradoxes are unsolvable of course – one must internalize them thoroughly and then move on —

quantum realm. Quantum reality is a collection of patterns that are observed in measuring instruments in ordinary physical reality.

Of course, even though physics tells us that ordinary physical reality may be portrayed as a set of patterns in quantum reality, it's very rare that we concretely perceive it as such. Mostly we conceive hypothetical quantum entities and then conclude that ordinary physical entities can be explained as patterns in these. For instance, we don't really perceive our hands as patterns among huge numbers of electrons, protons, neutrons and so forth – we hypothesize that all these particles exist and that our hand is a pattern of arrangement among them.

The perplexing aspect of the quantum theory of measurement is the fact that, apparently, quantum phenomena are affected in a fundamental way by the act of measurement. However, this means simply that the types of patterns we see when configuring our measuring instruments in a certain way, are different than what we expect based on what we see when we configure our measuring instruments in a different way. So from a patternist perspective, the intuitive bizarreness of the quantum world is interesting but not profound in a cosmic sense. (I'll have much more to say on this in Chapter 8.)

Next, bodily reality is a set of patterns in physical reality – but mostly in the same sense in which physical reality is a pattern in quantum reality, i.e. hypothetically. In experiential practice, we know our bodies as patterns of sense-percepts. Similarly, we know our minds as patterns of sense-percepts; but science helps us understand our minds as patterns of bodily phenomena, e.g. neuronal firings.

And organic bodies, properly configured, give rise to reasoning minds – which then define concepts like "organic bodies."   And ordinary physical reality may also be construed as consequent of the level of reasoning minds, and viewed as an experiential pattern – objects that we see are constructed by our minds as patterns among sensations that we receive. From this point of view, the elementary, unanalyzable things are the sense-percepts pouring into the awareness. This ties into the perspective of Buddhist psychology, which studies the role of consciousness in building patterns and structures and theories out of streams of sense-percepts.

Analytical minds, acting with sufficient freedom and power, give rise to abstract idea-spaces – intuitive realms. And these abstract idea-spaces, through shaping culture and language, have a formative impact on individual minds.

From the perspective of the collective intuition, ordinary physical reality is a social pattern – a "consensus reality" as it's sometimes called. There are a number of borderline phenomena – poltergeists and so forth – which many people aren't sure whether to classify as belonging to physical reality or not. But by and large, we agree that whatever is perceived by a whole bunch of people is probably really there.

Networks of interpenetrating intuitions give rise to (and emerge out of) a loosely-structured "remainder", in which there are no fixed realities and every form immediately subverts and transforms itself as soon as it arises. And this paradoxical, disconnected/continuous "realm of bliss" provides the stream of novel patterns needed to keep the collective intuition-field of more concrete patterns alive and growing.

Finally, in a sense, the highest/lowest level of the "formless void"/Brahman gives rise to all the other levels – including minds that have the capability to write essays and construct thoughts and intuitions trying to grasp it (and never quite succeeding –).

The Vedantic-hierarchical perspective is itself a meta-pattern, an interesting way of interpreting the overall patterning of the patterns in the universe. It's not useful for all purposes, and it doesn't explain everything, but it gives some unique insights, and pushes in some directions where the contemporary scientific view of the world is deficient.

Pattern Theology[18]

Now, I'll cap off this chapter of eccentric spiritualistic explorations by going even further out on a limb, and preaching to you my own peculiar brand of "theology." I said above that I'm a highly non-religious person – but I believe as deeply in patternist philosophy as I am able to believe in anything, and patternist philosophy has led me to a perspective on the universe that can be loosely classed as "theological."

The train of thought that leads to pattern theology begins as follows. We have patterns emerging from patterns emerging from patterns … and individual, embodied minds like human minds live somewhere in the middle of the universal pattern hierarchy. But what about the whole shebang – the whole heterogeneous collection of patterns that we call the cosmos? Can we think of this as a mind as well?

I have defined a mind as the set of patterns associated with a system that achieves complex goals in a complex environment. If we stick with this definition, then can we call the universe as a whole a mind? It's a bit of an existential long-jump, but I believe we can meaningfully do so. It makes a lot of sense to me to think of the universe as a system that tries to solve the problem of understanding itself. Its complex environment is itself, and its goal is complete self-understanding. The mind of the universe is then all the patterns in the universe!

Of course, calling self-understanding the "goal" of the universe begs some semantic questions. I don't mean by this to presume that the universe has a coherent, unitary self in the same approximative, yet meaningful sense that a human being does. Goals can be implicit or explicit – if a system acts as though it's trying to optimize some function, then we may say it has an implicit goal.

By "self-understanding" I mean nothing less than: Every part of the universe understands every part of the universe completely. This would be complete and utter self-understanding.

Now, the universe is pretty far from achieving this goal (a statement which is, in essence, just a complicated way of restating the classical "problem of evil"). How do we explain this? My personal favorite explanation is: *The universe just isn't all that smart!* Just because the universe is a mind doesn't mean it's an all-knowing, all-intelligent, AIXI-style mind. The universe is a peculiar sort of mind, quite different from human minds, with different strengths and different weaknesses. This is related to the notion – repeatedly put forth by Nietzsche – that the universe is finite. Of course, the universe could be infinitely computationally powerful and still have limited and finite intelligence, but then it would have to be somewhat perversely structured, since with infinite processing power it's fairly easy to display infinite intelligence. A finite universe implies a finitely intelligent universe implies a universe with necessarily imperfect self-understanding.[19]

Of course there is another explanation for why, if the universe has the goal of complete self-understanding, the level of self-understanding of the universe is currently so weak. The universe may have other goals as well. These goals may be too difficult for us to understand, with our paltry human mind/brains.

---

[18] The original title for this section was "Goertzellian Whacko Theology," but I decided to change it. Of course, the whole notion of "theology" in the context of this book and the philosophy it presents is highly tongue-in-cheek.

[19] And so, as I noted earlier, Bayes' Theorem suggests a finite universe!

I do believe the universe as a whole has a sort of consciousness – but, as will become clear in Chapter 5 when I discuss conscious experience in detail, this is definitely a different flavor of consciousness than the one we humans have.

If the universe is seeking self-understanding, but isn't all that clever, then that means we humans have an interesting potential role. It means we can fundamentally improve the nature of the universe by increasing its level of self-understanding. Of course, this kind of statement begs the question of "free will," which I'll discuss extensively in Chapter 9.

In brief: *a universal mind, conscious but with a different flavor of consciousness than humans, seeking the goal of universal self-understanding but not yet achieving it all that well due to its limited size and somewhat haphazard organization and consequent limited intelligence.* This is what happens when patternist philosophy meets the Perennial Philosophy.

This is not an entirely original vision: the idea of a Supreme Being with limited capabilities has arisen again and again and again throughout history, as a very obvious explanation for the "problem of evil." If you wish, you may view this "pattern-theology" as a simple reformulation of old theological platitudes in the patternist, systems-theoretic language that Ben Goertzel prefers using. However, I think I have added some new twists to the idea, which will become more apparent as successive chapters proceed.

But still, although this theological vision is far from unprecedented, it's definitely not the kind of God that the majority of religious people in the world want. So far as I can tell, this sort of Universal Mind is not very likely to answer your prayers for a new Toyota for Christmas, nor to give your mind a disembodied happy-ever-afterlife up in Heaven with everybody else who went to Church on Sunday and didn't violate too many of the holy dictums. On the other hand, this kind of theology is also not likely to motivate you to slaughter your neighbor, or the guy in the next country over – and rather than leading toward scientifically and intuitively implausible theories regarding the mental and physical world, it is harmonious with the deepest insights of scientific theory as well as spiritual and phenomenological experience.

Chapter 4
Intelligence[20]

I've talked a lot about "mind" so far, in these pages, but haven't yet given a detailed explanation of what I mean by the term. This is because, to me, "mind" is about the intersection of two other concepts, "pattern" and "intelligence." I've already said a bit about pattern so far; now I'll briefly address intelligence, a concept that is much less deep and foundational than pattern, but far from trivial nonetheless.

Of course, these are all ambiguous natural-language terms with a lot of different meanings, and my goal here is not to articulate "the correct meaning" of these or any other English words. Rather, I seek to articulate fairly precisely defined concepts that lie within the broad, fuzzy scopes of these natural language concepts, and then use more rigorous concepts to make specific philosophical points. Some of my colleagues have complained about this strategy — according to which I will, for example, define what I mean by "intelligence" in a manner narrower than the natural-language use of the term, and then proceed to use the term in accordance with my own definition. I have considered alternative strategies, such as coining my own terms ("cybernance" rather than "intelligence" was suggested by my friend and colleague Shane Legg), or using subscripts (e.g. $intelligence_1$, $intelligence_2$... — but it's just so ugly!). After a fair bit of reflection, I decided to stick with the strategy you find here.

The one thing that's most clear about the concept of intelligence is its ambiguity. "Mind" is a concept that many serious thinkers believe literally has no use at all – "intelligence" doesn't have that issue; basically everyone believes that such a thing exists. However, there is not even a conceptual consensus on exactly what it is, let alone on how to define it formally or measure it operationally. Anthropological studies have shown that different human cultures conceive the notion of intelligence in different ways (see Sternberg, 1988, for a good discussion of the issues in this and the following paragraphs). And theoretical psychology has struggled with issues related to the definition and measurement of intelligence, with no clear resolution.

To take just one issue with the measurement of intelligence, consider the problematic relationship between intelligence and genius. The IQ tests typically administered are basically measures of the ability to rapidly solve moderately hard logic puzzles in isolation and under pressure. But, clearly, this ability is not identical to the ability to solve extremely hard puzzles that integrate logical reasoning with other cognitive skills (such as information integration, strategic decision-making, the creation of new sorts of concepts, etc.), in a context of access to other humans and knowledge resources and without a high degree of time pressure or psychological pressure. It should be obvious that the cognitive capabilities required for these two different abilities are overlapping but quite different; and that while the IQ test basically measures

---

[20] The key ideas in this chapter, and some fragments of the text, are drawn from The Structure of Intelligence (Goertzel, 1993).

the former, much real-world intellectual achievement is based on the latter. But quantifying the latter by means of a test seems far more difficult. Any reasonable test of this sort of ability would have to last at least a few months, and it would be very hard to control for "cheating" of various types.

One of the key historical ideas in the psychology of intelligence is the "g factor" – a single numerical value quantifying the amount of "general intelligence" possessed by a person. And one of the ideas underlying the g-factor is that there is a general intelligence capability that underlies both of the abilities mentioned in the previous paragraph (solving moderately hard logic problems in isolation and under time pressure, or solving extremely hard problems that involve an integration of logical reasoning with other types of thinking, in a real world context).

On the other hand, the more recent "multiple intelligences" approach argues against there being any single g factor. The most common version of this approach decomposes intelligence into domain-specific components, e.g. mathematical intelligence, musical intelligence, literary intelligence, etc. (Gardner, 1993). General intelligence is then conceived as a combination of these different, semi-specialized intelligences, but the choice of combination function is acknowledged as arbitrary, and is generally avoided. In this perspective, both IQ-test-taking-ability and "real world extremely-hard-problem solving ability" may be viewed as a result of the intersection of multiple somewhat independent cognitive abilities, only some of which are measured in the IQ test.

There is clearly something to the multiple-components-of-intelligence idea. But even so, the g-factor approach does have some validity, at least within the limited scope of human intelligence. Among humans, there is a significant correlation between IQ and real-world genius — certainly IQ predicts genius better than other quantities such as physical strength, or short-term memory size, or the ability to do rapid arithmetic, or grades in a typical school. (In principle, grades in school could measure genius better than IQ does, but that would occur only if schools were intellectually challenging and students were motivated.). And of course, there is (among humans) a significant correlation between various multiple intelligences: logical intelligence, verbal intelligence and spatial intelligence, for example, are far from identical, but they co-occur much more often than would be expected by chance.

The multiple-intelligence approach could be ramified to an arbitrary degree; e.g. within mathematics it's well known that some mathematicians are smarter at abstract algebra, some are smarter at number theory, some are smarter at topology, etc. But the most common practice is to stop ramifying at the level of a half dozen to a dozen different intelligences, probably just because the human mind is so fond of the "magic number 7 +/- 2" (Miller, 1957). The difference between intelligence and genius may rely on a different sort of decomposition, into multiple intelligences that are defined not by different domains but rather by different cognitive abilities pertinent within a single domain; but this sort of idea has not yet been vigorously pursued within the psychology-of-intelligence literature.

It's very clear qualitatively that some people are much more generally intelligent than others. Yet it's also obvious that some who are not very generally intelligent have a high specialized intelligence in some particular sense; and it's also true that to compare the intelligences of two highly intelligent people, it's necessary to do a domain-by-domain and also specialized-ability-by-specialized-ability comparison ("X is better at abstract math and musical improvisation; but Y is better at complicated software programming and music theory"; etc.).

Much of my own thinking about the nature of intelligence has centered on my work in "artificial intelligence." It's only natural that anyone involved in AI should be a bit concerned about what the "I" in "AI" actually means. Some AI enthusiasts have said to me that the main reason AI hasn't yet been achieved is that no one has formulated the goal of AI work – the nature of intelligence – with adequate precision. I don't believe this at all. However, I do think the careful definition of notions of intelligence can be very useful for theoretical AI work.

Most contemporary work in the AI field involves software systems that are intelligent only in extremely narrow domains – such as chess playing, medical diagnosis, car-driving, mining patterns from biological data, etc. In terms of multiple intelligence theory, each of these "narrow-AI" systems possesses only a single extremely narrow form of intelligence – much narrower than any of the standard 7-or-so types of intelligence typically considered in human psychology.

Back in the 1950's, AI pioneer Alan Turing sought to bypass the problem of defining intelligence generally via his famous "Turing test," which states that a software program should be considered intelligent if it can precisely simulate human intelligence (Turing, 1950). Specifically, the Turing Test asks that an AI should be able to fool humans into thinking it's a human, in the context of a text-based conversational interchange." I think this criterion makes an excellent philosophical point, yet is not likely to be all that useful in practice. Turing was confronted with people who believed AI was impossible, and he wanted to prove the existence of an intelligence test for computer programs. He wanted to make the point that intelligence is defined by behavior rather than by mystical qualities, so that if a program could act like a human it should be considered as intelligent as a human. This is all very well, but in my view the development of AI is not that likely to follow a pathway in which AI's that strictly imitate humans come before AI's that combine refined self-awareness, radically superhuman intelligent capabilities, and a fundamentally nonhuman nature. Most likely, I think, the first obviously intelligent and self-aware AI's are going to drastically exceed humans in some dimensions (some components of "multiple intelligence") yet lag behind humans in others. My guess is that, by the time we have AI's that can pass the Turing test, there will already be AI's (perhaps human-imitating, perhaps not) that all reasonable people agree are highly intelligent and self-aware.

In the remainder of this chapter, I'll propose what I think is a sensible general way to approach the task of defining and measuring a system's intelligence. Most of the ideas given here are refinements of ideas I presented in 1993 in my first book, *The Structure of Intelligence*. The refinements have occurred mainly as a consequence of the years of practical and theoretical AI work I've experienced since writing that book. But even though my thinking about intelligence has evolved largely from an AI perspective, my conclusions are intended to possess general validity, applicable to any kind of complex system, not just intelligent computers.

Defining Mind

The key theme running through this book is to take pattern as a foundational concept – to view the universe as a network or system composed of multiple overlapping and interacting patterns. In line with this approach, it is natural to define mind in terms of pattern. But the best way I've found to do this so far seems to be to introduce an intermediate notion to intervene between pattern and mind – namely, the notion of "intelligence." Mind may be defined in terms of pattern and intelligence; and intelligence may then be defined in terms of pattern.

A mind, in my vocabulary, is a fuzzy set of patterns – in particular, the set of patterns associated with the intelligent behaviors of some system. By "associated with" the intelligent behaviors of some system, I mean patterns that are patterns *in* that system when it's carrying out intelligent behaviors or patterns emergent *between* that system and something else when it's carrying out intelligent behaviors. The fuzziness comes from the fact that some patterns are more intense patterns than others, and some patterns are more closely associated with the system's intelligent behavior than others – and, finally, from the fact that some behaviors are more intelligent than others. The maximum mind-ness patterns are those that are very intense patterns in a system, and very closely associated to that system's very intelligent behavior. This definition implies that, once one has figured out how to define intelligence, one can use this to define mind as well.

It's possible to mathematically quantify this definition of mind, and I've done so in prior publications as well as in an Appendix to this book, but there's not all much *immediate* value to this exercise other than to show it can be done or to point the way toward possible future science, because at this stage there's no way to calculate the mind of a system according to the sort of definitions one can formulate using contemporary scientific and mathematical language. What makes calculation difficult is the fact that recognizing all the patterns in a system is totally computationally intractable. It's an "uncomputable" problem according to standard formalizations of computation; and theory aside, in practical situations it's an unapproachably difficult problem. One can of course restrict the set of patterns under consideration (for instance to the set of patterns recognizable by some particular "observer system") and then look at the mind of one system relative to another observer system. This may be of some practical interest in the future when studying AI's, but for the moment AI "minds" (which barely deserve the name in any currently operational "AI" system) are sufficiently primitive that a qualitative understanding is sufficient for analyzing and engineering them.

As warned above, this is certainly not a comprehensive definition of the commonsense notion of "mind" in all its glorious and various senses. Eastern philosophy, for instance, tends to use notions related to "mind"[21] to refer to basically anything in the universe – any pattern or set of patterns, to use my own lingo. As should be evident from Chapter 3 above, I don't reject this usage, but it's not the emphasis I want to give to the "mind" concept. Saying "everything is mind" makes an important philosophical point (as in Peirce's proclamation "Matter is but mind hide-bound with habit") but also blurs the distinction between chimps and sea slugs – there's an important sense in which the former have more on their minds than the latter. My approach to defining "mind" agrees that everything is mind, but follows this up with the caveat that "some things are more mind-ful than others." Since some systems are more intelligent than others, it follows that some systems have more mind than others. The set of patterns associated with a sea slug is a mind, but its degree of mindness is much less than that of the set of patterns associated with a chimp.

A Working Definition of Intelligence

Now, getting to the meat of the chapter, I'll briefly state my own working definition of intelligence. My view is that intelligence is best understood as follows:

Intelligence is the ability to achieve complex goals in complex environments

The greater the total complexity of the set of goals that the organism can achieve, the more intelligent it is.

---

[21] Obviously, the various Sanskrit, Chinese, etc. terms used to denote "concepts related to mind" in Eastern philosophy are tough to translate into any English words. In spite of this difficulty, however, I think the point made in the text is valid.

Recall that I conceive complexity in terms of pattern: the complexity of something is the total amount of pattern in that thing. So this definition grounds the concept of intelligence in the concept of pattern, in accordance with the overall pattern-philosophy approach.

This is a subjective rather than objective view of intelligence, because it relies on the subjective identification of what is and is not a complex goal or a complex environment – which means that a system may potentially be unintelligent with respect to our own subjective measures, yet highly intelligent with respect to some other measure (e.g. its own).

From a pragmatic perspective, it seems that this definition of intelligence is most valuable for comparing different entities of like kind — i.e., different entities sharing largely the same goals, and comfortable in largely the same environments. This ties in with a more general and fairly obvious conceptual point — that if one were to formulate an IQ test for intelligent machines, it could not be used to compare Novamentes to humans or even to radically different types of AI programs. Really meaningful quantitative comparisons of intelligence, I believe, can only be done between similar systems. This is what Sternberg (1989) refers to as the *contextual* aspect of intelligence. In the patternist perspective, the context-dependence of intelligence is built right into the definition: it is unavoidable.

Another formalization of the notion of intelligence that I've found valuable is the one provided by my friend and sometime research collaborator Pei Wang[22] (1995, 2005, 2006). Pei posits, basically, that:

**Intelligence is the ability to work and adapt to the environment
with insufficient knowledge and resources.**

More concretely, he believes that an intelligent system is one that works under the Assumption of Insufficient Knowledge and Resources (AIKR), meaning that the system must be, at the same time:

- A finite system —- the system's computing power, as well as its working and storage space, is limited;
- A real-time system —- the tasks that the system has to process, including the assimilation of new knowledge and the making of decisions, can arrive at any time, and all have deadlines attached with them;
- An ampliative system —- the system not only can retrieve available knowledge and derive sound conclusions from it, but also can make refutable hypotheses and guesses based on it when no certain conclusion can be drawn;
- An open system —- no restriction is imposed on the relationship between old knowledge and new knowledge, as long as they are representable in the system's interface language;
- A self-organized system —- the system can accommodate itself to new knowledge, and adjust its memory structure and mechanism to improve its time and space efficiency, under the assumption that future situations will be similar to past situations.

---

[22] Although Pei and I have done a fair bit of collaborative AI research over the years, we have considerably different perspectives on a number of key issues in AI theory; and he has his own NARS AI system that is quite different from my Novamente approach.

It should be obvious that Pei's and my own definitions have a close relationship, in spite of their different orientations.

My "complex goals in complex environments" definition is purely behavioral: it doesn't specify any particular experiences or structures or processes as characteristic of intelligent systems. One could Pei Wang-ify it, by declaring that

Intelligence is the ability to achieve complex goals in a complex environment, using severely limited resources

However, I'm not sure that this is well-aligned with the intuitive notion of intelligence. Rather, it seems preferable to define a new concept, "efficient intelligence," as

Efficient Intelligence is the ability to achieve complex goals in a complex environment, using severely limited resources

Or equivalently:

Efficient intelligence is the ability to achieve intelligence using severely limited resources

This way, we can still say that a machine with an IQ of 500 running on 5000 machines is more intelligent than a machine with an IQ of 100 that runs on one machine. But we can also state in what sense the second machine is more intelligent: it has greater *efficient intelligence*.

The notion of efficient intelligence does not incorporate all the details of Wang's definition of intelligence. However, my choice not to incorporate these details into my working definition of intelligence shouldn't be taken to imply that I don't think they're important for the study of intelligent systems. Rather, I am just not sure that they should all be part of the *definition* of intelligence. As noted above, I suspect that certain structures and processes and experiences are necessary aspects of any sufficiently intelligent system. Perhaps the science of 2050 will contain laws of the form: *Any sufficiently intelligent system has got to have this list of structures and has got to manifest this list of processes.*

A full science along these lines is not, in my view, necessary for understanding how to design an intelligent system. But we do need some ideas along these lines in order to proceed toward human-level artificial general intelligence today, and Pei's definition of intelligence is a step in this direction. He posits that, for a real physical system to achieve complex goals in complex environments, it has got to be finite, real-time, ampliative and self-organized – and we suspect that this hypothesis is true. It might well be possible to prove this mathematically, but this is not the direction I have taken in my own research; instead I have taken this much to be clear and directed the bulk of my efforts toward more concrete tasks. I suspect that concrete AI work, as it progresses, is likely to lead to ample conceptual progress in the area of understanding the relationships between these concepts and general intelligence as here conceptualized.

What Are The Goals?

Everything I've said about intelligence has referred to some "complex goals" – but what are these goals? Many of the peculiar qualities of human intelligence have to do with the particular goals that drive human organisms; and of course, different software-embodied goal sets will lead to dramatically different AI's.

In the case of biological intelligence, the key goals have to do with the tight association that exists between minds and particular embodiments. These include the survival of the organism embodying the mind and its DNA (the latter represented by the organism's offspring and its relatives). These lead to sub-goals like reproductive success, status amongst one's peers, etc., which lead to refined cultural sub-goals like career success, intellectual advancement, and so forth. The external goal of survival gets internalized into personal goals that are believed to enhance survivability, and in social animals, into societal goals that enhance the chances of group survival.

An intelligent computer program, on the other hand, will almost surely emerge into an environment very different from the African veldt where humanity evolved, or any other familiar physical terrain. If powerful AGI is first achieved through robotics, then the analogy between its environment and the human one will be reasonably strong. But it's also quite possible that powerful AGI will first arise via software systems with radically different sorts of embodiments – ones involving simulated worlds, or "worlds" consisting of software agents browsing distributed quantitative and relational databases, etc. In the Novamente project, we are working toward an AGI capable of assuming a variety of embodiments, including a 3D simulation world, plus the textual information and scientific data available on the Internet. We are also exploring the possibility of moving in the direction of physical robotics[23] — but not as an exclusive mode of embodiment.

Human minds are specifically attached to individual embodiments, but this is both a strength and a weakness. It simplifies some aspects of cognition, via providing ready sensorimotor groundings of abstract concepts. But it also leads to a somewhat narrow view of the world compared to what will be possible for digital minds capable of shifting flexibly from one embodiment to another. The goals of an AGI system need not be tied to any particular embodiment, which is a major difference between AGI goals and the most common human goals.

There is also no particular reason why survival of self or offspring should be a particularly important goal to any AGI system. Of course, an AGI that didn't bother to maintain itself would quickly de-exist, and so there will be some natural selection among AGI's, encouraging those that value their ongoing existence and the propagation of others of their kind. However, one could also see the proliferation of AGI's whose goal systems involve values like:

V = "It's good to create other AGI's that are fundamentally and radically different from myself, especially if these other AGI's also share this value V."

This sort of goal system fits into the general framework of natural selection, yet nothing much like it has arisen through natural selection on Earth, and such a thing is very unlikely due to the particular nature of evolution using DNA. Self-propagating AGI goal systems may be drawn from a much larger class than self-propagating biological goal systems.

---

[23] The main reason we aren't pursuing a physical-robotics direction in the Novamente project right from the start is simply financial cost. Doing robotics right, in a manner flexible enough to be truly useful for AGI, would require a lot of expenditure on hardware and on robotics experts to tune and maintain it, to customize perception and action intelligence-components to the particular robotics hardware, etc. All this sort of work is extremely important and worthwhile, but we've judged that it's going to be possible to achieve powerful AGI without it.

The issue of AGI goals quickly becomes quite subtle, evoking some deep issues that will be raised in more depth in later chapters, and others that are not even raised in these pages (but see my recent book The Path to Posthumanity (Goertzel and Bugaj, 2006), which deals extensively with the ethical issues raised by massively superhuman, actively self-modifying AGI systems). What if the goal of an AGI is

G = "To modify myself in such a way as to make myself much cleverer, while still maintaining G as a goal."

or

G = "To modify myself in such a way as to make myself much cleverer and not hurt anybody, while still maintaining G as a goal."

All sorts of issues arise, such as "whose definition of clever? Whose definition of hurt?" The AGI, once it becomes vastly cleverer, may redefine "hurt" or "cleverness" in ways that its predecessors would hate. One can then look at goals like

G = "To modify myself in such a way as to make myself much cleverer and not hurt anybody, while still maintaining G as a goal, and where the terms in this goal are to be defined according to my best interpretation of what my creator meant by them."

In some of my prior writings on the future of AGI, I have explored ethics-oriented goal systems such as "Promote growth, joy and freedom". An AGI with an abstract goal system like this one will surely display a radically different sort of intelligence from one with a more human-like, individual-embodiment-focused goal system.

In the preceding paragraphs about AGI's, I've been speaking as if AGI systems have explicitly formulated goals that read like logical propositions – but, while this may be the case for some AGI systems, there may also be AGI's that don't have any explicitly formulated goals. An AGI, like a human, may be largely ignorant of many of its own major goals. The Novamente architecture is quite flexible and may in principle be used to build AGI's with abstract, explicit high-level goals, or else AGI's without any. The definition of intelligence given above doesn't say anything about how a goal is represented inside an intelligent system, nor about what the intelligent system (consciously, logically, explicitly or whatever) knows about its goals. Rather "achieving complex goals" is supposed to be interpreted as "acts as if it's achieving complex goals." It's intended to be interpreted as a behavioral pattern rather than an internal structural/dynamical pattern – even though, in many cases, behavioral patterns "happen to be" reflected by internal structural/dynamical patterns.

Above I mentioned the possibility that, for a wide variety of complex goals and environments (or even, for any complex goal/environment pair at all), there may be "universal structures and dynamics of intelligence" that are correlated with effective goal-achievement. Patternist philosophy suggests a number of possibilities in this regard: potential universal metapatterns of intelligence, many of which will be reviewed below.

Finally, it may also be that, given two goals of roughly equal complexity, one of them is intrinsically more amenable to intelligence than another – so that some goals are more "intelligence-friendly," depending on factors other than their raw "complexity" in the sense of the amount of pattern that they embody. It may be that some goals embody patterns more harmonious with the universal structures/dynamics of intelligence, and hence are more easily achievable (relative to their complexity) than others. This is an interesting possibility because it gives a way of assessing goals in a semi-objective way, using as a criterion the meta-goal of intelligence.

I'll return to these issues later in the context of moral philosophy – which obviously is all about "what the goals and values are, and/or should be" on the subjective identification of what is or is not a pattern. For example, if dolphins are bad at achieving goals that we think are complex, and operating in environments that we humans think are complexity-laden, this means that they are not very intelligent from our subjective view, but it says nothing about their intelligence from their own view, or from the view of other hypothetical beings with a perspective more comprehensive than that of either humans or dolphins.

Chapter 5
Experience[24]

Perhaps the trickiest issue in the domain of mind-philosophy is the problem of "consciousness," or more specifically the relation between the intuitive feeling of "being there / being aware" and the more physical, mechanistic view in which it's a particular assemblage of physical structures and dynamics that's doing the being there and aware. This leads directly into the issue that I'll discuss in the Chapter 9 – "free will", and the extent to which it's "illusory" versus "real."

In the Vedantic hierarchy discussed in Chapter 3, this "physical versus experiential views of consciousness" issue doesn't exist. One can't identify it with the relationship between the material and body levels (annamaya and pranamaya) and the higher levels (manomaya and vijnanamaya), because in Vedanta all of these are levels of awareness and experience. Basically, Vedanta, like other incarnations of the Perennial Philosophy, takes for granted that consciousness/awareness is the ground of being; the hierarchy of being is a hierarchy of progressively more abstract forms that this being-ground takes.

The issue also doesn't exist in neuroscience or physics or most other branches of science, the main exception being certain parts of psychology, because the scientific theories current in these fields speak of intelligence only from an underlying-mechanisms perspective. From these perspectives, conscious experience simply doesn't exist in any practical sense, because it can't be measured in any objective way.

The really interesting way to address conscious experience, however, is to look at it from both of these perspectives simultaneously. The essence of consciousness, I suggest, can only be seen in the light shed upon it via the intersection and interpenetration of the objective and subjective perspectives. In this chapter I will enlarge upon the relationship between objectivity and subjectivity, and the implications of this relationship for the nature and particularities of conscious awareness.

Subjectivity and Objectivity

So what is the right way to view the relationship between the subjective and objective worlds (or, to phrase it differently, the subjective and objective world-perspectives)?

---

[24] The ideas in this chapter were refined via discussions with Izabela Freire Goertzel, and also in long email dialogues on the SL4 e-mail list. Thanks are due to Izabela and the SL4 membership for prodding me to clarify my ideas on these very thorny topics, and pointing out some of the ambiguities and confusions in my prior writing and thinking on the topic.

In the objective-reality perspective, one views the objective world as defined by science and society as primary, and looks at the subjective worlds of individuals as approximations to objective reality, produced by individual physical systems embedded within physical reality.

In the subjective-reality perspective, on the other hand, one views the subjective, experiential world of the individual world (mine, or yours, for example) as primary, and looks at "objective reality" as a (very useful) cognitive crutch that the experiencing mind creates in order to make use of its own experience.

Both of these views are valid and interesting ones — they each serve valuable purposes. And, I contend, they don't contradict each other — because the universe (by which I mean the pattern-verse, not narrowly the physical universe) supports "circular containment". It's fine to say both "objective reality contains subjective reality, and subjective reality contains objective reality." The theory of non-well-founded sets shows that this kind of circularity is perfectly consistent in terms of logic and mathematics. (Barwise and Etchemendy's book *The Liar* (1988) gives a very nice exposition of this kind of set theory for the semi-technical reader. I also said a lot about this kind of mathematics in my 1994 book *Chaotic Logic*, and will review some of this material in Chapter 12 below as it is critical to the notion of autopoiesis which is the theme there.)

I will argue here that it's much easier to derive the existence of objective reality from the assumption of subjective reality, than vice versa. In this sense, I believe, it's sensible to say that the grounding of objective reality in subjective reality is primary, rather than the other way around.

On the other hand, I will also argue that it's probably easier to derive the details of subjective reality from the details of objective reality than vice versa.

These arguments give no clear answer regarding which one is "more fundamental." Rather, they illustrate the depth and subtlety of this particular circular containment relationship, a theme that will be enlarged upon in other ways as this chapter proceeds.

First, suppose one begins by assuming "subjective reality" exists — the experienced world of oneself, the sensations and thoughts and images and so forth that appear in one's mind and one's perceived world. How can we derive from this subjective reality any notion of "objective reality"?

Consider the example of a mirage in the desert — a lake of water that appears in the distance, but when you walk to its apparent location, all you find is sand.[25] This is a good example of how "objective reality" arises within subjective reality.

There is a rule, learned through experience, that large bodies of water rarely just suddenly disappear. But then, putting the perceived image of a large body of water together with the fact that large bodies rarely disappear, and the fact that when this particular large body of water was approached it was no longer there — something's gotta give. There are at least two hypotheses one can make to explain away this contradiction:

1. One may decide that deserts are populated by a particular type of lake that disappears when you come near it, or
2. One may decide that what one sees from a distance need not agree with what one sees and otherwise senses from close up.

---

[25] As it happens, many of the formulations in this chapter were conceived in conversations with my wife Izabela while backpacking in White Sands National Monument in southern New Mexico – one of the more surreal places on Earth, and a location where mirages are extremely difficult to avoid, especially in the overheated summer.

The latter conclusion turns out to be a much more useful one, because it explains a lot of phenomena besides mirage lakes. Occam's Razor pushes toward the second conclusion, because it gives a simple explanation of many different things, whereas explanations of form 1 are a lot less elegant, since according to this explanatory style, each phenomenon where different sorts of perception disagree with each other requires positing a whole new class of peculiarly-behaving entities.

Note that nothing in the mirage lake or other similar experiences causes one to doubt the veracity of one's experiences. Each experience is valid unto itself. However, the mind generalizes from experiences, and takes particular sensations and cognitions to be elements of more general categories. For instance, it takes a particular arrangement of colors to be a momentary image of a "lake", and it takes the momentary image of a lake to be a snapshot of a persistent object called a "lake." These generalizations/ categorizations are largely learned via experience, because they're statistically valid and useful for achieving subjectively important goals.

From this kind of experience, one learns that, when having a subjective experience, it's intelligent to ask "But the general categories I'm building based on this particular experience — what will my future subjective experiences say about these categories, if I'm experiencing the same categories (e.g. the lake) through different senses, or from different positions, etc.?" And as soon as one starts asking questions like that — there's "objective reality." Objective reality is a pattern that a mind constructs because it provides a useful simplified explanation of the long series of subjectively-perceived moments that it stores in its memory. (At least, this is what objective reality is *from a subjective perspective*.)

That's really all one needs in order to derive objective reality from subjective reality. One doesn't need to invoke a society of minds comparing their subjective worlds, nor any kind of rigorous scientific world-view. One merely needs to posit generalization beyond individual experiences to patterns representing categories of experience, and an Occam's Razor heuristic.

In the mind of the human infant, this kind of reasoning is undertaken pretty early on — within the first nine months of life. It leads to what developmental psychologists call "object permanence" — the recognition that, when a hand passes behind a piece of furniture and then reappears on the other side, it still existed during the interim period when it was behind the furniture. "Existed" here means, roughly, "The most compact and accurate model of my experiences implies that if I were in a different position, I would be able to see or otherwise detect the hand while it was behind the chair, even though in actual fact I can't see or detect it there from my current position." This is analogous to what it means to believe the mirage-lake doesn't exist: "The most compact and accurate model of my experiences implies that if I were standing right where that lake appears to be, I wouldn't be wet!" Notice from these examples how counter-factuality is critical to the emergence of objective from subjective reality. If the mind just sticks to exactly what it experiences, it will never evolve the notion of objective reality. Instead, the mind needs to be able to think "What would I experience if...." This kind of basic counter-factuality leads fairly quickly to the notion of objective reality. And once the notion has arisen, it's bound to stick around — because it's extremely useful for helping to predict what subjective patterns are going to arise in what sequences in the future.

On the other hand, what does one need in order to derive subjective reality from objective reality? This is a lot trickier! Given objective reality as described by modern science, one can build up a theory of particles, atoms, molecules, chemical compounds, cells, organs (like brains) and organisms — and then one can talk about how brains embodied in bodies embedded in societies give rise to individual subjective realities. But this is a much longer and more complicated story than the emergence of objective reality from subjective reality.

Occam's-razor-wise, then, "objective reality emerges from subjective reality" is a much simpler story than the reverse.

But of course, this analysis only scratches the surface. The simple, development-psychology approach I've described above doesn't explain the details of objective reality — it doesn't explain why there are the particular elementary particles and force constants there are, for example. It just explains why objective reality should exist at all.

The notion of deriving the detailed structure and dynamics of our physical universe from facts about subjective reality relates to deep issues of "grand unification physics", specifically John Archibald Wheeler's (1988) idea of "law without law" and "it from bit" —which hypothesizes that the laws of our physical universe are in some sense optimal, so that if one has an objective or physical world with unformed, indefinite laws, eventually the laws will settle into the optimal-law configuration (being the laws of our universe). Wheeler's idea was to assume some kind of pre-geometric, pre-lawful universe, and derive the laws of our universe from this. But the idea makes just as much (or little) sense if one places subjective reality in the place of Wheeler's primordial pre-geometric world.

On the other hand, starting from the assumption of objective reality – and furthermore assuming a complex, appropriately structured objective reality like our own, supporting the formation of complex, self-organizing systems – it seems possible in principle to derive the possibility of intelligent systems like humans and AI's, that will describe themselves as having subjective realities, and whose internal dynamics will be conveniently describable using the notion of "subjective reality." Contemporary science has not completed this derivation, but it seems easier and less far off than actualizing Wheeler's plan in a subjectivity-grounded way.

So we have an interesting asymmetry. While it's easier to explain the existence of objective reality based on subjective reality than vice versa, it seems like it's probably going to be easier to explain the details of subjective reality based on objective reality than vice versa. Admittedly, this statement is largely speculative at the moment, since right now we don't know how to do either — we can't explain particle physics based on subjectivist developmental psychology, but nor can we explain the nature of conscious experience based on brain function. However, my intuition is that the latter is an easier task, though both are probably possible. The provisional conclusion is that:

- At a coarse level of precision, "subjectivity spawns objectivity" is a simpler story than vice versa.
- At a higher level of precision, "objectivity spawns subjectivity" is a simpler story than vice versa.

And the patternist bottom line: Subjective reality may be derived as a pattern in objective reality (in that the assumption of minds having subjective realities may help one to explain the dynamics of intelligent systems within objective reality); and, objective reality may be derived as a pattern in subjective reality (in that the concept of objective reality arises in a mind as a way of organizing its subjective percepts/concepts).

Propositional Reality

An alternate way to think about these issues is, I suggest, to introduce a third kind of reality: propositional reality. By this I simply mean the universe of logical statements, or more properly: logical statements using a vocabulary of logical atoms that includes a set of pragmatic indicatives that is assumed understood by both speaker and listener (minimally, a logical atom serving the role of the indicative "this"). One can think about both objective and subjective reality by associating them with subsets of propositional reality.

Propositions may include, for example, "1+1=2" or "This [indicating a ball] is red" or "This [indicating the speaker] feels sad" or abstract statements like "Red things make people and donkeys feel happy", etc.

One can look at propositional reality as an abstraction of social reality and linguistic interaction. Any discourse or conversation can be viewed as an exchange of logical propositions among human minds. This is not a very useful perspective for us humans to take in most cases: when dealing with perception, actions, emotions or intuitions, there are better approaches to take than logicizing everything. But it has been clear since Leibniz and Boole that this sort of logicization is possible in principle, if one casts aside issues of computational efficiency – and so from the perspective of understanding the in-principle nature of things, it is reasonable to explore the light shed by logicization, while never forgetting the critical and foundational role of computational efficiency in the philosophy of mind and world.

Objective reality, in this perspective, boils down to a set of propositions that all or most of the participants in the dialogue can agree on.

We can also exchange propositions about our internal, subjective worlds – for instance, I can string words together evocatively describing a subjective feeling I've had, and you may read my words and map this approximately into some feeling you've had. So, subjective content also boils down to a set of propositions that multiple participants can understand. And in some cases, it boils down to propositions that multiple participants can agree on, such as, say, "Parting is such sweet sorrow."

We can't say that "parting is such sweet sorrow" is objectively real in the standard sense (thought we might find some neurological correlate of this subjective observation), but we can say that it's a proposition that is mutually comprehensible to members of a proposition-exchanging community, and that appears to very many of these individuals to be true.

This example shows that one of the main things distinguishing objective from subjective reality is that the former consists of propositions that nearly everyone in the dialogic community agrees to be true. Nothing in subjective reality elicits such widespread agreement. "Parting is such sweet sorrow" may be widely agreed on, but not quite as widely as "This rock is hard" or "The thermometer we are all looking at reads 20 at this moment."

In this context, the circular-containment relationship between objective and subjective reality boils down to mutual logical derivability. If we take a large enough set of propositions about an individual's subjective world, then through some complex derivations we should be able to derive important propositions about objective reality – and vice versa. However, these derivations may be long and complex. We may view subjective and objective reality as "islands" in proposition-space, separated by long and difficult chains of derivation.

The propositional point of view certainly has its shortcomings – one may argue that projecting into the space of propositions loses something of the essence of both objective and subjective reality. However, this is largely the same sort of projection we make when we discuss these matters via intellectual discourse like this one. (Though of course, sharing propositions is not the only way for humans to share knowledge: there is wordless transmission of feeling and intuition, and there is purely physical interaction in the context of the objective world.)

One reason I like the propositional reality idea is that it places the idea of scientific validation in a proper (i.e. sociological) perspective. Some people like to say that objective reality is somehow more real than subjective reality because statements about objective reality can be scientifically validated. But what does the scientific method really come down to, in practice? It's a complex thing (which I'll address in Chapter 13), but part of it has to do with the replication of experimental results by different experimenters. In other words, it has to do with the formulation of propositions that multiple individuals will agree are true. In this sense scientific validation is not really so different from the creation of agreeable abstractions regarding subjective experience.

The relation between propositions and patterns has to do with the notion of truth. Most propositions of relevance to humans or other complex systems are not "crisp" but rather "uncertain" – they have truth values that are better modeled as probabilities than as Boolean 0/1 quantities[26]. The probabilistic truth value of a proposition is closely related to the intensity with which that proposition is a pattern in the world. A simple proposition that has proved true over a large body of observations is a way of simplifying the observations, and thus constitutes a pattern in these observations. A body of interrelated logical propositions regarding a body of data may allow one to reproduce most or all of the data within a high degree of approximation, but may be much more compact, and may then constitute a highly significant pattern in the set of data. Minds construct propositions that individually and collectively form patterns in themselves and their environments; and in mutual discourse minds construct propositions that collectively form patterns not present in any of the individual minds participating.

Finally, on reading some notes created in preparation for writing this chapter, my wife Izabela pointed out that one can think about propositional reality as the declarative content of the "social mind" consisting of the collective of humans. I like this observation. In this sense, grounding subjective and objective reality in propositional reality takes on a hierarchical flavor. One can envision a hierarchy of minds, where the subjective and objective realities of minds are defined in terms of the subjective reality of the overmind in which they're contained. This becomes particularly poignant in the case where this overmind is a mindplex and explicitly contributes its own content to the social proposition-space.

The Problem of Qualia

Now I turn from subjectivity/objectivity to a closely related but more difficult topic: the "problem of qualia." I discuss this issue from several views and at the end present a unified summary perspective.

Let me first explain what I think the problem is.

First of all, a "quale" is defined (by dictionary.com) as "a property, such as whiteness, considered independently from things having the property." In discussions of consciousness, the term is normally used to refer to something that is perceived by a conscious, feeling, aware mind – as opposed to something that is conceived as having an absolute existence in an objective world. We experience the whiteness of the wall (this is a quale); but we, as subjectively experiencing

---

[26] This is a fact that was understood by Leibniz in his original formulation of what is now called "Boolean logic", but was not sufficiently emphasized by Boole in his own work.

minds, infer the "objective" existence of the wall, via recognizing patterns of relationship among multiple qualia.

So what is the "problem" of qualia?

Sociologically, the problem is that there are two very distinct camps out there, dividing the set of educated humans who think about these things at all.

Some of us think "qualia" is a meaningful concept, and that it makes sense to think in terms of things like "Ben's subjective experience of whiteness."

Others believe that "qualia" is a nonsensical conceptual construction. They make arguments like "The whole idea of qualia is meaningless, because I can never measure a brain and tell if there are qualia in there or not." Or: "Not being George Bush myself, how could I ever tell the difference between George Bush, and a version of George Bush that had no qualia?"

How can we reconcile these two perspectives?

This sociological distinction reflects a basic conceptual problem, which may be cast as a linguistic problem: How can we meaningfully connect the language of subjective experiences (which we all use on an everyday basis) with the language of empirical observations, in a way that ecognizes each as intrinsically meaningful rather than reducing one to the other?

Clearly there *is* a connection, because when I say "I experience that tree as tall", you or I can then go measure the tree and see if it's tall or not. So, from a subjective point of view, I can detect correlations between my qualia and scientific measurements. But, from a scientific point of view, I can never detect such correlations because there is no such thing as a quale-ometer. I can detect correlations like "There is a link between a tree being tall and people saying they have the experience of perceiving the tree as being tall." But this kind of correlation introduces an unsatisfying level of indirection: one is then dealing with talking rather than with experiencing directly — one is dodging the qualia problem rather than confronting it.

I think the quale-skeptics are correct that qualia are not measurable. However, I don't think this means "qualia" is a meaningless concept. I will give two analogies now to explain the sense in which I think non-measurable entities can be very meaningful.

The first analogy is one that will be discussed extensively in Chapter 8 below: quantum phenomena. Quantum reality contains many phenomena that are known to be unmeasurable, but that scientists still find very useful to talk about. Because positing and analyzing these particular unmeasurables proves very useful for explaining various measurable results.

The second analogy is simpler: time. How can we prove time exists? We can't. The statement that time exists is not a falsifiable hypothesis, because the very concept of doing a scientific experiment involves the notion of time (how can we talk about "replication" of experiments without assuming the existence of time?). Similarly, I suggest, the statement that qualia exist is not a falsifiable hypothesis, but the very concept of doing a scientific experiment involves qualia: we only accept the results of an experiment if we experience them ourselves. "Seeing is believing" basically means that we want to observe the outcome of an experiment on a measuring instrument, in order to definitively accept it. In short, individual experience, like time, is part of the language that is used to define scientific experimentation, rather than something that can be measured via scientific experimentation. Asking "can the existence of qualia be scientifically verified?" like asking "can the existence of time be scientifically verified?", is making a category error.

But as with time and quantum reality, the concept of qualia can be judged in a human sense via its usefulness. Is the concept good at producing interesting, surprising, valuable, ideas? As we'll see in Chapter 13, a careful analysis of the history and philosophy of science reveals that this is ultimately the way scientific research programs are validated anyway.

My view is that qualia are very useful as a way of explaining various aspects of human experience – for instance, I wouldn't want to try to describe an LSD trip or a meditative experience or a love affair without referring to various qualia. On the other hand, they've proved less useful so far for explaining certain aspects of human cognition. But I suspect that is because the theory of qualia (in contrast to e.g. the theory of quantum phenomena) has been extremely poorly developed. With this in mind, in the next section of this chapter, I will sketch some ideas indicating what I think some aspects of a theory of qualia might look like. (Of course, this is a very big topic and I'll barely scratch the surface.)

Pertinent to Chapter 3 above, I note in passing that Buddhist psychology has a lot to say about the nature of qualia. However, its vocabulary is obscure and tends to tangle up descriptive ideas with normative ones. I have been conceptually inspired by Buddhist psychology in my thinking on these topics, but in this instance I prefer to take a fresh start and introduce a new vocabulary and set of concepts more closely tied to cognitive science than Eastern philosophy.

Qualia and their Properties

Okay, so…suppose "qualia" exist in some meaningful sense, then what else can we say about them? What general propositions can be made about this "subjectivity-reflecting propositional content?"

First, a comment about networks of qualia. When thinking about qualia like "whiteness", it's hard to see how it could ever be possible to infer things like walls, trees, electrons and people from qualia. But the trick is that qualia like "whiteness" are not the only kind — there are also more abstract properties that present themselves as raw, immediate perceptions. There are properties of relationship, such as "on-ness", "beside-ness" and so forth. One can experience beside-ness, independently from the things that are beside each other. This is the quale of besideness. But then one can piece the beside-ness together with the things that are beside each other, which is a matter of building networks of relationship among qualia.

This — networks of relationships among qualia — is what the subjective world is made of.

As noted above, from a subjective perspective, there is no such thing as a "quale versus a non-quale." Every subjective entity has a quale aspect — it's "being-in-and-of-itself" — but then subjective entities may also have relational aspects as well (e.g. whiteness vs. whiteness-of-the-table, where "table" itself is a network of relationships among qualia; or besideness vs. besideness of whiteness and blackness, or besideness of the white wall and the black table).

Qualia, which are properties, can themselves have properties. For instance, some important properties of qualia are the ones I call arity, centrality, intensity, solidity and historicity.

Arity has to do with whether a quale is applicable to one thing ("whiteness") or two things ("besideness") or more ("give-ness" relates three things). Qualia with different arities have different subjective feels to them. Qualia of arity one may be called "elementary"; those of higher arity may be called "relational."

Centrality has to do with how much focus is on a given quale. Some things seem to be at the fringe of awareness — say, a vague sense of unease or confusion, or an idea that one can sense forming but doesn't quite grasp yet. Other things are right at the center of awareness.

Intensity has to do with how vivid a quale is, how much attention it demands. This is different from centrality — because, for instance, sometimes the center of one's awareness can be occupied with something quite pale and calm, other times by something exciting and demanding-of-attention.

A key hypothesis I make is that quale-intensity is correlated with pattern-intensity. Quale-intensity is subjective; to use a language to be introduced in the following chapter, it is a matter of Peircean Firstness (raw experience, raw beingness). Pattern-intensity is about evaluations and comparisons and is hence about Peircean Thirdness (the Peircean realm of relationship). The proposed relation between quale-intensity and pattern-intensity is thus a link between the domains of First and Third, and as a subcase a link between subjective and objective reality, because our theories of objective reality are systems of relationships and therefore part of the domain of Third.

Solidity has to do with, for example, the difference between qualia that appear to be perceived and those that appear to be imagined. A tree in the outside world has a different "feel" to it than a tree imagined inside the mind. This is not a matter of intensity or centrality; it's a different dimension. Normally non-solid qualia have less detail to them than solid qualia but this is not a hard-and-fast rule.

Relationship qualia can connect qualia with different degrees of solidity. For instance, if I believe there is (in objective reality) a fly behind my computer monitor, but I can't see the fly (only the monitor), then I can experience the (somewhat solid) relationship between the non-solid quale of the imagined fly and the solid quale of the perceived monitor.

Finally, **historicity** has to do with time — with, basically, with whether the quale has ever been Present or not. This quality gives us our innate sense of whether a quale is a memory or not.

There are empirical laws relating qualities of qualia; for instance, central qualia tend to be more intense than peripheral ones, but this is not a universal law. One may distinguish and analyze those contexts in which peripheral qualia become unusually intense.

Building objective reality from subjective has to do (in a "proposition space" framework) with the formation of hypothetical relationships. The blackness of the fly behind the monitor is not directly experienced as a solid quale, but the hypothesis is made that if I were to get up and look in back of the monitor, I would then experience it as a solid quale. Now, getting up to look in back of the monitor itself involves a bunch of different qualia, including plenty of relational ones — so the hypothesis of the fly behind the monitor is basically a set of implications of the form "If these qualia, then those qualia." Once the ability for this kind of abstract implication emerges in a mind, then the capability to construct a working concept of "objective reality" is there. Specifically, objective reality has to do with abstract implications whose conclusions involve solid, elementary qualia.

These various descriptors of qualia are somewhat useful for discussing the standard "paradoxes" of conscious experience. For instance, Tennessee Leuwenberg, on the SL4 email discussion list, recently told the parable:

> *An intelligent scientist in the future is born on, and living in a spaceship. The inside of the spaceship is not devoid of light, but the colouring of all the internal surfaces happens to be black-and-white in appearance. However, she has a huge amount of information about physics. In this experiment, she is not capable of reproducing anything that is coloured for her to see, but she is able intellectually to fully understand the nature of light, its effects on the human eyeball, brain, nervous system etc.*
>
> *One day she lands on Earth at the end of her mission. Upon opening the hatch, she casts her eyes first on an enormous bunch of red roses which have been given to her.*
>
> *'Oh', she says, 'so that's what it's like.'*

(Similar parables have been told by others before, and discussed in the philosophy literature.)

Tennessee then asked: "Has she learnt anything new about colour? If you accept that she has, then qualia must be real, because she already knew everything that science could inform her about the world and about colour. There must, therefore, be something real about colour which is not addressed by science."

Daniel Dennett (1992) has presented a reasonable counterargument against this sort of argument, which is that we don't really know what it would be like to have a truly comprehensive understanding of all the phenomena related to red roses. So our intuition about the experience of the alien in Tennessee's parable is bound to be flawed and biased based on our experience being far more limited creatures. Dennett's counterargument shows that our intuitive reaction to this sort of parable cannot logically be considered decisive evidence about the reality of qualia – yet I still find that the parable is valuable.

What the parable reflects, to me, is the difference between elementary and abstract qualia, and between solid and non-solid ones. The scientist in the spaceship could understand roses using abstract, non-solid qualia. Once on Earth, Tennessee's alien could understand them using elementary, solid qualia. The key point of the story, then, in terms of the present discussion of qualia, is that qualia are differentiated by (among other qualities) arity and solidity.

Future science will give us the capability to manipulate and explore the various parameters of qualia in a way that is barely conceivable now. For instance, consider the notion of the centrality of qualia, which has to do with the experience of a "stream of intense consciousness" that we all have, at least in our more ordinary states of mind. It is a particular property of human minds (and probably any vaguely humanlike AI minds as well) that centrality is closely connected with verbalizability – because our verbalization processes are hooked up to verbalize the things that are most globally spread across our brains, most globally significant to us. In future, I project, we will be able to manipulate the "reporting mechanism" of intelligent systems (in humans, this means mainly but not exclusively verbalization) to probe the relationship between "actual qualia" and "reported qualia." Suppose, for instance:

- We connect reporting mechanisms to various parts X of a real or simulated brain and then find that reported qualia often occur in X in this circumstance (when various co-factors are met, probably), whereas they rarely occur in X otherwise.
- We find that some physical correlate of qualia:
    - occurs in X even when X is not connected to the reporting mechanism;
    - occurs in X *and* the reporting mechanism when X is connected to the reporting mechanism, and ensuing qualia are reported.

This will provide some interesting circumstantial evidence that maybe qualia are present in X all the time but just aren't being reported.

I feel sure a lot of other, even more interesting ideas will come up once experimental tools advance appropriately, allowing us to more fully explore centrality as well as other qualia parameters. Unfortunately, with experimental tools at the stage they are now, it's hard for me to see how to make real progress in these areas – but fortunately, science progresses fast.

Summary: Qualia and Subjective/Objective Reality

By way of summary, I will reiterate here how my position on qualia differs both from the standard anti-quale position and the more extreme pro-quale positions.

Materialist/positivist types, arguing against the meaningfulness of the "qualia" concept, will often suggest that quale-fans like me don't really appreciate the full power of materialism to explain complex systems like brains. "Once science advances a little further and you see the detailed neurophysiological explanation of your precious feeling of love," the argument goes, "then these mystical notions will disappear, just as the mystical notion of Biblical Creation disappears once one really understands the evolutionary explanation of the origin of species."

I don't buy this argument. Because, unlike some other qualiacs, I *do* think there is an empirical, scientific, physics-based explanation of the behaviors and utterances of beings like humans (and future AI's) that claim to be conscious and have qualia. I think one will almost surely be able to formulate this explanation – once future science achieves it — in terms that don't involve qualia, awareness, and the like. Modern science doesn't have all the details worked out yet, but I'm pretty confident they will be.

However, this does *not* indicate to me that qualia are meaningless, nonexistent figments of mystical imaginations.

I predict that, as science develops further, we will find that theories involving qualia form a far *simpler* way of explaining the behaviors and utterances of beings-that-claim-to-be-conscious-and-have-qualia than theories that don't involve qualia.

Thus, I think that as future science unfolds, even die-hard materialists will have to admit that qualia have at least the same sort of "reality" that electrons do. (No one has ever directly experienced electrons. They exist purely as theoretical constructs, for helping us explain various macroscopic physical observations within a complex theoretical framework.)

Also, I think that future science will allow us to begin with some plausible assumptions about qualia, and then derive the mathematics of the physical world therefrom. I.e., to explain apparent objectivity in terms of subjectivity; just as future neuroscience will allow us to explain subjectivity in terms of objectivity.

Overall, then, it should be clear that my belief in the reality and interestingness of qualia is not due to a lack of careful thinking about the potential power of mechanistic explanations to account for subjective phenomena. I just don't think these mechanistic explanations, even when complete, will tell the whole story in a maximally simple way. Qualia-incorporating explanations will always provide greater compactness, I project.

And, as noted above, while my hypothesized "explanation of physics in terms of qualia" and "detailed explanation of cognitive systems' behavior in terms of qualia dynamics" are not yet actualized – neither is the detailed explanation of the behaviors and utterances of intelligent systems based on physical law. On all sides we are dealing with speculations about what future science may provide: a chancy game, though a fun and a necessary one.

In terms of current science, we have no way to explain subjective experience neither in terms of empirical science nor vice versa. To assume that future science will yield an explanation in one direction (subjectivity in terms of objectivity) but not the other direction (physics from qualia, a variant of John Wheeler's "it from bit") is wholly unjustified. I am indulging in speculation biased by my own metaphysical assumptions, but I admit it, whereas positivist/reductionist types seem not to: they seem to accept the detailed explanation of the subjective in terms of the objective as a done deal even though it isn't, and dismiss out of hand the possibility of a detailed reverse explanation.

I've explained how my views differ from those of materialist/positivist/ reductionist types, but how do they differ from those of others who find qualia interesting? Consider for example Mitchell Porter's hypothesis (made on the SL4 list in 2005) that an AI engineered based on purely reductionist theories of cognition would probably not have qualia. Well, I really doubt it. I think that if we engineered an AI to reflect on its own structures and dynamics in such a way that it believed, based on its own self-study, that the simplest hypothesis for explaining its own behavior was a hypothesis involving "I have qualia" — then this AI would have qualia in exactly the same sense that human do. Not because qualia are somehow an epiphenomenon of material phenomena, but because subjectivity and objectivity are different perspectives on pattern-sets — meaning they are pattern-sets that often have significant intersection, and the intelligence-associated patterns in intelligent self-reflective systems lie in this intersection.

I do think, however, that it might be possible to engineer an AI (like the sort of hypothetical AI that Eliezer Yudkowsky has called a "non-sentient highly powerful optimization process") that lacked the kind of self-reflective cycles correlating with the human use of qualia for self-modeling, and hence had a very qualitatively different sort of qualia than we do. Not an AI with no experience at all, but an AI with such a different sort of experience that we would have a hard time classifying its experiences as "experience" on an intuitive basis (in a vaguely similar way to how most people in Western civilizations have a hard time intuitively classifying a plant's experience as "experience"). Hutter's infeasible AIXItl system would fall into this category – but there might also be some pragmatically feasible AI's in this category as well. This is the kind of possibility that will be a lot easier to explore once we have a philosophy and science of qualia and associated phenomena going far beyond the relatively simple ideas presented here…

Subjectivity and Objectivity in the Experience of Time

As a coda to this chapter, I will now briefly discuss how the above ideas on subjectivity and objectivity relate to the philosophy of time. As it happens, the subjectivity/objectivity distinction as pursued here matches up very naturally to a twofold distinction that has arisen in modern philosophy, as a reaction to certain paradoxes observed with the commonsense notion of time. And a careful study of the common human experience of time reveals it to be a fascinating amalgam of subjective and objective elements – a case where subjective reality incorporates an objectivist model of reality into itself so fundamentally that it creates a kind of objectivity-fueled subjective delusion.

Perhaps the most significant idea in the modern philosophy of time is McTaggart's paradox (McTaggart, 1908), which shows that the standard "folk psychology" view of time as consisting of "past, present and future" is self-contradictory. McTaggart begins with the distinction between two ways or organizing temporal events: A-series and B-series. An A-series is a series of time-markers assigned to an event indicating its "flow" through time – an event begins as distant past, then becomes near past, then becomes present, then near future, then distant future. A B-series, on the other hand, is a series of events ordered in terms of the relationships of simultaneity and precedence. In a B-series of events, each event occurs before some others, after some others, and simultaneously with some others. A B-series may also involve time-markers assigned to events indicating their location in objective time – one event may occur at 1:04 AM on October 12 2007, another may occur at 3:33 PM on January 9 1995, etc. In this case the precedence and simultaneity relationships may be defined in terms of the explicit, objective time-markers. These two types of series correspond to two different ways of talking about time – A-expressions like "The train crashed four days ago" or B-expressions like "The train crashed before the bus exploded" or "The train crashed before 9/11."

McTaggart observed that, while B-series can be derived from A-series, the reverse is not possible. Given A-series information such as "The train crash was in the recent but not immediate past; the bus crash was in the immediate past", one can derive the B-series relationship "The train crash was before the bus crash." But given B-series information alone, there is no way to determine which moment is the present moment, ergo no way to derive A-series information. A-series information derives only from B-series information plus the assumptive identification of a particular moment as the present one.

However, McTaggart argued, the whole concept of an A-series is intrinsically untenable. His argument goes as follows:

1. Different A-series positions are mutually incompatible, so no event can have more than one of them (in an A-series, e.g., the same event can't be both "recent past" and "far future").
2. But: Assuming there is an A-series, then, since A-series labels change, each event will assume all A-series labels (e.g. the event of the train crash will at one point assume the label "recent past" and at another time the label "far future").
3. Thus: The idea of an A-series is contradictory, since it requires each event to have exactly one label, but also requires each event to have all labels.

It's a very simple argument, but subtler than it first appears. Intuitively, it seems at first that the argument is based on a misinterpretation or a "game with words": the sense in which different A-series positions are incompatible is that no event can hold both of them *relative to the same present moment*. And the sense in which A-series labels change is that they change when the present moment changes. But the problem is that to formalize these differences in sense requires some formalization of the notion of change – and then, to formalize the notion of change, one seems to have two choices: A-theory or B-theory. If we formalize change in terms of B-theory, then we need to define a B-theoretic precedence relationship between various present-moments, and we no longer have any way to specifically identify which moment is the present one. We are back to the notion of "B-theory plus a special marker indicating which moment is the present one." On the other hand, if we formalize change in terms of A-theory, then we are saying things like "The train crash is present from the *present* present, but future from the *past* present," and one doesn't escape the problem – one then has the same contradiction regarding the relationship between the different "present moments" (the ones that are distant past, recent past, future, etc. with respect to the original present present). One can construct a regress of present present presents, {future "recent past" "far future"}'s, etc., but one doesn't escape the problem.

The two reasonable solutions to this elementary but irritating problem seem to be: Assume the B-theory of time is correct; or, adopt a "presentist" position in which the past and future are not real, only the present is. Various philosophers have adopted each of these positions, and some have adopted other positions seeking to work around McTaggart's paradox.

In the terms discussed here, the presentist position corresponds to subjective reality. Presentism is basically the mystical notion that "time is unreal; it's all one big fluid moment." According to presentism, the past and the future exist only in the form of memories and conjectures that arise during the (implicitly) present moment. Assumption 2 of McTaggart's argument, above, is rejected, because it is denied that the train crash that is considered "present" at 3PM on July 12 2007 is the same as the train crash that is considered "past" a year later. The idea is that the train crash in Jim Smith's "July 12 2007 present"'s direct experience is not the same as the train crash in Jim Smith's "July 12 2008 present"'s memory. Of course there is a relationship between these two train crashes but it's not correct to say that this is an identity relationship. The relationship itself only exists in some particular present moment, experienced by some particular temporally localized mind.

On the other hand, the B-theoretic position corresponds to the standard, scientific, objective-realistic view of time, in which there is a time axis that exists independently of any observer, and each event has a particular position on the time axis. This view of time is conceptually unproblematic but its relationship to experience is unclear. From a subjective view, a B-theoretic time axis is constructed only within a particular present moment, as a way of organizing remembered, inferred or predicted events. A mind may construct a B-theoretic time axis at one moment, and then another one at another moment – but one can't assume that the events on one moment's B-theoretic time axis are identical to the events on another moment's B-theoretic time axis.

Now, what one can say, however, is that if a mind, at a particular subjective moment, constructs a B-theoretic time axis, then in the context of this time axis it can make A-theoretic constructions. However, to do this, it requires something more than just the B-theoretic time axis, it also requires an intuitive notion of "pastness" – it has to be able to distinguish present experiences from memories. The intuitive notion of "pastness", together with a constructed B-theoretic model of its memory, allows the mind to construct an A-theoretic understanding.

What is perplexing here from a psychological point of view is that, much of the time, the A-theory feels phenomenologically primary – that is, we feel a sense of the "flowing" and "moving" of time. However, the conclusion one comes to is that this is a case where the human mind deludes itself. We will see a similar case in Chapter 9 when we study the issue of free will. In free will, the mind feels (during ordinary states of consciousness) like a certain high-level "decision" process is causing a cascade of lower-level processes, but more careful reflection (e.g. during meditative states of consciousness) and objective neuroscience analysis both suggest that this is not the case, and in fact the lower-level processes are causing the higher-level one. On the other hand, in temporal experience, the mind feels like its experience of the world is A-theoretic first and B-theoretic second. But what's really happening (unconsciously, at least with respect to the ordinary everyday human states of consciousness) is that B-theory is being used to organize events, and then the intuitive subjective sense of the present moment is being used to turn this B-theoretic understanding into an A-theoretic understanding.

Phrased in terms of objective and subjective reality, what this means is that our ordinary experience of time is a peculiar and fascinating amalgam of reality-types. It is a consequence of the subjectively experiencing mind creating a theory of objective reality and then using this theory to structure its experience – so that subjective experience relies fundamentally on implicit, "unconscious" objectivity for its nature. A more thoroughly reflective and self-aware intelligence would not make this sort of mistake; it would understand the primal nature of the B-theoretic and presentist perspectives and would not have a need to experience time as "flowing."

The next question becomes why this delusion of A-theoretic time-flowing ever arose. The answer of course is that it provides a compact and useful way of describing and reasoning about the world. Many propositions about experience are more compactly expressible using A-theoretic language, and this is the case to such an impressive extent that the human mind has come to use this language implicitly and automatically. For this reason, it may be expected that a more thoroughly self-aware transhuman intelligence might still use A-theoretic constructs in its thinking about time – but it would be unlikely to use them as frequently or as fundamentally as humans do, because our tendency is to use A-theory even in many cases where B-theory would be more useful.   So in spite of the sometime utility of A-theoretic constructs, it seems likely that transhuman minds will not experience time as flowing in anything near the same sense we do.

Chapter 6
Four Levels of Mind
with Ted Goertzel

The Vedantic hierarchy is one important application of the hierarchy meme to the philosophy of mind, but it's not the only one. In this chapter I will explore a different approach to the hierarchical categorization of mental phenomena, which I call the *fourfold model of mind* — a model of mind as existing on four different "levels of being." This concept is not entirely an original one; it has many precursors, and I will present it here as an elaboration of late nineteenth century American philosopher Charles S. Peirce's categorical metaphysics, giving mentions to other relevant thinkers such as Jung and Buckminster Fuller along the way. Following the discussion of Peirce, I will briefly present some ideas from the thinking of Friedrich Nietzsche, which make the same conceptual points that Peirce made, but present them in a different language and with different emphases.

In short, the four levels to be discussed here correspond to:

1. **First**, raw experience;
2. **Second**, physical reaction;
3. **Third,** relationship and pattern;
4. **Fourth**, synergy and emergence.[27]

Each of these levels constitutes a different perspective on the mind; and many important mental phenomena (consciousness being a prime example) can only be understood by considering them on several different levels.

The relation between these categories and two category systems proposed in previous chapters is worthy of brief discussion. Firstly, in the previous chapter I presented a dichotomy between subjective and objective perspectives on reality. I believe the Peircean hierarchy is partially aligned with, and partially orthogonal to, this distinction. Both subjective and objective realities exist at the levels of Second, Third and Fourth. However, First is a level that objectivity does not recognize: it belongs purely to the subjective.

The relation between the Peircean levels and the Vedantic hierarchy is subtler, and is loosely indicated in the table below. Here again the category systems are partially aligned and partially orthogonal; but here we can see that the intersections between particular Vedantic categories and particular Peircean ones have specific (and interesting) semantics. The table is not intended as complete in any sense, but merely as an evocation of a few things associated with a

---

[27] I am indebted to my friend and fellow philosopher Kent Palmer for pointing out to me the critical nature of the Fourth category, and for many interesting discussions on this topic (among many others), mostly during long e-mail dialogues with myself, Onar Aam and physicist Tony Smith during the period 1994-1997.

few of the category-intersections. Cells that are left blank are not hypothesized to be senseless concepts, but merely not to correspond to any well-known human concepts that occurred to me at time of writing!

	Firstness: Being	Secondness: Reaction	Thirdness: Relationship	Fourthness: Synergy
Quantum	The subjective experience of hypothetical "quantum minds"	Uninterpretated interactions of uninterpreted quantum phenomena	Laws of quantum theory	Wave function of the universe
Empirical		Uninterpreted interactions of classical physical objects	Science	Complex systems science; the holistic unity of scientific theories
Body / Life	Awareness of plants	Habitual reflexes of protoplasm and higher biological systems	Traditional medicine	Holistic medicine; systems biology
Mind	Awareness of animals and humans	Cognitive and motor reflexes; hard-wired perceptions	Logicist and mechanist views of mind	Complex systems view of intelligence (e.g. patternist view)
Intuition	Creative inspiration		Patternist model of creativity	Beauty and unity of creative works
Bliss	Psychedelic experience; extremely intense creativity		"Pattern-space" as a model of the universe transcending subjective and objective reality	"Oceanic feeling" of oneness with the universe
Atman	Enlightenment!			

Peirce's Four Categories of Being

Peirce believed that on the most fundamental level, the universe was organized numerically, and he divided the universe into categories according to the first three integers, called First, Second and Third. He believed that the small integers - particularly one, two and three - were not just arbitrary human creations, but fundamental organizing principles of the universe. In this section we will follow Peirce and then extend him a bit, introducing a new category called Fourth.

Peirce's concept of fundamental categories is closely related to psychologist Carl Jung's notion of *numerical archetypes*. In general, Jung used the term "archetype" to refer to patterns which are pervasive and recurrent in the mind and universe, and which seem to express the fundamental nature of phenomena. Although Peirce did not use the word "archetype", he observed the archetypal nature of the small integers in fairly explicit terms:

> *Three conceptions are perpetually turning up at every point in every theory of logic, and in the most rounded systems they occur in connection with one another. They are conceptions so very broad and consequently indefinite that they are hard to seize and may be easily overlooked. I call them the conceptions of First, Second, Third.*

He explained the significance of his three archetypal categories as follows:

> *First is the conception of being or existing independent of anything else. Second is the conception of being relative to, the conception of reaction with something else. Third is the conception of mediation, whereby a first and second are brought into relation... (1935, p.25)*

In psychological terms, on the other hand, he described the categories as:

> *In psychology, Feeling is First, Sense of reaction Second, General conception Third...*
> *Chance is First, Law is Second, the tendency to take habits is Third. (p. 26)*

While Peirce focused on the numbers 1, 2 and 3 as representative of archetypal categories, in the mid 20th century Jung extended the numerical archetype notion to other numbers such as 0, 4 and 5, with interesting results. Zero, for instance, he viewed as the archetype of absolute nothingness, emptiness – basically, though he didn't use precisely such language, the Formless Void of Zen Buddhism. Four he viewed as the archetype of wholeness and unity.

The numerical names of these categories have an interesting double meaning. On the one hand, he is simply naming fundamental categories in order of complexity. First is the simplest category; Second, the second-simplest category, builds on First, etc. On the other hand, each category is fundamentally associated with the number that names it, in that the *minimal examples* of category named "Nth" are always examples involving N elements.

For instance, First minimally requires only one entity. With only one entity, there is nothing but raw experience. There is no interaction of any kind, because interaction requires two things.

With two entities, on the other hand, one can have interaction. Or one can have change, the two entities then reflecting before and after states.

With three entities, one can have relationship. One can take a reaction between two things, and create a third thing, a symbol that stands for this reaction. This is critical to intelligence, as we'll see very explicitly in Chapter 15 when we discuss the Novamente AI design: Novamente, in essence, is a large collection of relationships (Thirds), implemented in a programming language that allows these relationships to be enacted in physical reactions (Seconds) among electrons flowing through pathways on a silicon chip.

Jung incorporated a category of Fourth, without a direct analogue in Peircean philosophy. The Jungian Fourth corresponds to wholeness or synergy – to the phenomenon whereby several relationships are interwoven into a network, in which each relationship relates the others. Every collection of relationships is not a Fourth, only a relationship that has some intrinsic wholeness to it.

In terms of pattern theory, a pattern itself is a Third, a relationship. The network of emergent patterns existing in a complex system is a Fourth – a coherent whole, emergent from a web of interrelated patterns.

These general concepts are very simple but they possess significant power to cut through complex issues in cognitive science, as we will see as we proceed. In complex systems terms, one may make the mappings:

- First is pure experience, "raw consciousness;"
- Second is the physical structures and dynamics underlying complex systems;
- Third is the domain of patterns ("habits" or "relationships" is what Peirce called them);
- Fourth is the domain of emergent patterns, arising from systems of simpler patterns; and of cooperativity among pattern-generating processes.

In Peircean terms, the Novamente design focuses on engineering things at the level of Second and Third, so as to cause other things to emerge at the levels of Third and Fourth. First, the domain of conscious experience, is not explicitly addressed in our Novamente work, although (as will be discussed below) we do believe that a sufficiently complex intelligent system will necessarily possess its own intense conscious experience.

I will now review the philosophical categories of First, Second, Third and Fourth in more detail, with an emphasis on their implications for patternist philosophy, cognitive science and artificial intelligence.

First: Raw Being

The Peircean category hierarchy begins with First: *"The conception of being or existing independent of anything else."*

In physics, First corresponds to chance behavior, apparent randomness — what Peirce called "the swervings of Atoms" and we today call the quantum indeterminacy of matter. The random choice of an electron whether to spin up or down is independent of everything else in the universe.

In psychology, on the other hand, Firstness is:

feelings, comprising all that is immediately present, such as pain, blue, cheerfulness, the feeling that arises when we contemplate a consistent theory, etc. A feeling is a state of mind having its own living quality, independent of any other state of mind ... an element of consciousness which might conceivably override everything.

In modern psychological language, Firstness corresponds closely to *consciousness* – the raw, experiential aspect of consciousness, rather than the structured aspect of consciousness that is the subject of scientific psychology.

Perhaps the clearest statement Peirce made on Firstness as consciousness was in his essay on "The Logic of the Universe" (1935):

> *The sense- quality [First] is a feeling. Even if you say it is a **slumbering** feeling, that does not make it less intense; perhaps the reverse. For it is the absence of **reaction** - - of feeling **another** – that constitutes slumber, not the absence of the immediate feeling that is all that it is in its immediacy. Imagine a magenta color. Now imagine that all the rest of your consciousness – memory, thought, everything except this feeling of magenta - - is utterly wiped out, and with that is erased all possibility of comparing the magenta with anything else or of estimating it as more or less bright. That is what you must think the pure sense- quality to be. Such a definite potentiality can emerge from the indefinite potentiality only by its own vital Firstness and spontaneity. Here is this magenta color. What originally made such a quality of feeling possible? Evidently nothing but itself. It is a First.*

Here the connection with consciousness is made quite explicit. Consciousness, when separated from the apparatus of memory and cognition, is a First. Consciousness-as-First is pure apprehension of subjective qualities, of what Peirce calls qualia:

> *The **quale**- consciousness is not confined to simple sensations. There is a peculiar **quale** to **purple**, though it be only a mixture of red and blue. There is a distinctive **quale** to every combination of sensations so far as it is really synthetized - - a distinctive **quale** to every work of art - - a distinctive **quale** to this moment as it is to me...*
> *Each **quale** is in itself what it is for itself, without reference to any other. It is absurd to say that one **quale** in itself is considered like or unlike another... (p. 152)*

And a connection between consciousness and chance is drawn quite explicitly:

> *[There] is no check upon the utmost variety and diversity of **quale**- consciousness as it appears to the comparing intellect. For if consciousness is to blend with consciousness, there must be common elements. But if it has nothing in itself but just itself, it is **sui generis** and is cut loose from all need of agreeing with anything. Whatever is absolutely simple must be absolutely free; for a law over it must apply to some common feature of it...*
> *And thus it is that that very same logical element of experience, which appears upon the inside as unity, when viewed from the outside is seen as variety. It is **totus, teres, atque rotundus**. (p. 154)*

In Peirce's vocabulary, variety is synonymous with chance - - he speaks of "the infinite diversity of the universe, which we call chance." Chance ensues from elemental freedom. Raw consciousness appears in the world of structure as the random. And what regulates these chance eruptions is the law of mind, the tendency to take habits. Consciousness and habituation work together to produce the structured diversity of the mind.

It is important to note that Peirce's notion of chance is not identical with the modern mathematical notion of algorithmic randomness. Rather, what Peirce meant by chance was something more subjective. He meant that consciousness, from the perspective of the mind experiencing it, consists of the emergence of entities that the mind did not expect – the emergence of new stimuli or actions or patterns into the mind "as if out of nowhere."

Peirce is in essence an animist; he believes that every entity in the universe has a little spark of consciousness – i.e., everything exists in the realm of First, as well as in other realms. But some entities, he notes, have more intense consciousness than others. "More intense" is a Third concept, not an aspect of Firstness; Firstness exists at a level where there can be no comparisons. In this vein, he differentiates between two kinds of consciousness:

- "quale-consciousness", pure Firstness, raw experience;
- "consciousness intensified by attention," which has to do with Third as well as First.

This perspective agrees with the one taken in the previous chapter, where we concluded that, in a sense, quale-consciousness comes along for free with every system in the universe, although particular structures and processes (such as some embedded in Novamente) are better at amplifying and intensifying it than others.

As for which entities display this intensified consciousness, Peirce does not give a complete and coherent theory, but he makes some interesting observations:

> *And now I enunciate a truth. It is this. In so far as **qualia** can be said to have anything in common, that which belongs to one and all is **unity**; and the various synthetical unities which Kant attributes to the different operations of the mind, as well as the unity of logical consistency ... and also the unity of the individual object, all these unities originate, not in the operations of the intellect, but in the **quale**- consciousness upon which the intellect operates...*
>
> *Perhaps it may be thought that hypnotic phenomena show that subconscious feelings are not unified. But I maintain on the contrary that those phenomena exhibit the very opposite peculiarity. They **are** unified so far as they are brought into one **quale**-consciousness at all; and that is why different personalities are formed. Of course, each personality is based on a "bundle of habits"... But a bundle of habits would not have the unity of self- consciousness. That **unity** must be given as a centre for the habits.*
>
> *The brain shows no central cell. The unity of consciousness is therefore not of physiological origin.... I say then that this unity is logical in this sense, that to feel, to be immediately conscious, so far as possible, without any reaction nor any reflection, logically supposes one consciousness and not two nor more...*
>
> *In **quale**- consciousness there is but one quality, but one element. It is entirely simple...*
>
> *Thus consciousness, so far as it can be contained in an instant of time, is an example of **quale**- consciousness. Now everybody who has begun to think about consciousness at all has remarked that the present so conceived is absolutely severed from past and future...*
>
> *So I might express my truth by saying:*
> *The Now is one, and but one. (p. 153)*

Among the observations Peirce makes in this passage, there is one that puzzles the neuropsychology community even today. Neuropsychologists have proven what Peirce suspected, that "the brain has no central cell," no Cartesian Theater of consciousness. But they have not yet fully come to grips with the function of this distributed phenomenon of consciousness. Somehow, modern neuroscience suggests, "focused attention" is an emergent phenomenon, coming out of the interaction of various parts of the brain. And somehow this emergent pattern of brain dynamics is correlated with intense quale-consciousness, intense conscious experience. But how? Neither Peirce nor modern neuroscience has proposed a reasonable, reasonably complete answer. We have our own hypothesis, which was hinted at in the previous chapter, and which we will discuss a little differently below, in the context of the Fourth archetype.

Second: The Reacting Object

The passage from First to Second is the passage from the subjective moment of experienced reality, to physical reality. First is the now which is one and but one; whereas second involves the movement of time, the recognition of now versus then. When one particle strikes another, or one unit of electrical charge enters into a neuron, one has a Second, a reaction.

As Peirce puts it, *"Second is the conception of being relative to, the conception of reaction with, something else"*. It relates to *"sensations of reaction, as when a person blindfold suddenly runs against a post, when we make a muscular effort, or when any feeling gives way to a new feeling."*

In physics, the laws which describe the relationships between different phenomena are Third. But the reactions between individual phenomena, which the laws describe – these are Second.

Psychologically, Secondness, the feeling of reaction or being- in- the- world, is most closely related to touch and kinesthesia, senses which are direct in the sense of admitting very little representation. By means of paintings or photographs one can give a false impression of looking at sand, but using current technology, to give someone a false impression of *feeling* sand one has to touch their skin with something very similar to sand. Experiential Secondness is exemplified by the feeling of absent- mindedly tracing a finger across an object. As soon as one compares what one is feeling to a memory store of objects, one is involved with Thirdness, and one can ask whether what one is feeling contains recognizable patterns or else is random. But the mere sense of sensory difference, of change, is freedom making itself felt as reaction, as Second.

In terms of AI design, the Second level does not often require explicit attention, but it is implicitly there, within numerous considerations on the Third level. All the complex workings of a software program like Novamente "bottom out" in physical reactions within machinery. We orchestrate these reactions based on our knowledge of relationships that exist among them; but it is the reactions themselves that allow the software to enact itself in the world.

Third: Relationship, Pattern, and the Law of Mind

With Thirdness we get to the essence of Peircean philosophy. *"Third is the conception of mediation, whereby a first and second are brought into relation."* Third is habit, relationship — pattern. It is the crux of abstract thought, as well as of complex perception and orchestrated action. The number three occurs here because, to have a relationship, one needs two items to relate, and a third item doing the relating.

In Peirce's view, Thirdness is the inevitable product of the human mind:

When we think, we are conscious that a connection between feelings is determined by a general rule, we are aware of being governed by a habit...the one primary and fundamental law of mental action consists in a tendency to generalization. Feeling tends to spread; connections between feelings awaken feelings; neighboring feelings become assimilated; ideas are apt to reproduce themselves.

The Thirdness of mind is what Peirce referred to, in an important passage, as the "one law of mind":

Logical analysis applied to mental phenomena shows that there is but one law of mind, namely, that ideas tend to spread continuously and to affect certain others which stand to them in a peculiar relation of affectability. In this spreading they lose intensity, and especially the power of affecting others, but gain generality and become welded with other ideas.

This is an archetypal vision of mind that I think of as "mind as relationship" or "mind as network." Although Peirce articulated it in the late 1800's, it sounds a lot like a verbal rendition of modern connectionist cognitive science. In modern terminology Peirce's "law of mind" might be rephrased as follows: "The mind is an associative memory network, and its dynamic dictates that each idea stored in the memory is an active agent, continually acting on those other ideas with which the memory associates it."

Peirce took his vision of mind as Thirdness a little further than we have articulated here. For example, he articulated three different kinds of inference: deduction, induction and abduction, which play a central role in Novamente's inference engine.

Peirce's emphasis on Third was closely related to his philosophy of pragmatism, which consisted basically in the proposal that ***the only real aspects of an entity are those aspects which can be measured***. However, his definition of measurement was a bit eccentric: To measure an entity, in his view, is to recognize a pattern in it, or a pattern between it and other entities in the world – in short, to find a relationship involving it.

The communicable, comprehensible reality of an entity, according to Peirce, consists of the relationships that one can point out regarding that entity. Peirce did not deny the "existence," in some sense, of the unmeasurable – Firsts, in his philosophy, are unmeasurable. But the unmeasurable is not the domain of science; science has to do with relationships. The notion of "objective reality," reality that goes beyond a single mind's subjective perspective, is tied to the notion of measurement, of patterned relationship.

Fourth: Emergence, Synergy, and the Unity of Consciousness.

Peirce never talked about Fourthness. He had a detailed mathematical argument for his decision not to go beyond three fundamental categories, but the essence of his argument was quite simple. Three, he argued, was the minimum number of entities you needed to express a relationship. Everything more complex can be treated as combinations of groups of three. But while mathematically this is indeed possible, we feel that it is not necessarily the most *convenient* way to understand the phenomena around us.

In my view, there is a certain arbitrariness about stopping at three. Peirce gives a logical, relational argument that Fourth, Fifth and other such categories aren't needed because, mathematically, all n-ary relationships can be reduced to ternary relationships via usage of ternary relationships. On the other hand, there is no mathematical way to reduce ternary relationships to binary relationships without using ternary relationships.[28]   However, the problem with this argument is that it exists entirely in the domain of logical relationships, i.e. of Third. Similarly, one could argue that there is no need to go beyond 2 in a category scheme, because Third and higher don't introduce any physical reactions besides the ones existing in Second. And one could argue that there is no need to go beyond 1 in a category scheme because as far as raw experience is concerned, First sums things up totally. In the end, all Peirce's argument demonstrates is that ternary relationships are the be-all and end-all where relationship are concerned — but this doesn't argue that it isn't meaningful to look at a level that goes beyond relationship in the same general sense that relationship goes beyond reaction and reaction goes beyond being.

Jung and other thinkers, such as Buckminster Fuller, have recognized the importance of the Fourth archetype, as an archetype of wholeness. If Third is pattern and relationship (there are two things being related, and a third thing doing the relating), Fourth is synergy. Fourth is when several relationships lock together into a whole. The archetypal pattern of Fourthness is a web of relationships which support and sustain each other so that the whole is greater than the sum of the parts. This occurs in the brain and also in complex AI systems like Novamente. In complex systems terms, Thirdness has to do with pattern, whereas Fourthness has to do specifically with *emergent pattern.*

There is also a fascinating connection between Fourthness and consciousness, which I hinted at above. In the comments quoted above, Peirce mentioned the unity of consciousness, and its relevance to the notion of First.   He also mentioned the existence of some systems in which consciousness is "intensified". This is the dilemma faced by all animist philosophies. If one declares that amoebas and even atoms are slightly conscious, then one is liberated from having to explain why human brains are conscious and why AI programs might be conscious – because, after all, consciousness is just part of the universe. But one still has to explain why and in what sense humans are *more conscious than* amoebas or atoms.

Part of the answer to this question has to do with the particular structure of human intelligence — an issue that I'll take up in later chapters. But the Fourth category gives us an important clue, meaningful in itself. Each First has a certain unity, as a consequence of its monadic, indecomposable, nature. But Fourths also have unity. A Fourth is a coherent whole – a unity forged out of diversity, out of a collection of underlying reactions and relationships. And the "intensified awareness" of human mind/brains (and one day, AI mind/brains) is connected to the emergent unity of Fourth.   Consciousness has a lot to do with the creation of coherent wholes, with the building of perceived unities out of disparate sensory or cognitive information. And, as a feature of subjective experience, the more coherent and intensely emergent Fourths – the more powerful unities built by the mind – are more vividly and frequently experienced as intense Firsts. A mind's stream of consciousness is a series of Fourths built up via its attempts to understand the world and itself, on a real-time basis, each one vividly presenting itself as a First, but with interior structure (irrelevant to the level of First) constructed via the process of pattern-emergence.

---

[28] For instance, Peirce points out, one can reduce (a,b,c) to (a,(b,c)) – thus apparently reducing a ternary relationship to two (nested) binary relationships. But in Peircean terms the latter expression (the nested binary relations) is still Third because it is equivalent to {(a,x), x = (b,c)}, and the expression {x=(b,c)} involves three terms: x, b and c. On the other hand, one can represent (a,b,c,d) using an expression like (a, (a,b), (a,(b,c)), (a,(b,(c,d)))), which that doesn't require any relationship involving more than three terms to articulate, since e.g. (a,(b,(c,d))) can be written {(a,(b,x)), x=(c,d)}.

Nietzsche on Mind

Writing at around the same time as Peirce, Friedrich Nietzsche presented a similar view of the mind in a very different language, with a very different emphasis. It is interesting to see how such different thinkers, writing within different traditions, expressed similar insights into the mind so very differently.

One of Nietzsche's major points was the illusory nature of "objective, physical reality." The world that a mind perceives, he pointed out tirelessly, is the creation of that mind. In this he was following the lead of earlier German philosophers such as Kant and Schopenhauer, who had both argued – in very different ways — that the world we see is a *mind-constructed* world rather than an objective world. Kant (1990) spoke of the n*oumenal* world of imperceptible true realities and the *phenomenal* world that we construct and observe, approximating noumena. Schopenhauer (2005) spoke of the "world as will and representation," will being similar to Kant's noumena and representation being similar to Kant's phenomena.

Nietzsche followed up on these ideas in a more concrete and psychologically realistic vein. Consider, for instance, the following passage:

> *Just as little as a reader today reads all of the individual words (let alone syllables) on a page — rather he picks out about five words at random out of twenty and "guesses" at the meaning that probably belongs to these five words — just as little do we see a tree exactly and completely with reference to leaves, twigs, color and form; it is so very much easier for us to simply improvise some approximation of a tree. Even in the midst of the strangest experiences we still do the same: we make up the major part of the experience and can scarcely be forced **not** to contemplate some event as its "inventor." All this means: basically and from time immemorial we are — **accustomed to lying**. Or to put it more virtuously and hypocritically, in short, more pleasantly: one is much more of an artist than one knows.*

This is a very important point from an AI perspective. It reminds us that Novamente will build its own subjective perspective, its own view of the world, which may be very, very different from our own. If most of what we think we "perceive" is really our invention (an insight that Nietzsche achieved purely philosophically, but which contemporary neuroscience has validated in spades).

The point where Nietzsche differed from Kant, Schopenhauer and most of his other predecessors was as to whether, when our perceptual systems construct a "fake" tree, they are constructing some approximation to a *real* tree ... or whether there are *only* "approximations" and no reality. Schopenhauer viewed the perceived world as a collection of representations, but his *representations* were still representations of *something* — of Ideas that, though generally inaccessible, were nonetheless objectively "real." Nietzsche, on the other hand — like Peirce — wished to throw out Plato's Ideas altogether, to keep *only* observable, measurable forms and patterns. Like Peirce, he recognized that, while entities more basic and simple than patterns might be said to exist, that these entities then could not be considered "objectively real" in any sense. Peirce called these pre-relational entities Firsts and Seconds; Nietzsche, as we will see in a moment, called them "dynamic quanta."

Schopenhauer spoke of the Will, a mysterious force identified with ultimate reality, with Ideal essence to which observed forms approximate. Nietzsche replaced this Will with a "will to power" which animates each form in the world to *exceed itself* and become what it is not, to overcome other forms and incorporate them into its own. Instead of being the essence of what each thing *really is,* the will becomes each thing's *drive to expand itself.* Thus, the focus is on dynamics, on change, on the constant competition between forms to dominate one another.

Nietzsche rejected Kant's *noumenal* world, the world of things-in-themselves. Instead, he proposed, there is no deeper world — the world is *only surfaces.* This is parallel to the Zen Buddhist statement that nirvana (noumena) and samsara (phenomena) are the same thing. However, Nietzsche taught that the nature of the world — the "surface" world — had been consistently misunderstood by Western philosophers. The absolutely real world of Newton and Kant was an utter illusion. Instead, immediate reality was a non-objective, non-subjective universe, full of teeming relationship and competitive flux. A world of entities which are relations between each other, each one constantly acting to extend itself over the other, while blending in with its neighbors harmoniously. To quote from Nietzsche's notebooks (1968):

> *The mechanistic world is imagined as only sight and touch imagine a world (as "moved") — so as to be calculable — thus causal unities are invented, "things" (atoms) whose effect remains constant (— transference of the false concept of the subject to the concept of the atom)*
>
> *The following are therefore phenomenal: the injection of the concept of number, the concept of the thing (the concept of the subject), the concept of activity (separation of cause from effect), the concept of motion (sight and touch): our eye and our psychology are still part of it.*
>
> *If we eliminate these additions, no things remain but only dynamic quanta, in a relation of tension to all other dynamic quanta: their essence lies in their relation to all other quanta, in their "effect" upon the same. The will to power, not a being, not a becoming, but a* **pathos** *— the most elemental fact from which a becoming and effecting first emerge —*

This last paragraph is perhaps the purest, most elegant distillation of Nietzsche's final world-view. The world, he declares, is nothing but relations among each other, constantly struggling to subsume each other — this is so simple and so profound that there is almost no way to say it. A non-objective, non-subjective universe of relationship and competitive flux. A world of entities which are *relations between each other*, each one constantly acting to *extend itself over the other*, in accordance with the will to power, which is its essence. Each "thing" is known only by its effect on other things; by the observable regularities which it gives rise to. But this web of interrelationships is *alive,* it is constantly moving, each thing shifting into the others; and the way Nietzsche chose to express this dynamic was in terms of his principle of the "will to power," in terms of the urge for each relationship to extend over the others.

Nietzsche's "dynamic quanta" wrap up Peirce's Firsts and Seconds. They are moments of experience, but they also react with each other. They form relations amongst each other, leading to Nietzsche's comment that "all the world is morphology and will to power."

Nietzsche's will to power, the basic elemental energy that causes dynamic quanta to interact and form patterns of interrelationship, is a kind of combination of Peirce's "chance," governing the swervings of Atoms, with Peirce's "law of mind," that causes entities to spread over and influence and transform other entities to which they relate.

Numbers, objects, space, causality, societies, minds and ideas, in Nietzsche's view, are all epiphenomena – essentially, attractors or emergent patterns in the web of interreacting, interproducing, intercreating dynamic quanta. Everything emerges. The perceptions and concepts that we have are not approximations to some ideal realm, but are rather self-organizing constructs from a dynamical underlayer, driven by chance and by the tendency of each entity to spread over other related entities.

A Multilevel Perspective on Mind

And so, to sum up this voyage through Peirce and Nietzsche: Mind may be viewed on four different levels — experience, physical reaction, pattern, and emergence. All the levels are important, and mixing them up often leads to conceptual confusion. The numerical archetypes provide a useful, though highly abstract, conceptual tool for structuring the complex subjective/objective system we call the universe.

There is the point of view of First, of the stream of raw consciousness – from this point of view mind is unanalyzable, simply present. This view of mind lives outside the domain of science. This is the perception of a "dynamic quantum of being/becoming." It simply is. This exists only in subjective reality. We can never know this, any more than we can know the subjective experience of any other human being – or of our own selves a few hours, days or years in the past. (Of course, we have memories of the subjective experiences of our past selves, but that is not the same thing as really knowing the experience.)

Then there is the physical point of view, the perspective of Second – of reactions in brains and silicon chips and the like.

There is the point of view of Third, of relationship. From this perspective, mind is a web of relationships, of patterns. Peirce views mind as a network of habits, each one extending itself over the other habits that it related to. Nietzsche views mind as a field of dynamic quanta, each one extending itself over other quanta to which it is related. It's basically the same thing, in different language. The point is that mind is a web of patterns – a dynamic web, continually rebuilding itself by a dynamic in which each component, each pattern, continually modifies the other patterns that it's related to. This very simple dynamic is the essential dynamic of mind. Novamente embodies this dynamic in one way, the human brain in another; and other AI systems may embody it in yet other ways.

And Fourth, synergy, is what makes all the relationships comprising a mind come together into a hierarchy of coherent wholes. Synergy is what allows practical intelligence to occur – it allows a large number of useful relationships to be packed into a relatively small system. It also allows the "unity of consciousness," ultimately emanating from the level of First, to display itself on the level of system patterns. Fourth binds the disparate patterns Third generates together into coherent wholes, which can be Firsts on their own. Pattern-theoretically, Fourth is all about *emergence* – a phenomenon definable in terms of pattern, but constituting something qualitatively different from pattern. Emergence is the pattern of wholeness. It can be viewed as a relationship, but it can also be viewed as a qualitative entity fundamentally different from relationship.

Chapter 7
Complexity[29]

The theme of this book is patterns, but patterns don't live in isolation – patterns come bundled in systems. Complex systems, more often than not. This brings up the question whether there is something relevant to be learned from the emerging interdisciplinary field of science that has been called "complexity science" or "complex systems science." Does complexity science tell us something about complex systems of intercreating patterns?

The answer seems to be: sort of. Complexity science currently provides no hard facts or rigorous theorems allowing the derivation of definite and nontrivial conclusions about minds or complex pattern systems in general. However, the "complexity science way of thinking" seems to be a valuable one, and leads to a number of interesting hypotheses regarding complex pattern systems and intelligent systems in particular. Rather than complexity science providing knowledge to patternist philosophy, it seems that these two young, actively evolving approaches to understanding the world may benefit from a certain amount of unification and cross-pollination. In this chapter I'll present some of my own views on complexity science, with a focus on concepts that will arise in later chapters.

A Brief History of Complexity Science

The "complexity perspective," construed broadly, is not new at all. The basic concept of understanding the world as a collection of systems with holistic properties has been around for quite some time. In fact, in a sense this is the original way of understanding the world, far predating modern scientific reductionism and having a firmer foundation in everyday intuition. E.g. the Vedantic perspective shows no lack of appreciation of the complexity of the world.

Carrying out holistic, systemic understanding in a scientific and mathematical way, however, is a somewhat more recent endeavor. Serious efforts in this direction have been going on for three-quarters of a century, but progress has increased exponentially in recent years, as a direct result of advances in computing power.

The scientific area now known as "complexity science" is, in large part, an outgrowth of the "systems theory" discipline that began in the middle of the last century and still flourishes today. The main thing complexity science has added to old-fashioned systems theory is computer technology. Computers allow the simulation of complex systems of various types with a level of accuracy not possible with any prior technology, and this has led to the development of a variety of new ideas regarding complex systems and their dynamics, including for instance a far fuller development of the notion of "chaos" (deterministic dynamics that qualitatively emulate

---

[29] Much of the content in this chapter is drawn from my prior books *Chaotic Logic, Creating Internet Intelligence,* and *From Complexity to Creativity.* But there are also some significant modifications and refinements.

randomness) than was previously possible. A full understanding of complex systems of patterns requires the insights of the old general systems theorists, the new chaos and complexity theorists, plus other insights not common in either of these traditions.

The first major landmark in the development of systems theory was probably Norbert Wiener's book Cybernetics (1965), which came out in the 1930's and was the first systematic attempt to use mathematics to explain computational, biological and cognitive systems in one fell swoop. Following up on Wiener's early ideas, in the 1940's and 50's, a fairly large amount of work was done under the name of "general systems theory." This body of work dealt with engineering, biological and psychological systems, and involved many of the same people who laid the foundations for what we now call computer science. Among the various successes of this research program were Gregory Bateson's (1979) psychological theories, Ashby's (1960) work in cybernetics, McCulloch's (1965) groundbreaking work on neural networks, and a variety of ideas in the field of operations research. As molecular biology reduced more and more of human life to mechanism, von Bertallanfly (1993) and others were tirelessly demonstrating what many modern biochemists and geneticists still forget: that the essence of life lies in emergent properties of whole systems, not in individual mechanisms.

The general systems theorists understood that the whole is more than the sum of the parts – that in a complex system, behaviors and structures emerge via cooperative processes that you can't easily predict from looking at the parts in isolation. Furthermore, they realized that many of these cooperative phenomena didn't depend on the details of the parts, that the same essential phenomena occurred for many different systems. They thought about brains, robots, bodies, ecosystems, and so on and so on. But they failed to articulate a general systems theory that was really useful at solving problems in particular domains, and because of this, as the 1960's progress, General Systems Theory faded. The brilliant work of the early systems theorists was absorbed into various disciplines: neural network theory, nonlinear physics, computer science, neurobiology, operations research, etc. The systems theorists of the forties, fifties and sixties recognized, on an intuitive level, the riches to be found in the study of complex self-organizing systems. But, as they gradually realized, they lacked the tools with which to systematically compare their intuitions to real-world data. And we now know quite specifically what it was they lacked: the ability to simulate complex processes numerically, and to represent the results of complex simulations pictorially.

The general systems theorists focused on emergent structures, on properties that may be observed in a wide variety of systems with differing underlying construction. The resurgence of complex-systems-thinking has focused substantially on the notion of "chaos": apparent unpredictability in a system that nevertheless is known to follow predictable rules. But after a lot of talk about chaos in the 1980's and early 1990's, modern complexity science has in recent years come to focus more on emergence as well. In fact, the two concepts are closely linked. The balance between chaos and emergence is crucial to intelligence and is continually observable, for instance, in human mind and behavior. There is no way to predict what an individual neuronal group in the cognitive cortex is going to do; but it is not hard to make some predictions about what general cognitive structures and behaviors will emerge in a particular human brain in a particular situation.

Mathematical chaos theory focuses on simple dynamical systems that can be proved (via analytical math or detailed computer simulations) to have strong chaotic properties, in the sense that their trajectories are statistically indistinguishable from series of random events. But, while mathematically and conceptually interesting, these extreme cases have little directly to do with real-world complex systems. Pragmatically speaking, the really interesting point is that complex, self-organizing systems, while unpredictable on the level of detail, are *interestingly predictable on the level of structure*. This is what differentiates them from simple dynamical systems that are almost entirely unpredictable on the level of structure as well as the level of detail. What most mathematical chaos theorists are currently doing is playing with simple low-dimensional "toy equations" displaying pure and extreme chaos; but on the other hand, what most popular expositors of chaos and complexity are thinking about is the more interesting topic of *the dynamics of partially predictable structure.*

The interesting thing about intelligent systems, dynamically, is not that the details of their dynamics are hard to predict, but rather that their unpredictable microdynamics will give rise to macrodynamics with definite cognitively-meaningful patterns, and emergent macrostructures with important cognitive functions. And an important part of intelligent cognitive function is its ability to recognize patterns in its own dynamics and structure: to tease out the predictable patterns from the low-level chaos, and then embody these patterns as explicit ideas in its own mind, allowing the formation of yet more complex and subtle patterns out of the embodied representations of previously recognized patterns.

The dynamics of any intelligent system will almost surely involve many "strange attractors" in the chaos theory sense – I have already seen this in simple experiments with my research AI systems, and researchers like Walter Freeman (2001) have found this to be true of the human brain. But, the structure of these attractors need not be as coarse as that of the Lorenz attractor, or the attractor of the logistic map, or the other toy examples dealt with by mathematical chaos theorists. The structures of a mind's attractors contain a vast amount of information regarding the transitions from one patterned system state to another. And this, not the chaos itself, is the interesting part – although chaos does play an important role in terms of generating new patterns.

Unfortunately, there is no apparent way to get at the structure of the strange attractors of a dynamical system like an AGI program, or the human brain. Such a system presents hundreds of millions to hundreds of billions of interlinked variables. To understand such systems, we believe, it is necessary to shift one's attention up from the level of physical parameters, and take a "process perspective" in which the mind and brain are viewed as complex *networks of interacting, inter-creating processes*. And of course, this brings one back to more of a general systems theory perspective. Admitting we cannot understand all the details of such complex systems, the question becomes what aspects of their overall structure and dynamics we can understand without understanding all the details. This is the key question preoccupying the modern study of complex systems.

The work done over the last few decades in the area of complex dynamical systems is diverse and defies a simple summary. Example subfields include:

- Neural networks
- Cellular automata
- Artificial life
- Evolutionary programming (GA's, GP's)
- Evolutionary/ecological models of immune systems and ecosystems
- Distributed agent systems
- Artificial economies

In each of these cases, intricately detailed work has been done, exploring the emergent properties that arise due to the interactions of a large number of simple elements. This is not the place to review the successes of the complex systems approach in detail, but a brief list of results include:

- An understanding of which variants of neural net learning rules allow neural nets to learn attractors that store memories (Amit, 1991);
- An understanding of how to use abstract principles of evolution by natural selection to solve mathematical optimization problems (Koza, 1991; Goldberg, 1988);
- The creation of digital "artificial life forms" displaying emergent social behaviors like flocking and group problem-solving (Langton, 1997);
- The explanation of complex and chaotic economic activity as a consequence of rational activities of individual economic agents, acting based on the knowledge available to them (Batten, 2000);
- The modeling of hugely complex fluid dynamics problems using simple discrete dynamical systems (Wolfram, 2002);
- The modeling of the mammalian immune network, leading to the conclusion that much immune function and many immune disorders come out of complex network phenomena (Perelson, 2002);
- A huge body of work on pattern formation in chemical and physical systems, e.g. the Benard cell, the Belousov-Zhabotinsky reaction, etc. (Gao, 1994).

But still, in spite of all these interesting developments, there is still no well-organized "complex systems science" with general laws of complexity. The extent to which such laws will ever exist is not yet clear, but my main goal in this chapter is to present some well-grounded speculations regarding what some of these laws might potentially look like.

In addition to lacking general laws, something else complex systems science lacks is a guiding mathematical formalism, to do for it what, say, differential equations does for physics. Various authors have proposed various mathematical formalisms to serve this role, including (but not limited to):

- Prigogine's (1994) thermodynamics-inspired nonlinear differential equations;
- Time-discrete dynamical systems, as presented e.g. in Devaney's (1989) books, and relating closely to the field of fractal geometry launched by Mandelbrot's (1982) classic work;
- Cellular automata and other time-and-space-discrete dynamical systems (Wolfram, 2002);
- "Brownian logic," a special kind of formal logic invented by G. Spencer-Brown (1994) and further developed by Francisco Varela (1978) and Louis Kauffmann (1996).

None of these has proved fully satisfactory, however. One of my motivations in developing mathematical pattern theory (as briefly described in Appendix A) was a hope that it could perhaps help with this situation and provide part of a sound and useful mathematical foundation for the study of complexity. Whether pattern theory will ever fulfill this promise, however, remains to be seen.

Anyway, promising possibilities notwithstanding, right now complex systems science does not include any robustly, rigorously demonstrated "general laws," nor a generally accepted formalism in which such laws might be compactly and usefully expressed. Instead, all we have are:

1. Some general conceptual principles that intuitively seem to be broadly true;
2. A lot of interesting results about particular kinds of natural and artificial system, inspired by these general conceptual principles;
3. Some mathematical and computational tools that seem to be useful for studying many different kinds of complex systems.

Even if it fails as a general formalism of complex systems, the mathematics of "pattern theory" may come to play a role in Category 3; it's intended to be a framework useful for studying emergent patterns in many different types of complex systems.

In the rest of the chapter, I'll discuss a number of items that belong in Category 1: general conceptual principles of complex systems science, that are conceptually important for understanding the nature of mind. I will give a series of potential principles of complex system dynamics. Each of these so-called "principles" is stated here as a hypothesis, and a rough idea is given of the relationship between the principle and the structures and dynamics of human and artificial intelligence. If one could turn these proposed principles into actual theorems, with assumptions not overly restrictive, then one would have — lo and behold! — a start toward a real science of complexity. We suspect this may be possible, but, time will tell.

Optimal Connectivity

My first complex-systems-words-of-wisdom principle is a simple one, which I call the "optimal connectivity principle." This is, quite simply, the observation that, in many complex systems, both *optimal functionality* and *maximum structural complexity* are obtained when system components are, on average, connected neither too densely or too sparsely.

In the computational domain, this concept has been most rigorously demonstrated in Stuart Kauffmann's (1993) NK models: a particularly simple, though not very general, formalism for modeling complex systems. However, it has not been formulated in a really general way before because of the lack of a general definition of connectivity. The principle is as follows:

Optimal Connectivity Principle: **There is an optimal range of connectivity in a complex system. When the connectivity is too small, or too large, the system's functioning in its environment is relatively unintelligent.**

The subtle issue in formalizing this principle is clarifying what is meant by "connectivity." I believe the correct approach to connectivity is to ground it in terms of causality. I'll say that *component X is connected to component Y if changes in the state of X are likely to cause changes in the state of Y, in such a way that the path of causation is clearly traceable by an observer system with modest computational resources.* This is a relativist definition in the sense that it relies on an observer system, but there seems to be no viable alternative approach lacking this attribute. The definition of causality itself is a very subtle matter, which is treated in depth in

Chapter 17 – and, as reviewed there, I do not think causation is an objective and rigorous concept, but rather a subjective concept used by minds to understand the world. The subjectivity of causality implies an additional subjectivity for connectivity, beyond the one explicit in the notion of the observer system mentioned above. I believe this kind of subjectivity is one of the things that has made complex systems science so slippery to formalize in a rigorous way. Complex system science is mostly about physical and biological rather than mental systems, but as compared to other kinds of physical science, it is more explicitly about properties of these systems *as perceived by minds*.

Viewed in terms of causality, what the optimal connectivity principle means is that, for effective functionality in a complex system, there should be clear causal pathways from each component to only a moderate number of other components. If each component can easily cause significant changes in too many other components, then the system's dynamics will either become so complicated as to be unmanageable, or else may fall into some simple repetitive global pattern. On the other hand, if each component can easily cause significant changes in too few other components, then the system's dynamics will lack flexibility and not much complexity will be achieved.

This perspective has more subtlety than may be apparent at first, because the notion of "causality" is itself a very subtle one. As will be detailed in Chapter 17, causation is a messy, qualitative, subjective concept, rather than something crisp and mathematical. This implies that "connectivity" as defined above is also subjective in nature, being relative to the pragmatic causal judgments of some perceiving, judging mind. I think this is unavoidable: complexity is subjective, and this is just one more manifestation of this fact.

Pursing the connectivity theme a bit further, one may consider an intelligent system as being broken down into very rough "clusters" of data-objects dealing with different functions and/or different types of information. For instance there may be a cluster of data-objects dealing with the perception of color, one dealing with the perception of pitch, one dealing with the conjugation of verbs, one dealing with self-monitoring of high-level statistics of the overall system state, etc. The optimal connectivity principle suggests that the most effective functionality will be achieved when there is neither too much nor too little connectivity between these subsets of the system's overall mindspace.

If the connectivity between these clusters is too low, then each cluster will bring other parts of the system to bear on its own problems only very occasionally. This is not optimal because in a complex environment, there are many patterns emergent from different inputs, so that the parts of the system dealing with different inputs will need to communicate with each other.

On the other hand, if the connectivity of clusters is too high, then each cluster will interact with the others nearly all the time. This is not optimal because it destroys modularity: each cluster has no independent functionality, which means that adaptive learning is more difficult. The breaking-down of complex processes into somewhat independent clusters each performing parts of the process, is important, because it means that a lot of learning can be done on the smaller scale of individual clusters.

Intelligent systems, like evolving complex systems in general, are caught between the real-time learning efficiency of exploiting emergence, and the evolutionary efficiency of a modular structure. The optimal connectivity, among other things, represents a balance between these two factors.

The Edge of Chaos

Following up on the same conceptual theme as the optimal connectivity principle, the "edge of chaos" idea states that complex structures and dynamics seem often to reside "between simple periodic/constant behavior, and completely crazy chaotic behavior." This concept has been much discussed in the complex systems literature over the years (Langton, 1990), though its validity has never been demonstrated in a general rigorous way, largely because of the many different ways of defining and measuring complexity and chaos in practice.

Edge of Chaos Principle: **In a complex system, there tend to be large regions of parameter space leading to chaotic behavior, as well as large regions leading to repetitive behavior. The regions leading to complex behavior tend not to be embedded entirely in the chaotic or repetitive regions but rather to border both.**

This principle has often been vividly demonstrated in the domain of 1-D cellular automata, where it's easy to visually see the difference between periodic, chaotic and complex behavior (Wolfram, 2004).

The edge of chaos relates conceptually to the optimal connectivity principle, in the following sense. Suppose one has a very complex system with many different states, and that one considers a "coarse-grained state space," each element of which corresponds to a number of related basic system states. One may then construct a graph corresponding to the system's dynamics: the nodes are coarse-grained states, and the directed links indicate which states lead to which other ones. Links may be weighted to indicate probabilities of transition. This graph may be called a "coarse state graph."

In this terminology, we may observe that:

- The coarse state graph of a chaotic system is often an overly richly connected graph, in which each state can lead to a whole host of other states.
- The coarse state graph of a stable system has one node; the coarse state graph of an oscillatory system is in the shape of a ring.
- A system whose coarse state graph is too sparse, will tend to have no rapid way to get from any one state to any other state – its cognitive transitions will be sluggish.

Intuitively, it seems that the most desirable thing for a system that has to act in the real world is to have a coarse state graph permitting fast transitions between any two states, but not richly connected enough to lead to fully unpredictable behavior (which is difficult to adapt and to model).

The (Dubious) Principle of Computational Universality

Carrying the system categorization underlying the "edge of chaos" principle a little further, Stephen Wolfram, in his book *A New Kind of Science* (2002), proposed a principle called "computational universality." This principle says that there are basically four kinds of dynamical systems:

1. Those that lead to stable, unchanging behavior;
2. Those that lead to periodic behavior;
3. Those that lead to chaotic, quasi-random behavior;
4. Those that are universal, in the sense that they can lead to any possible finitely-describable behavior.

Wolfram understands that this is an overstatement, but his view is that it holds for the vast majority of dynamical systems. He has done a lot of detailed work with cellular-automaton-like systems to back this up, though nothing resembling a rigorous proof exists.

However, there is one disturbing fact that that makes Wolfram's computational universality principle perhaps not quite as powerful as he would like. In practice, *some complex, universal dynamical systems* can lead to *some types of behavior* much more easily and efficiently than others. One could create a thinking machine by programming it as an initial condition for the 1D CA rule 110, but this would be a very awkward way to implement such a thing, and getting acceptable runtime behavior on current or near-future hardware would probably not be possible.

For instance, my Novamente AI system certainly does have universal computing ability; this is guaranteed by inclusion in its internal procedure vocabulary of elementary operators called "combinators", which have been known to have universal computing power since the 1930's (i.e. since before the notion of "computing power" was fully formalized). But in the course of practical AI development, I've found that what's important is that the kinds of behaviors needed for practical intelligence in the world, are *relatively simple* within the Novamente computing framework, as opposed to for example within CA rule 110, or within a Turing machine, or the register machine underlying a contemporary von Neumann computer. It doesn't really matter what's computable in principle, it matters what's computable in practice given available computational resources.

MetaSystem Dynamics

Our next would-be complex systems principle has to do specifically with intelligent systems. Like the previous principles, it is closely related to the notion of "connectedness":

Principle of MetaSystem Dynamics: **The most intelligent systems are those in which it is a pattern that the major subsystems cause each other to act more intelligently**

Philosophically, this one goes hand in hand with the connectivity principle. Complex systems that are broken down into subsystems will tend to be easily adaptable; hence, they will tend to be more intelligent than non-hierarchical systems. But the relationship between subsystems needs to be the right one. Subsystems need to help each other, not hurt each other.

The Chaos Language Hypothesis

The Chaos Language Hypothesis, which I proposed in *From Complexity to Creativity*, states intuitively that there is a set of dynamical behavior patterns that occur very broadly in a wide variety of complex systems.

To state it more formally, one must introduce the notion of a "formal language" implicit in a dynamical system. Suppose one takes the space of possible states that a certain system can achieve, and divides this space into a finite number of regions. Suppose one denotes each region by a certain symbol. Then the evolution of the system over time can be used to generate a series of symbols, which can be thought of as a kind of linguistic expression. If, whenever the system has a state in the region R1, it never immediately transitions into a state in the region R7, then we have a grammatical rule "R1 cannot precede R7" or "R1 R7 <0>" (where the <p> is used to denote the probability of a rule). If, half the time that the system has a state in the region R1, it next moves into a state in region R4, then we have a rule "R1 R4 <.5>".

Jim Crutchfield and Karl Young (1990), among others, have studied the grammars that emerge from simple dynamical systems, and arrived at some fascinating conclusions, for instance observing that as a system's parameter values are brought close and closer to chaos-inducing values, the implicit languages in the system become more and more complex.

In *From Complexity to Creativity* I presented the notion of an optimal partitioning of the state space of a system into regions. The most natural partitions are of the same size and shape, but this isn't always optimal. Sometimes another partitioning will lead to more powerful grammatical rules. The quality of a partitioning is measured in terms of the quality of the set of grammatical rules describing system trajectories. Rules with probabilities that differ strongly from chance expectation are considered higher-quality.

Next, I introduced there the "Chaos Language Algorithm," meaning the process of finding an optimal partition of a certain size, and then producing a grammar describing a system's dynamics based on this partition.

The hypothesis made in *From Complexity to Creativity* is that if one follows this process, then in many cases, similar grammars will emerge from very different systems. In other words, it is suggested that real-world systems may demonstrate a small number of "archetypal" attractor structures, and that the attractor structures observed in real systems approximate these archetypal attractor structures. Mathematically speaking, the approximation of these archetypes should reveal itself as a clustering of inferred formal languages in formal language space. Thus one obtains the following formal hypothesis:

1. The formal languages implicit in the trajectories of real-world dynamical systems (using optimal partitionings) show a strong tendency to "cluster" in the space of formal languages.
2. The formal languages implicit in psychological systems show an even stronger tendency to cluster.

The Structure-Dynamics Principle

The Chaos Language Hypothesis postulates similarity of structure across different systems. Our next "complex systems hypothesis", the Structure-Dynamics Principle, also postulates a similarity of structure, but in a somewhat different context: not between different systems, but between the purely static and purely dynamic aspects of the *same* system. This is a principle that has a particularly close connection to the Novamente architecture.

The essential idea of the Structure-Dynamics Principle is that, in many cases, specific *components* of a system are able to place the *entire system* in well defined states. Consider a system S which possesses two properties:

1. The components S_i are each capable of assuming a certain degree of *activation*.

2. There are perceptible *pathways* between certain pairs (S_i, S_j) of system components, indicating a propensity for activation of S_i to lead to activation of S_j.

The paradigm case here is obviously the brain: activation has a clear definition on the neural level, in terms of neural firing, and pathways are defined by dendritic connections and synaptic conductance. Similarly, if, following Edelman and others, one takes the fundamental components to be neuronal groups, one finds that the neuronal definitions of activation and pathways naturally extend up to this level.   However, this picture also applies entirely to Novamente, in which case the components are Atoms, and the pathways are links and chains of links.

Now let Sys_i denote the collection of global system states that immediately follow high-level activation of system component S_i. Then the question is whether the sets Sys_i will emerge as regions in the Chaos Language Algorithm, i.e. whether they will emerge as "natural categories" for understanding the trajectory of the overall system S.

Suppose this is the case: then what we have is a grammar of overall system states corresponding to the "grammar" of individual system components that indicates the order in which system components will be activated. But the latter grammar is, by the second assumption above, perceptible from purely *structural* information, from the observation of pathways between system components. Thus one has a grammar which is both a purely structural pattern (a pattern emergent in the collection of pathways in the system at a given time) and a purely dynamical pattern (a pattern emergent in the symbolic dynamics of the trajectory of the system). This, when it occurs, is a most remarkable thing.

At this stage, there is no hard evidence that the Structure-Dynamics Principle is obeyed by naturally occurring complex systems. However, the hypothesis is a plausible one, and, especially in the context of cognitive neuroscience, it would go a long way toward solving some vexing problems.   As we will see in Chapter 15, the Novamente AI system manifests the Structure-Dynamics Principle vividly, by design.

Chapter 8
Quantum Reality and Mind[30]

What pertinence does quantum physics have for the study of mind?

A lot has been written about quantum theory and mind, and some of it is best categorized as well-intended nonsense. However, when one considers the issues carefully, the truth turns out even more interesting than the various misunderstandings and fabrications that have been propounded. Quantum physics does seem to yield significant insights into the nature of both physical and mental reality.

Before explaining what I mean by this I'll briefly clarify what I *don't* mean.

First of all, it's obvious that if we consider intelligence as emergent from physical systems, then quantum physics is generally relevant to human and artificial intelligence. So far as we can tell based on the physical observations made by humans so far, the entire world is apparently based on quantum physics (in some as yet undetermined fusion of quantum theory with general relativity and perhaps other ingredients). Brain chemistry relies fundamentally on quantum dynamics, as does the flow of electricity through silicon chips. But, along with this observation, one must also observe that many parts of the world don't explicitly display the freakier properties of the quantum domain – indeterminacy, nonlocality and so forth. Humans owe their existence to weird quantum effects, in the sense that our proteins and other component molecules behave the way they do because of quantum effects, but nevertheless (except in sci-fi novels, e.g. (Egan,1995)) we tend not to tunnel through the walls of our houses in the same way that electrons tunnel through barriers. This has to do with the notion of decoherence, which I'll discuss a little later on.

Some theorists argue that human brain dynamics relies on macroscopic quantum phenomena (Jibu and Yasue, 1995; Penrose, 1996, 2002; Hameroff, 1987), but this remains an ungrounded speculation. More frustratingly, this has sometimes been used as an argument why non-quantum-computing-based AI software can never be truly intelligent! My own opinion – to be enlarged upon below — is that, while a macroscopic-quantum-effects basis for brain dynamics is possible, even if it's true it doesn't tell us anything about the viability of creating non-quantum AI systems. The theory of quantum computation tells us that quantum computers can't compute anything special that's beyond the domain of ordinary computers – their only benefit is that they can make use of quantum nonlocality and "multiversality" to compute some things a lot faster than ordinary computers (Deutsch, 1997, 2000). If the brain does use some quantum effects as a "cognitive accelerator," this is interesting – but it doesn't mean that quantum-based acceleration is necessary for intelligence. There are plenty of other acceleration techniques that are accessible to software programs on modern computer architectures but not to human brains. And the way the

---

[30] This chapter owes a great deal to conversations with others, primarily Steve Omohundro, Jesse Mazer, and Izabela Freire Goertzel

brain uses quantum computing is evidently not all that powerful, since the brain can't carry out the feats of rapid factorization achievable via certain known quantum computing architectures, and it is famously bad at intuitively predicting the dynamics of quantum systems (ergo the famous subjective-weirdness-to-humans of basic quantum theory).

Whether or not the brain is a macroscopic quantum system in the strong sense that some theorists would like, however, it seems to me that quantum theory offers an insight into the nature of the universe that ultimately does give us some insights into the nature of mind and brain. I will not present any novel insights into quantum physics here, but after reviewing some familiar and not so familiar aspects of physical theory and interpretation, I will make some new suggestions regarding the relationship between quantum theory and mind. After reviewing the notion that decoherence is the essence of quantum measurement, I will take the idea one step further, and make an argument that ***decoherence is essentially the physical correlate of conscious experience***. However, most decoherences are not associated with particular complex systems – they involve the distribution of information widely throughout regions of the universe. In terms of consciousness, they may be said to represent the experience of the world; whereas decoherence that results in the generation of memories concentrated within individual systems may be associated with the consciousness experiences of those systems. This is not presented as a complete theory of conscious experience – I've already given my views on the more subjective and experiential aspects of consciousness – but merely as a new way of looking at the connection between conscious experience and material reality.

Elementary Quantum Weirdness[31,32]

I'll begin by reviewing some of the basic concepts of quantum theory, with a focus on philosophically significant aspects rather than technical details. A little later I'll dig deeper into recent quantum physics research (specifically, a very profound experiment called the "delayed choice quantum eraser"), but I need to lay some groundwork first.

Quantum theory began relatively unadventurously as a modern incarnation of ancient Greek atomistic theory. The idea of the "quantum" is that there is a minimum size to the stuff that happens in the universe. Specifically, according to quantum physics, there is a quantum of "action" – action in physics is defined as energy multiplied by time, and the quantum is Planck's constant, $6.6260755 * 10^{-34}$ joules-second.

The existence of this minimum quantum of action, however, has some very peculiar consequences. The key point is that, if there is a minimum amount of action one can exert, then there is no way to measure very small things without disturbing them somewhat. If some phenomenon exists on the Planck scale, then measuring it necessarily also involves action on the Planck scale, which will therefore cause a significant perturbation. This means that what happens to a particular quantum system when we're looking will not necessarily be all that close to the

---

[31] This section, and the later section on evolutionary quantum computing, is basically cribbed from an article I wrote for the Frankfurter Allgemaine Zeitung on quantum computing.

[32] If you're not familiar with the strange properties of the microworld, this chapter isn't really going to fill you in adequately. There are loads of books on the subject; two favorites from my youth are Gary Zukav's *The Dancing Wu Li Masters* (2001) and Fred Alan Wolf's *Taking the Quantum Leap* (1981; though some of Wolf's other books are full of what seems to me like New-agey nonsense, that book is a good one). I'm grateful to my college physics professor George Mandeville for introducing me to Zukav's book, and to the literature of physics-philosophy interpenetration generally.

same as what would have happened if we hadn't looked.

What is really odd about quantum systems, however, is that when *in principle* there's no way to know what happens in a system – because of the limits on observability posed by the minimum quantum of action – then the logic of what happens in the system becomes different. The in-principle knowability affects the system dynamics itself. This was an unexpected discovery, and I think it may well rank as the most surprising scientific discovery ever made.

One common way to describe the situation is to say that, in the microworld of particles and atoms, an event does not become definite until someone (or something) observes it. An unobserved quantum system remains in an uncertain state, a superposition of many different possibilities, defined by a mathematical object called a "wave function." Observation causes the wave function to "collapse" into a definite condition, which is chosen at random from among the possibilities the wave function provides. This notion of indeterminacy coupled with collapse is known as the "Copenhagen interpretation" of quantum physics as it was originated by the famous Dane Niels Bohr, who invented some important parts of quantum theory in the early 1900's.

Counterintuitively, in quantum physics, the dynamics of systems in unobserved form is different from the dynamics of systems in (what the Copenhagen interpretation calls) collapsed form, after observation. Observation affects the logic of the physical world in a concrete way. Of course, these statements beg the question of the meaning of "observation" – an important issue that I will take up a little later in this chapter.

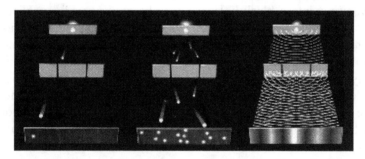

The classic double-slit experiment[33]

To better illustrate what all this peculiar verbiage means, consider the classic double-slit experiment. A particle passes through one of two slits in a barrier and leaves a mark on a plate on the other side of the barrier, indicating which slit it passed through. If one observes each particle as it passes through the barrier, the marks on the plate will be consistent with one's observations: "Fifteen hundred through the top slit, six hundred and ninety through the bottom slit," or whatever. But if the particles passing through the barrier are not observed (meaning: are not observed by any human, or gerbil, or any macrosopic recording instrument – more on this later), then something very strange happens. There are marks on the plate where there shouldn't be any -- marks that could not have been made by particles passing through either slit. Instead of passing through the slit like a good particle should, the particle acts as if it were a wave in some mysterious medium, squeezing through the slit and then rapidly diffusing. The key point is whether the particle was "observed" or not. If not, then mysterious superpositions-between-states take the place of the definite states we are accustomed to. If so, then the wave function defining

---

[33] This figure was taken from http://www.fortunecity.com/emachines/e11/86/qphil.html

superpositions between definite states is the true reality, and what one sees on the screen is consistently only with the "wave function" point of view, not the "everything must be in one or another definite state" point of view.

But what is this "observation" process? What counts as observation, from the point of view of "collapsing the wave function"? When a human looks at it? A mouse? A recording instrument hooked up to a printer? What if it's just a single molecule that registers the behavior of the particle in its dynamics? It turns out that, due to the particular mathematics of decoherence (to be discussed below), even a molecule can be considered as an "observer" in this context. To obey the peculiar logic of quantum reality, a system has to really be undisturbed to an extreme degree, avoiding coupling with any other systems that themselves possess significant internal decoherence. Otherwise the quantum beauty shatters and you're left with plain old classical-type reality.

A related issue is that, in unobserved quantum reality, time doesn't have the same significant as it does for us big classical beings. As the great physicist John Archibald Wheeler – the inventor of the term "black hole" and a leading developer of Einsteinian gravitation theory — pointed out, this even works if the choice is *delayed*, i.e. if the choice whether to record the photon's trajectory or not is made after the photon has already hit the screen. One then arrives at the phenomenon of the "quantum eraser" – a subtle and complex phenomenon that I'll describe in detail a little later.

Richard Feynman, one of the quantum pioneers, once said: "He who to tries to understand quantum theory vanishes into a black hole, never to be seen again." [34]   Niels Bohr, one of the founding fathers of quantum physics, said in 1927: "Anyone who is not shocked by quantum theory does not understand it."[35] The reason for these declarations of bafflement is clear. If the dynamical equations of quantum theory are taken literally, nothing is ever in a definite state; everything is always suspended in a superposition of various possibilities. This is what we see on the screen in the double-slit experiment, when no one observes the paths of the photons. But yet that's not what we see in the world around us - - neither in the physics lab nor in everyday life. When then does the superposed world become the actual world? When it is recorded by a machine? When it is recorded by a person? What about an intelligent machine ... or an observant chimpanzee, dog, mouse, or ant?

An added, related peculiarity is provided by the phenomenon of nonlocality. The classic example of this (the Einstein-Podolsky-Rosen or EPR experiment) involves two particles that are at one point joined together, but are then shot off in different directions until they're far distant. One supposes that the particles are, for instance, electrons, each of which has a property called "spin" that takes one of two values: either Up or Down. One may know, since the two particles were initially joined together, that only one of the particles may have spin Up, the other having spin Down. But which is Up, which is Down? This is random. There's no way to predict this.

Now, when you observe one of the particles, it automatically tells you something about the other particle – no matter how far away the other one is. If the first particle is observed to have spin Up, then the other particle is known to have spin Down, even if it's 10 quadrillion light years away. But the funny thing is that, because of the critical role of observation in quantum measurement, this act of observation in some sense causes a physical change. By observing the

---

[34] I haven't been able to track down the source of this quote, though I have found a number of other authors quoting it. The Web page referenced in the following footnote gives some other quotes by Feynman with similar semantics.

[35] http://en.wikiquote.org/wiki/Quantum_mechanics

one particle to have spin Up, the state of the other, far-distant particle is then in a way caused to have spin Down. Its state is caused to collapse from uncertainty into definiteness.

When Einstein, Podolsky and Rosen discovered this phenomenon in the 1930's they thought they had proved quantum theory false. It seemed to contradict Einstein's special relativity theory, which says that no information can travel faster than light speed. But there's no contradiction, because it's not classical information that's traveling instantaneously, it's just bizarre counterintuitive quantum collapse-into-definiteness. Although Einstein was himself a pioneer of quantum theory, he never liked its indeterminacy: he said, famously "God doesn't play dice." But it turns out that the universe at a very basic level does play dice — and this dice-playing is not only empirically verifiable, but useful as the foundation for a new generation of computing technology.

And, just as critically, it turns out that this dice-playing has far weirder consequences than one would naively think dice-playing would have. Randomness is one thing, but the observation-dependency we see in the quantum world is something else entirely, and in fact goes beyond playing dice. To me, the odd thing isn't that God plays dice; it's that God makes the dice obey different rules when we're not looking at them versus when we are. I'll return to this point a little later.

Quantum theory is conceptually weird enough that it has spawned a large number of interpretations. I have introduced one already, the "Copenhagen interpretation" which says that observation (a concept not precisely defined in the Copenhagen interpretation) collapses the wave function and reduces a superposition to a definite value. The saving grace of this approach is that, mathematically, it doesn't matter where one says the collapse happens. If one has a scientist watching a human watching a mouse watch a recording device monitor a molecule reacting to a particle, one can say that:

- The particle is in a superposition of states and the molecule "collapses the wave function" via having one or another definite reactions to the particle.
- The molecule is in a superposition of states and the recording device "collapses the wave function" via having one or another definite records of what the molecule did.
- The recording device is in a superposition of states and the mouse "collapses the wave function" via having one or another definite image of the recording device.
- The mouse is in a superposition of states and the human "collapses the wave function" via having one or another definite image of the recording device.
- The human is in a superposition of states and the scientist "collapses the wave function" via having one or another definite image of the recording device.

Wherever you decide the collapse occurs, it doesn't matter – the probability distribution of the outcomes the scientist will see comes out the same. That is why the Copenhagen interpretation can get away with being so vague about the notion of observation, because it doesn't really matter where the observation occurs – all that matter is that, by the time *you* observe something, it's already collapsed, because after all what you're seeing is a definite reality and not a superposition.

Another leading interpretation of quantum theory, the Many-Worlds Interpretation (MWI), argues that the reason is doesn't matter where you place the wave-function collapse is that no such collapse ever occurs. A naïve description of the MWI is that there are many possible universes, and different values for observations are to be viewed as different "branches" into different parts of the multiverse. In the double-slit experiment, for instance, when the photon passes through the slits, there is a branching between universes in which the particle goes to the right and universes in which it goes to the left. Basically, the MWI involves accepting that quantum reality is reality, and there is no mysterious collapse operation.

The immediately apparent problem with the MWI is that it doesn't explain why our everyday world doesn't look much like quantum reality. But the classic answer to that is because each of us lives on only one branch. For instance, consider Schrodinger's famous cat, which is trapped in a box with a gun aimed at his head, and the gun is rigged to fire only if an electron (whose spin is quantum-random) is measured to spin UP rather than DOWN. From an observer outside the box, according to the Copenhagen interpretation the cat exists in a superposed state until observed, at which point it collapses into a definite state. This creates a peculiar image of a cat in a superposed state, neither fully alive nor fully dead. But in the MWI, there are only dead cats in cat-is-dead universe-branches and living cats in cat-is-alive universe branches – there are no superposed cats.

The MWI deserves a little more elaboration than this, so I'll give it a few more paragraphs here; but I hasten to add that there are many different variants of the MWI, and I'll only lightly touch a few of them. The original name for the MWI was the "relative state" interpretation (1957). The rough idea underlying the name is that after two systems are entangled there is no meaning to discussing their states independently of each other; their states only have meaning relative to each other. For instance after a measuring device interacts with the system, it is no longer possible to describe either system by an independent state; the only meaningful descriptions of each system are relative states, such as the relative state of the system given the state of the measuring device or the relative state of the measuring device given the state of the system. The "branching" aspect of the MWI can be introduced into this relative state picture in several different ways; e.g., in Bryce deWitt's (1972) popular version, it comes in via hypothesizing that the state of a system after measurement is given by a quantum superposition of alternative histories of the system – the "many possible worlds" in the past of the system that could potentially explain its present. For instance, the double slit experiment, when it shows quantum interference, can be viewed as showing a quantum superposition of two possible past universes: one where the particle went to the left and one where it went to the right. The idea is that both universes exist and when the observer is not entangled with the particles then, relative to the observer, they exist in superposition. On the other hand, when the observer becomes entangled with the particle, then the state of the particle exists only relative to the observer, and it exists in one of the two possible worlds (right slit or left slit).

Of course, one could look at histories of observers in classical physics too. The novelty in the MWI is that the various complete alternative histories are supposed to be superposed to form new states. (Or, in the language of exotic probabilities to be introduced below: that the various complete alternative histories are supposed to be reasoned about using complex-valued rather than real-valued probabilities.)

Measurement, in the MWI, involves no special collapse operation, but is simply modeled by applying the ordinary laws of quantum physics to the system comprising the observer and the observed. Each observation can be thought of as causing the universal wavefunction to change into a quantum superposition of two or more non-interacting branches, or "worlds"; and since many observation-like events are constantly happening, there is a continual process of branching in which universes are distinguished from universes are distinguished from universes.....

The MWI is simple enough on the surface – though of course wildly counterintuitive relative to ordinary experience – but when you scratch below the surface subtleties arise and different thinkers present different approaches. For instance, according to DeWitt, a "world" in MWI is more precisely defined as "a complete measurement history of an observer." On the other hand, according to the Stanford Encyclopedia's definition of the MWI:

> *A world is the totality of (macroscopic) objects: stars, cities, people, grains of sand, etc. in a definite classically described state.*

I have seen at least a couple dozen different interpretations of the MWI, all of which are philosophically about the same, and all of which are intended as English paraphrases of the same quantum mathematics.

It is conceptually desirable to say that some universe-branches in the MWI are more likely than others, but this interpretation requires some care. It's hard to give a "frequency interpretation" to the probability of a universe occurring (introducing multiple multiverses doesn't really help because then one has to know how to weight the different multiverses). On the other hand, there is an alternate philosophy of probability in which the notion of probability is divorced from the frequency interpretation, and probabilities are considered simply as "measures of plausibility." In this interpretation, it makes perfect sense to say that some branches are more plausible/possible than others.

But which ones? Physicist Juergen Schmidhuber (1997) has proposed an interesting answer, which is to use "speed of computation" as a prior probability here. He proposes that:

> *whenever there are several possible continuations of our universe corresponding to different wave function collapses, and all are compatible with whatever it is we call our consciousness, we are more likely to end up in one computable by a short and fast algorithm. A re-examination of split experiment data might reveal unexpected, nonobvious, nonlocal algorithmic regularity due to [this phenomenon].*

The idea is that the probability of a "next branch" in the multiverse could have something to do with the speed of computing that branch conditional on the present universe-state. And there may be a tie-in with decoherence here, in that in the case of macroscopic systems the possible future universes not involving interference effects are probably much algorithmically simpler.

If Schmidhuber's hypothesis is correct, then one can envision the universe as a kind of "universal mind" which is constantly making choices using computational simplicity as a guide. And, as I'll argue in later chapters, our individual minds are also constantly making choices using computational simplicity as a guide. However, the difference is that the universal mind gets to make every choice at once; it just has to weight them based on computational simplicity. On the other hand, we individual minds, when we make choices, have to choose just one possibility or a limited number of possibilities. Because we are "classical systems" with limited resources. Loosely speaking, the universe then comes out looking something like an infinite-resources mind, which reasons using complex probabilities. On a huge quantum computer, one might be able to "simulate" a mind that worked more like the universe as a whole —- a kind of middle level between individual and universal mind.

The Decoherence Approach to Quantum Measurement

The most popular take on the quantum-measurement-interpretation problem these days seems to be the "decoherence" approach, which focuses on the interaction of quantum systems with their environments. Generally, the decoherence approach is considered together with the MWI, though the two are not necessarily coupled. Dieter Zeh (see Joos et al, 1993) and Wojciech Zurek (1991) originated this line of thinking via developing mathematics demonstrating that, as soon as a quantum system interacts with an environment, it very rapidly "decoheres", meaning roughly that the various portions of its quantum wavefunction stop interfering with each other, because they get so wrapped up in their interactions with the wavefunctions of entities in the environment. This gives the impression of an "almost-collapse" of the wave function of the quantum system, and makes it appear essentially like a classical system. The striking thing is the small amount of environmental coupling that is required in order to induce massive decoherence.

This is an elegant and practical observation about quantum systems, with obvious and dramatic implications for measurement. However, from the orthodox Copenhagen-interpretation approach, there is still something missing here conceptually: the gap between almost-collapse and collapse. The decoherence program doesn't supply the kind of "collapse" that the Copenhagen interpretation wants. It almost does, but not quite. In the decoherence approach, everything is really a quantum system, but some things don't act that way because of the peculiar and special condition into which their wave functions have evolved. This matches fine with the MWI, however.

Going back to the double-slit experiment, the decoherence approach tells us something about the nature of that mysterious "observation" process. Let's begin with a question: What if, instead of just not observing which way the electron goes, you make an apparatus that prints out which way the electron went on a piece of paper – but then don't look at the paper. What if you print out the information on the paper and then burn the paper? Then is there interference or not?

The answer is, there is not interference, and the reason is a subtle one. The problem is that, if we print which way the electron went on a piece of paper, then it is possible *in principle* for someone to figure out which way it went. Even if we burn the paper, the act of printing on the paper disrupted a whole bunch of air molecules, in such a way that a sufficiently clever observer could study their perturbations and figure out what was printed on the paper and thus which way the electron went. And this "in principle observability" of the which-path information is enough to make the quantum interference go away! Peculiar, indeed, but that's the way the universe appears to work (so far as we can tell based on current data).

In the words of Anton Zeilinger (Rev.Mod.Phys., 1999, p.S-288):

The superposition of amplitudes [i.e. the appearance of quantum interference] is only valid if there is no way to know, even in principle, which path the particle took. It is important to realize that this does not imply that an observer actually takes note of what happens. It is sufficient to destroy the interference pattern, if the path information is accessible in principle from the experiment or even if it is dispersed in the environment and beyond any technical possibility to be recovered, but in principle 'still out there.'

An example of this principle is that if you try to do the electron version of the double-slit experiment in open air rather than in a vacuum, the electron's interactions with the air molecules will destroy the interference, even though in practice it would probably be impossible for human experimenters to reconstruct which slit the electron went through by measuring all the air molecules. According to quantum physics what is observed by us has to do with what is knowable by us in principle rather than what is known by us in practice.

Next, consider the plight of Schrodinger's poor cat from a decoherence perspective. The point of the Schrodinger's cat experiment is one about the differences between different subjective perspectives and how the elegant mathematics of quantum theory makes them all consistent though different. The many-worlds interpretation of quantum tells us that things look very different from "inside" the system than from outside – but that the inside and outside perspectives are ultimately mathematically consistent. From the perspective of an outside viewer the cat becomes a superposition of states which observe different outcomes. From the perspective of the cat living *in* the system, on the other hand, he sees a definite answer to measurements and it appears to him that the wavefunction of the system being observed has collapsed into a state consistent with his measurement. And amazingly the quantum math works out so that the different subjective realities (some of which involve the cat being superposed and some of which involve it being collapsed) are consistent with each other even though different.

But this still leaves the question of why the alternate-universe cats don't interfere with each other later. The answer, according to the decoherence approach, is that because the cat has many degrees of freedom the different branches decohere and can no longer interfere with each other (barring a very, very unlikely coincidence).

To put it a little differently: Because of the impossibility of isolating the cat from its environment (including the human observer), the (quantum) reality is that the state of the cat+observer system looks like a sum of a state in which the cat is alive and the observer is in some state (knowing or not knowing its state) and a state in which the cat is dead and the observer is in some other state (knowing or not knowing its state). Regardless of whether you know or not, any neutrino affected by both of you or any air molecule mildly perturbed by the thud of the cat as it falls will decohere the the two branches. So in fact the observer is never in a single state with the cat in a superposition relative to the observer. The two entities, cat and observer, are in a superposition together.

Another, more technical question arises here (non-mathematical readers may wish to skip ahead a couple paragraphs), which arises if one formulates the Schrodinger's cat situation in terms of density matrices and basis vectors. In this case, the key question becomes why the universe uses the basis:

```
|alive> and |dead>
```

(|alive> is standard quantum theory notation for the state of the system in which the cat is alive) rather than say:

```
|alive> + |dead>  and |alive> - |dead>
```

for the matrices modeling the Schrodinger's cat situation. Both of these bases span the space of possible worlds – any possible state of the system can be written as a linear combination of |alive> and |dead>, but also as a linear combination of |alive> + |dead> and |alive> - |dead>. So why do we model the situation as involving a collapse into "alive" versus "dead" instead of "alive+dead" versus "alive-dead"?

This seems like a simple question but in the book *The Nature of Space and Time*, which is a dialogue/debate between Stephen Hawking and Roger Penrose (Hawking and Penrose, 2000) these two great physicists manage to disagree on the issue. Penrose argues that quantum physics doesn't provide any means to make the choice but Hawking argues for a decoherence-style perspective. Reading that book, I found myself wanting to agree with Hawking but not quite understanding his point. In a fascinating e-mail conversation, Stephen Omohundro[36] provided me with an explanation clearer than Hawking's, though conceptually in the same vein. What Omohundro wrote to me was:

> *The problem with the alive+dead basis is that an observer in that relation to the cat would quickly split into two branches, one which sees alive and one which sees dead and the two branches would quickly be incoherent with one another and so they stop affecting each other.*
>
> *Thus the operation of the viewer's brain is only coherent when viewed in the alive vs. dead basis...*
>
> *I think that the ultimate answer to your question of why the dead/live cat basis instead of the dead+live/dead-live comes down to the fact that physical interactions (primarily electrostatic) occur at spatial points. For multi-particle objects, this prefers the basis which is a tensor product of spatially localized wavefunctions as the one in which eigenstates stay most coherent. This leads to the dead/live basis because dead+live isn't spatially localized. The very same cat atom is in a superposition of two locations corresponding to its live position and its dead position.*
>
> *Then that leads to the even more fundamental question of why physical laws are spatially localized. It occurs to me that physical space might only be apparent "from the inside". Abstractly we just have this Hilbert space with a vector spinning around. I wonder if there is an extreme version of Lakoff's embodied intelligence where space itself requires an embodied observer making measurements within the system.*
>
> *As a thought experiment to think more about this I've been thinking about an intelligent entity running on a cellular automaton Life board (do you know Conway's argument that a random infinite life board will evolve intelligent creatures that can move around, reproduce, and make decisions even though the underlying rules are deterministic?). The question is how much of the "physics" (i.e. the Life cellular automaton rule) can a Life creature figure out from its perceptions (e.g. certain kinds of glider collisions, etc.). I think it would definitely develop an internal model of space as a pair of integers based on the movements it can make. It would probably have to do some serious searching to go from its "particle phenomenology" (e.g. rules about its perception of glider crashes etc.) to find the simple "laws of the universe" that so succinctly explain them.*

Omohundro's latter point reminds me of an article I read recently by Ross Rhodes, entitled "A Cybernetic Interpretation of Quantum Mechanics" (Rhodes, 2001). Rhodes' argument is that the weirdness of quantum physics can be interpreted as evidence that we actually live in a computer simulation. Basically, he argues, quantum physics resembles the "code layer" underlying our physical reality, similar to the cellular automaton rules that Omohundro alludes to. Quantum nonlocality makes a lot of sense if you assume that all processing for the universe-machine is being done by some central CPU, for example. I find this an amusing speculation, at any rate! Along similar lines, Omohundro also suggested that special relativity (which says nothing in the

[36] Steve is an AI theorist with a physics PhD and a diverse research background; see http://home.att.net/~om3/

physical universe can travel faster than light) is a method one might well use in a universe simulation to enable comparisons to be made between objects traveling at similar speeds with many fewer bits of precision than would be required in a Newtonian simulation.

The Delayed Choice Quantum Eraser

To illustrate more clearly the peculiarities of the quantum world, in this section I'll run through a particular quantum theory experiment in more detail. This section is a bit more technical than the prior ones, and readers with very little physics background may want to skip it over. However, I present it here because I think the points raised aren't widely enough understood and are actually critical to the understanding of the nature of quantum reality. Frankly, these are issues that confused me somewhat for a long time and only became entirely clear to me in the course of discussing them with quantum-theory experts in the context of finalizing this chapter.

The specific situation I want to discuss here is a "quantum eraser" experiment that was proposed by Marlan Scully in 1982, and finally carried out in reality in 1999 (see Kim, Kulik, Shih and Scully, 1999). The idea of the experiment is to combine quantum entanglement (as in the EPR experiment) with the basic logic of the double-slit experiments, to enable a kind of "quantum erasure." Put loosely, the idea is that the quantum interference in the double-slit experiment disappears when the "which-path" information is obtained, but then reappears when this information is "erased." However, it is easy to interpret this loose formulation in misleading ways, which is why it's important to dig a little deeper.

The experimental setup used to explore this idea is illustrated in Figure 1.

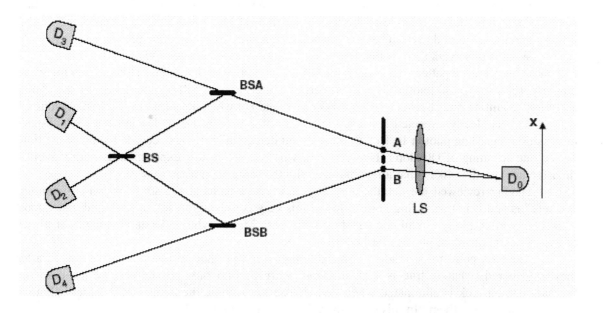

Figure 1. *A pair of entangled photons is emitted from either atom A or atom B by atomic cascade decay. Clicks at D3 or D4 provide which-path information, and clicks at D1 or D2 erase the which-path information (Kim et al, 1999).*

It is described in Kim et al as follows[37]. Two atoms labeled by A and B are excited by a laser pulse. A pair of entangled photons, photon 1 and photon 2, is then emitted from either atom A or atom B by atomic cascade decay. Photon 1, propagating to the right, is registered by a photon counting detector D0, which can be scanned by a step motor along its x-axis for the observation of interference fringes. Photon 2 (the "idler" photon), propagating to the left, is injected into a beamsplitter.

If the pair is generated in atom A, photon 2 will follow the A path meeting BSA (beam-splitter A) with 50% chance of being reflected or transmitted. If the pair is generated in atom B, photon 2 will follow the B path meeting BSB with 50% chance of being reflected or transmitted. Under the 50% chance of being transmitted by either BSA or BSB, photon 2 is detected by either detector D3 or D4. The registration of D3 or D4 provides which-path information (path A or path B) of photon 2 and in turn provides which-path information of photon 1 because of the entanglement nature of the two-photon state of atomic cascade decay. Given a reflection at either BSA or BSB, photon 2 will continue to follow its A path or B path to meet another 50-50 beamsplitter BS and then be detected by either detector D1 or D2, which are placed at the output ports of the beamsplitter BS. The triggering of detectors D1 or D2 erases the which-path information.

The point is that either the absence of the interference or the restoration of the interference can be arranged via an appropriately contrived photon correlation study. The experiment is designed in such a way that L0, the optical distance between atoms A, B and detector D0, is much shorter than the optical distance between atoms A, B and detectors D1, D2, D3, and D4, respectively. This means that D0 will be triggered much earlier by photon 1. After the registration of photon 1, we look at these "delayed" detection events of D1, D2, D3, and D4 which have constant time delays relative to the triggering time of D0. From the experimental setup, it follows that these "joint detection" events must have resulted from the same photon pair.

What is observed, then, is that the "joint detection" rate between D0 and D1 will show interference pattern when detector D0 is scanned along its x-axis. This reflects the wave property (both-path) of photon 1. However, no interference will be observed in the "joint detection" counting rates between D0 and D3 or D0 and D4 when detector D0 is scanned along its x-axis. This is clearly expected because the receipt of signals at D3 or D4 indicates which-path information regarding photon 1. But all these "joint detection" rates are recorded at the same time during one scanning of D0 along its y-axis. So what we see in this experiment is both wavelike (interference) particle-like behaviors, obtained with the same apparatus.

Having reviewed the details, it is important to understand the sense in which quantum "erasure" is and is not occurring here. I found this very confusing myself at first, and understood it fully only after a long e-mail conversation with Jesse Mazer, who understands the practicalities of experimental quantum physics far better than I do.[38]

The first point to be made is that, although it is true that no records are destroyed after they're gathered, what is true is that an *opportunity* to find out retroactively which path the "signal" photon took is eliminated when you choose to combine the paths of the "idler" photons instead of measuring them. In this sense there is a genuine "erasure."

---

[37] The next few paragraphs are paraphrased from Kim et al, with some sentences directly quoted.

[38] The ideas in the next few paragraphs were largely obtained via an e-mail dialogue with Jesse, and much of the text is paraphrased or quoted from his e-mails. However, I am the one responsible for the final form the present reportage of the discussion has taken; don't blame Jesse!

If the idler photon is detected at D3, then you know that it came from atom A, and thus that the signal photon came from there also; so when you look at the subset of trials where the idler was detected at D3, you will not see any interference in the distribution of positions where the signal photon was detected at D0. This is similar to how you see no interference on the screen in the double-slit experiment when you measure which slit the particle went through. Likewise, if the idler is detected at D4, then you know both it and the signal photon came from atom B, and you won't see any interference in the signal photon's distribution.

But if the idler is detected at either D1 or D2, then this is equally consistent with a path where it came from atom A and was reflected by the beam-splitter BSA or a path where it came from atom B and was reflected from beam-splitter BSB. So, in this case, you have no information about which atom the signal photon came from – and you'll get interference in the signal photon's distribution, just like in the double-slit experiment when you don't measure which slit the particle came through.

A very subtle point arises here. If you removed the beam-splitters BSA and BSB, then you could guarantee that the idler would be detected at D3 or D4 – in this situation the path of the signal photon would be known. Similarly, if you replaced the beam-splitters BSA and BSB with mirrors, then you could guarantee that the idler would be detected at D1 or D2 and thus that the path of the signal photon would be unknown. By making the distances large enough you could even choose whether to make sure the idlers go to D3 and D4 or to go to D1 and D2 *after* you have already observed the position at which the signal photon was detected, so in this sense you have the choice whether or not to retroactively "erase" your opportunity to know which atom the signal photon came from, after the signal photon's position has already been detected.

This seems at first to be a violation of causality, because it seems to imply that your later choice determines whether or not you observe interference in the signal photons earlier. But there is a trick here, which is observable if one looks at the graphs in the Kim et al paper, reproduced here exactly as Figures 3, 4 and 5. This trick is an important one and gets at the essence of quantum erasure and quantum measurement.

Figure 3 shows the interference pattern in the signal photons in the subset of cases where the idler was detected at D1, and Figure 4 shows the interference pattern in the signal photons in the subset of cases where the idler was detected at D2 (the two cases in which the idler's "which-path" information is lost). Both of these graphs show interference — but if you line the graphs up against each other, you see that the peaks of one interference pattern line up with the troughs of the other. So the subtle "trick" here is that if you add the two patterns together, you get a non-interference pattern just like if the idlers had ended up at D3 or D4. What this means is that even if you did replace the beam-splitters BSA and BSB with mirrors, thus guaranteeing that the idlers would always be detected at D1 or D2 and that their which-path information would always be erased, you still wouldn't see any interference in the total pattern of the signal photons; only after the idlers have been detected at D1 or D2, and you look at the subset of signal photons whose corresponding idlers were detected at one or the other, do you see any kind of interference.

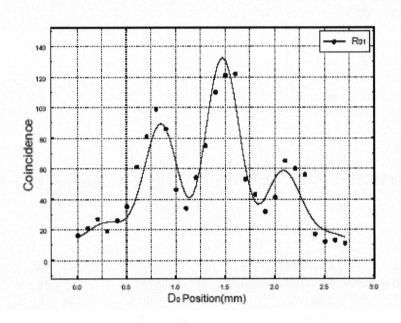

Fig. 3 from Kim et al

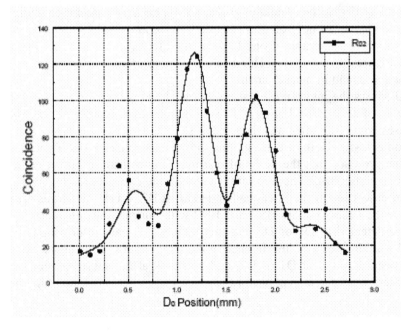

Fig. 4 from Kim et al

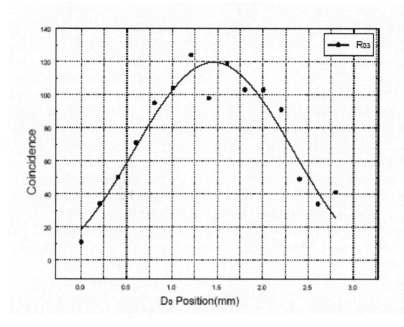

Fig. 5 from Kim et al
(the sum of Figures 3 and 4, with no interference)

After Jesse Mazer explained all this to me, I asked him to apply the same reasoning to a more complex situation. I wanted to insert a human inside the quantum eraser, instead of just a bunch of machinery involving little particles. I wasn't sure exactly what apparatus I wanted to use here, but Jesse came up with a good suggestion: "Imagine," he wrote:

> *something like the double-slit experiment with an electron, except that the slits are on one side of a closed box whose insides resemble a cloud chamber, with the electron gun on the inside of the box on the opposite side. Imagine that this box is an idealized one that can perfectly isolate the inside from any interactions with the outside world, along the lines of the box in the Schrodinger's cat thought-experiment (perhaps the only way to realize this practically would be to simulate the insides of the box on a quantum computer). Now, if an electron comes out of the slits and hits a screen, then if we immediately open the box and look inside, we'll probably still be able to see the path the electron took through the cloud chamber, and thus we'll know which slit it went through. On the other hand, if we wait for a long time before opening the box, the insides will have gone back to equilibrium and we'll have no way of telling which slit the electron went through.*
>
> *In analogy with the quantum-eraser experiment, no matter which we do, I don't think you'll see an interference pattern in the total pattern of electrons on the screen. But, again in analogy with the quantum-eraser experiment, if you were to look at all possible outcomes of measuring the exact quantum state of the inside of the box in the case where you wait a large time t to open it (and the number of possible distinct quantum states would be huge, because of the number of particles making up a cloud chamber), and then you performed the experiment an equally huge number of times so you could look at the*

subset of trials where the box was found in a particular quantum state X, then in that subset, my guess is that you'd see an interference pattern in the position that the electron was detected on the screen. But if you add up all the different subsets involving each possible quantum state for the inside of the box, my guess is that just as in Fig. 3 and Fig. 4, the peaks and troughs of all the various subsets would add together so the total distribution of the electron's position would show no interference. And on the other hand, if you opened the box immediately instead of waiting a large amount of time, then most of the exact states you would find when you open the box would show clearly which path the electron took, so that even if you looked at a subset of trials where the box was found in a particular quantum state Y, you'd still see no interference pattern in the electron, just like you don't see an interference pattern in Fig. 5 of the paper which looks only at the subset of trials where the idler ended up at D3 and thus was known to have come from atom A.

And of course, instead of just having the inside of the box contain a cloud chamber, you could have it contain some even more complicated macroscopic recording device, like a cloud chamber and a little man who can see the condensation track in the cloud chamber and remember it, and then you could choose whether to open the box and measure the state of the system inside while the man's memory was still intact, or after a bomb had gone off inside the box and made the information impossible to recover even in principle from a measurement of the state. The basic idea here should be the same — you'll never see interference in the total pattern of electrons, but if you repeat this experiment some vast amount of times and look only at the subset of trials where the inside of the box was found in a particular precise quantum state, then you may see interference in that subset, in the cases where the information has been erased (in this example, the cases where you wait until after the bomb has blown the man and his memories to smithereens).

As Stephen Omohundro pointed out to me, this is related to a simpler paradox involving quantum physics and faster-than-light causation. Suppose one does the EPR experiment of creating two particles with opposite spin. Suppose one sends one particle far away to Alpha Centauri and sends the other through a Stern-Gerlach magnet (a device that measures whether the spin of the particle is up or down), and lets the spin up and spin down outputs interfere in the context of a double slit experiment. If the faraway particle is measured up vs. down, the local particle must definitely go through the up hole or the down hole and we get no interference pattern. If he measures the faraway particle sideways we get a superposition of states and we get interference. Thus by rotating his measurement one should be able to communicate to us faster than the speed of light. We should see our pattern blinking between interference and not. What's wrong with this argument?

The problem, of course, is that there is no "change" to what a particular particle does when you observe its coupled pair in a certain way. Its individual results do not visibly change from non-interference to interference. (If that did happen, you'd have the basis for a faster than light communicator.) Instead, when you observe some of the particles sideways and some vertically, you are merely creating "correlational information" that exists only statistically as a correlation between that's happening in Alpha Centauri and what's happening locally.

The specific manifestation of this is another "miraculous" cancellation, like in the quantum eraser experiment. If one restricts attention to the cases where the faraway particle is measured "right" then interference is seen, and if one restricts attention to the cases where the faraway particle is measured "left" then interference is seen; but if one looks across all cases where the faraway particle is measured sideways, then the peaks and troughs of the different cases cancel out and no interference is seen.

Kim et al, in the conclusion of their paper, observe that "The which-path or both-path information of a quantum can be erased or established by its entangled twin even after the quantum has been registered." But the details work out so that this cannot be used to violate the forward causality of time on a macroscopic level. In fact, the phrasing they chose in their conclusion is telling. It is the entangled twin that can erase or establish the which-path information, not the human experimenter. The state of the entangled twin, when observed by a human, is random – and in the subset of quantum states where this random observation comes out a certain way, one observes quantum interference. But one cannot cause this random observation to come out a certain way. Random events occur which erase prior quantum events, but the patterns of these erasures.

Philosophically, what this means about causality would seem to be as follows. The past is not necessarily determined by the time the future occurs. And random events in the future can "cause" random events in the past, in the sense that one can make a setup guaranteeing that once they are observed, they will correlate with past events. But if there is any "backwards causation" going on here, it is restricted to the ability of random events to create backward causation.

And so – insofar as we can tell presently — quantum reality winds up to be almost mystical, but not quite. The quantum world violates locality and temporality, but always in a subtle way that doesn't violate causality because it doesn't allow faster-than-light information transmission – only backward-going "causation" triggered by *random* events. The unidirectionality of informational causality and the definitiveness of the macroscopic world are preserved in spite of the nonlocality and uncertainty underneath. The underlying quantum reality is exactly as "mystical" as it can get away with without violating the solid and causal nature of the humanly observed world.

Psychokinesis and Time Travel

Even though the quantum eraser experiments don't allow true "backwards causation," this doesn't prove that such a thing is impossible. It just proves that there is no way to do it within the commonly accepted constraints of physical law. There is at least once concrete possibility for how currently known physical law may be breakable, in a way that would allow backward causation (and, as an effective consequence, time travel – since being able to cause events in the past would mean being able to create an exact replica of oneself in the past, including a brain-state possessing the feeling of having just been quantum-magically transported into the past). This possibility is "quantum psychokinesis" – a notion which sounds bizarre, but is apparently supported by a variety of experiments done by respected scientists at various institutions including Princeton University.[39]

The simplest of these experiments involve people trying to influence, by the power of concentration, random events such as the direction of an electron's spin. A long list of experiments shows that, after some training, people have a weak but real ability to do this. Over tens of thousands of trials people can make electrons spin in the direction they want to 51% of the time or so, whereas chance would dictate merely 50%. This is a small difference but over so many trials is highly statistically significant.

---

[39] http://www.fourmilab.ch/rpkp/strange.html

Hooking this kind of PK experiment up to a quantum eraser apparatus, one would obtain a practical example of reverse causation. If this kind of PK actually works, then in the context of the above "paradox" situation, for example, it really would be possible for someone on Alpha Centauri to send messages faster than light to someone back home, via biasing the direction of spin of the coupled twin particle observed on Alpha Centauri. The rate of information transmission would be extremely low, since all that PK has ever been observed to do is give a slight statistical bias to events otherwise thought random. But with an appropriate code even a very slow rate of information transmission can be made to do a lot. And hypothetically, if this sort of PK phenomenon is actually real, one has to imagine that AI's in the future will find ways to amplify it far beyond what the human brain can do.

The (Possibly) Quantum Mind/Brain

The peculiar role of observation in quantum theory has caused some theorists to associate quantum theory with consciousness. This idea goes back to the 60's and quantum pioneer Eugene Wigner, and has more recently been adopted by a veritable army of New Age thinkers. Basically, Wigner (1962) proposed that the mysterious "collapse" in the Copenhagen interpretation may be associated with the activity of consciousness. This is an appealing idea, but as we'll see, this connection as commonly drawn runs into some serious conceptual problems.

As the above discussion of decoherence emphasized, quantum theory does not support any unique role for human observes as opposed to observers that are mice or experimental devices or molecules. So if there is a link between quantum theory and consciousness, it must be explicitly made within a panpsychist perspective, in which all these other things are assumed to have consciousness just like humans do. This is not necessarily philosophically problematic – in fact it is a view that I advocate – but it is not what Wigner intended when he introduced the consciousness/quantum-wave-function-collapse correspondence.

Relatedly, some "quantum consciousness" proponents have sought to establish a crucial role for quantum phenomena in the human brain. This is not as outlandish as it may seem; while quantum theory is generally considered a theory of the very small, it is not firmly restricted to the microworld. There are some well-known macroscopic systems that display elementary-particle-style quantum weirdness; for example, SQUIDs, Superconducting Quantum Interference Devices, which are little supercooled rings used in some medical devices (Gallop, 1991). A SQUID is around the size of a wedding ring, but its electromagnetic properties are just as indeterminate, and just as able to participate in peculiar nonlocality phenomena, as the spin of an electron.

Is the brain a macrosopic quantum device, like a SQUID? It's not impossible. If it's true, then all of contemporary cognitive neuroscience potentially goes out the window. Neuroscientists now commonly think in terms of "neural networks." The brain cells called neurons pass electricity amongst each other along connections called synapses, and this constant to-and-fro of electricity seems to give rise to the dynamics of mind. Thoughts and feelings are considered as patterns of electrical flow, and synaptic conductance. But if the brain is a macroscopic quantum system, all this electrodynamics may be epiphenomenal – the high-level consequence of low-level quantum-consciousness-based cognitive magic.

It's a wonderful story; at present, however, there's basically no evidence that the brain's high-level dynamics display quantum properties. The Japanese physicists Jibu and Yasue, in their book *Quantum Brain Dynamics and Consciousness* (1995), put forth a seductive hypothesis regarding quantum effects in water megamolecules floating inbetween neurons. But the jury's still out; in fact, from the point of view of the scientific mainstream, there's not even enough cause yet to convene the jury in the first place. We don't currently have enough evidence to verify or refute such theories.

It's quite possible that there are quantum effects in the brain, but without the dramatic consequences that some theorists associate with them. Perhaps the quantum effects aid and augment the neural network dynamics traditionally studied; perhaps they're just one among many factors that go into our experience of consciousness. Or perhaps the skeptics are right, and the brain is no more a quantum system than an automobile – which is also made out of tiny particles obeying the laws of quantum physics.

Quantum Theory and Consciousness

My own view of the relationship between quantum reality and consciousness is a little different. Unlike what Eugene Wigner suggested back in the 1960's[40], we can't quite say "consciousness is the collapse of the wave function" because in the decoherence approach the wave function does not collapse – there are merely some systems that are almost-classical in the sense that there is minimal interference between the different parts of their wave function. We can always say "everything is conscious" but this doesn't really solve anything – even if everything is conscious, some things are more conscious than others and the problem of consciousness then is pushed into defining what it means for one thing to have a higher degree of consciousness than another.

The analogue of "consciousness is the collapse of the wave function" in the decoherence approach would seem to be "consciousness is the process of decoherence." I propose that this is actually correct in a fairly strong sense, although not for an entirely obvious reason.

Firstly, I suggest that we view consciousness as "the process of observing." Now, "observation," of course, is a psychological and subjective concept, but it also has a physical correlate. I suggest the following characterization of the physical substrate of observation: *Subjective acts of observation physically correspond to events involving the registration of something in a memory from which that thing can later be retrieved.*

It immediately follows from this that observation necessarily requires an effectively-classical system that involves decoherence. But what is not so obvious is that all decoherence involves an act of observation, in the above sense. This is because, as soon as a process decoheres, the record of this process becomes immanent in the perturbations of various particles all around it – so that, in principle, one could reconstruct the process from all this data, even though this may be totally impractical to do. Therefore every event of decoherence counts as an observation, since it counts as a registration of a memory that can (in principle) be retrieved.

Most events of decoherence correspond to registration in the memory of some fairly wide and not easily delineated subset of the universe. On the other hand, some events of decoherence are probabilistically concentrated in one small subset of the universe – for example, in the memory of some intelligent system. When a human brain observes a picture, the exact record of the picture cannot be reconstructed solely from the information in that brain – but a decent approximation can be. We may say that an event of registration is approximately localized in some system if the information required to reconstruct the event in an approximate way is contained in that system. In this sense we may say that many events of consciousness are

---

[40] I suspect Wigner's perspective would be a bit different these days if he were still alive. Most likely he'd be extremely excited by the work on decoherence done during the last couple decades, and would revise his views accordingly. Unfortunately we'll never know exactly how.

approximately localized in particular systems (e.g. brains), though in an exact sense they are all spread more widely throughout the universe.

So, just as the Copenhagen-interpretation notion of "wave function collapse" turns out to be a crude approximation of reality, so does the notion of "wave function collapse as consciousness." But just as decoherence conceptually approximates wave function collapse, so the notion of "decoherence as registration of events in memory as consciousness" conceptually approximates "wave function collapse as consciousness."

How is this insight reflected in the language of patterns? If a system registers a memory of some event, then in many cases the memory within this system is a pattern in that event, because the system provides data that allows one to reconstruct that event. But the extent to which a pattern is present depends on a number of factors: how simple is the representation within the system, how difficult is the retrieval process, and how approximate is the retrieved entity as compared to the original entity. What we can say is that, according to this definition, the recognition of a pattern is always an act of consciousness. From a physics point of view, not all acts of consciousness need to correspond to recognitions of patterns. On the other hand, if one takes a philosophical perspective in which pattern is primary (the universe consists of patterns) then it makes sense to define pattern-recognition as identical to consciousness.

Of course, none of this forms a solution to the "hard problem of consciousness," which may be phrased as something like "how does the feeling of conscious experience connect with physical structures and dynamics?" I have already given my views on this philosophically subtler issue, in a previous chapter. But an understanding of the physical correlates of consciousness is a worthwhile thing in itself, as well as an important component of the intelligent discussion of the "hard problem."

Quantum Probability Theory

One of my favorite ways of thinking about the relation between quantum reality and ordinary reality is the notion of "quantum probability theory," as developed in a series of papers by the physicist Saul Youssef (1995). (This is another somewhat technical section which the less math-and-physics-savvy reader may want to skip or skim over lightly.)

In ordinary probability theory, one assigns probabilities to events and then manipulates these probabilities using special arithmetical rules. Probabilities are numbers between 0 and 1, as in "there's a .5 chance of rain tomorrow" or "there's a .9 chance that Ben will say something obnoxious in the next 55 minutes." Probabilistic reasoning plays a large role in my Novamente AI system, and in this context I'll talk about it more in Chapter 15.

Ordinarily probabilities are real numbers, but one can also develop an alternate probability theory, which is also completely mathematically consistent, in which probabilities are complex numbers.[41] So we might say "there's a .5 + .6i chance that the electron will pass through this portion of the diffraction grating." This seems intuitively rather peculiar – but Youssef has shown that, if we consider probabilities as complex numbers, then after some fairly natural

---

[41] If you're not familiar with complex numbers, it would be too much of a digression for me to review the basic facts here, but you really should get a primer. There are plenty of good online tutorials, such as http://www.clarku.edu/~djoyce/complex/ . The basic idea is to introduce a new "number" called i, which represents the square root of negative one. This simple idea lets you solve all polynomial equations and leads to all manner of amazingly beautiful mathematics.

mathematical manipulations, the basic laws of quantum theory fall out as natural consequences.

Quantum probability theory provides a beautiful crystallization of the true weirdness – from a human-intuition perspective – of quantum reality. One must reason about the probabilities of observed phenomena using real probabilities. But, one must reason about the probabilities of unobserved phenomena using complex probabilities. Now, of course, one can never verify directly that these unobserved phenomena are actually obeying the conclusions of complex-probability inference. But assuming this allows one to make very simple explanations of the real-probability-tables one creates from tabulating the results of observations. So the simplest known explanation of the real probabilities one observes via one's laboratory equipment, is that when one isn't looking, the world is acting in a way that can only be predicted via complex probabilities! Somehow, complex probabilities are the logic of the unknowable, whereas real probabilities are the logic of the knowable.

Youssef's mathematics also leaves room for yet more exotic probability theories – involving not just the complex numbers, but abstract "numbers" drawn from higher algebras called quaternions and octonions. As Geoffrey Dixon (1994), Tony Smith (2006), John Baez (2002) and others have observed, the octonionic algebra has all sorts of subtle relationships with modern physics, including the Standard Model of strong and electroweak forces as well as general-relativistic gravitation. It seems quite possible to me that octonionic probabilities are part of the key to the long-quested unified model of quantum theory and general relativity – though I stress that this is purely a speculation; I haven't had the time to work through the copious mathematics that would be needed to try to prove (or disprove) such a thing.

Now, Youssef's approach presents an interesting "correspondence principle" issue, in that in the classical macroscopic world, probabilities are more accurately thought of as real numbers than complex numbers. The "wave function collapse" issue then becomes one of why complex probabilities tend to collapse into real probabilities. This of course ties in with decoherence. What happens is that, when interference terms disappear, complex quantum probabilities collapse onto the real line. What this means is that, from the perspective of a macroscopic observer, like a person, the directly perceived world will always appear to obey real probabilities. However, in order to explain why certain aspects of the directly perceived world behave as they do, it is necessary to model the unperceived portion of the world (e.g. the microworld) using complex rather than real probabilities. The principle is, in sum:

- Ordinary probability theory is the way to reason about uncertainty in the directly perceived physical world, i.e. in the portions of the subjectively perceived world that are considered as objective reality.
- Complex probability theory is the way to reason about uncertainty in the context of the unperceivable physical world, i.e. in the portions of the hypothesized objective world that are not in principle perceivable by any observer (because perceiving them would fundamentally alter their nature).

Real probabilities are for reasoning about the perceivable; complex probabilities are for reasoning about the imperceivable, which may nevertheless have perceivable implications. And, the simplest explanation we know for important aspects of the perceivable world (quantum theory) implies the hypothesis of an imperceivable world that has radically different properties from the perceivable world.

Quantum Theory, Time, and Complexity

A different way of thinking about the relationship between quantum theory, minds and brains is to introduce the notion of "complexity", as reviewed in some depth above. Qualitatively, what we mean by a complex system is one that has a lot of different patterns in its dynamics. As noted above, this usually comes along with a certain unpredictability in the details of system dynamics ("chaos"), for reasons that become clear when one studies mathematical dynamical systems theory.

In classical physics complexity is generally tied to "irreversibility" – this point was made beautifully by Ilya Prigogine in his classic work on nonlinear, far-from-equilibrium thermodynamics (1984).   This ties into chaos theory, because the defining characteristic of chaotic dynamical systems is that they lose information so quickly that it's pragmatically impossible to compute their past from their future. Quantum theory presents some difficulties for this perspective on complexity because the underlying equations of quantum theory are linear, so there is no irreversibility. There is no chaos. There is a burgeoning field of "quantum chaos" (Stockmann, 1999) but what is studied here is actually not chaos – it's the quantum systems that, in their "classical limit" (an ill-defined notion) — converge to chaotic systems.

It turns out, however, that in quantum mechanics the role of irreversibility is played by complex wave-function entanglements. The transition from a pure state to an entangled one plays the role of irreversibility. For a decohered system, there is a pragmatic irreversibility even though not a fundamental one. The memory of the system's past is there in the system according to quantum theory, but, extracting this memory may take huge amounts of computing power. Relative to any finite observing system (any observing system with a remotely humanly comprehensible amount of computing and observing power), decohered systems are effectively irreversible.

This has an interesting implication regarding the directionality of time – an implication that has been drawn out in beautiful detail by Julian Barbour in his book *The End of Time* (1994), though his work was more inspired by general relativity than quantum theory.   Since the basic equations of quantum theory are time-symmetric, it would seem that we could have macroscopic systems operating either forward or backward in time depending on the particular configuration of systems in the universe).   Thus we could have observers whose memories go in either direction. The question then becomes, is it possible for those of us going in one direction to encounter minds going in another direction? Or (my suspicion) will a single connected universe always go in one direction only, locking into a single pattern just as some crystalline structures may lock into one of several possible patterns but not two at once? As far as I can tell, the mathematics of decoherence doesn't answer this yet, due to unresolved issues regarding the interconnection of quantum theory and general relativity.

Anyway, even though there is no fundamental irreversibility in the universe according to quantum theory (or general relativity), if one views the universe at a suitably abstracted level, then there *is* irreversibility. If you assume any reasonable degree of coarse-graining (as is the case in the perspective of any real observer), then you are seeing decohered systems, and so you see irreversibility, because decoherence involves complex dynamics that requires massive computing power to track in detail.   Any intelligent system's subjective view of the objective physical world will display irreversibility — but at the quantum level, if we get rid of the inevitable coarse-graining, the irreversibility is seen to actually bottom out in theoretically-reversible (though not plausibly reversible) entanglements.

The result of this is that the world of macroscopic objects and events perceived by an ordinary macroscopic observer operates within a realm of real probabilities – and in order for some quantum event to register as observable within our real-probability realm; it must transform itself into a real-probability event as well, via entangling with us and therefore decohering. And this is related to the way we experience *time* – we experience it as moving forward in one direction; whereas in the quantum domain, causal effects can move backwards as easily as forwards, as illustrated by Wheeler's delayed-choice double-slit experiment, discussed above. Complex numbers are used in relativity theory to represent time; and in quantum theory, complex probabilities allow us to reason accurately about events that have impact both forward and backward in time. Real-probability inference, on the other hand, seems to work only when one lacks the nonlocal quantum interdependency that allows time-symmetric interaction. This is because entanglement forces an effective single-directionality. Even in a decohered system, one can infer the past from the future according to the laws of quantum physics – but it becomes incredibly computationally difficult compared to inferring the future from the past. In a coherent quantum microsystem, on the other hand, computing the past from the future and computing the future from the past are about equally difficult.

Thus, the "arrow of time" seems to come about through computational complexity, through information loss – it exists because it's easier to predict the future of a complex[42] dynamical system than its past. A chaotic dynamical system is one whose dynamics are so unpredictable as to look random; a complex dynamical system – the more interesting kind – is one that has some chaotic aspects, but which has some predictable patterns to its dynamics even though it's not predictable in detail. In the quantum realm, there are no strictly chaotic dynamical systems, but there are dynamical systems that appear chaotic to all macroscopic (decohered) observers due to the complex entanglements within them.

Viewed from the perspective of any actual observer, each moment of time that a chaotic or complex system evolves, it loses information, in a sense. This loss of information is the arrow of time; it is the information-theoretic version of the Second Law of Thermodynamics.[43] Of course, this information loss is often compensated by a pattern gain. Complex systems lose information about their initial conditions, which makes reversibility impossible; but they often do so by converging to "attractor" states, which embody interesting emergent patterns.

The logic of the observed physical universe, then, seems to be as follows. Complex number probabilities describe events that can propagate effects forwards or backwards in time. But complex dynamical systems (from the perspective of any actual observer) lose information about their initial conditions, and therefore in a complex dynamical system (as observed by any actual observer), one can no longer propagate effects backwards in time. Thus, where complex dynamical systems are concerned, one can no longer reason using complex probabilities, one has to reason using real probabilities. Therefore, when a quantum system becomes correlated with a complex dynamical system, it "collapses," meaning that it is only explicable in terms of real probabilities.

This point of view doesn't "explain" everything – it doesn't explain why reality is time-reversible and complex-probabilistic in the first place. What it does explain, I think, is why, if fundamental reality is time-reversible and complex-probabilistic, our observable reality is time-irreversible and real-probabilistic. It's because of the chaotic aspect of our complex dynamics.

---

[42] The word "complex" in the context of dynamical systems does not denote complex numbers; rather, it denotes a system whose dynamics are subtle and difficult to understand. This polysemy of the word "complex" in science is confusing, but it's there, so you may as well get used to it if you're not already.

[43] This point was first gotten across to me by the physicist Michel Baranger

But the chaotic aspect of complex dynamics appears to come along with attractor formation – i.e. with the formation of emergent patterns in complex systems. Emergent patterns require chaos which entails irreversibility which collapses wave functions. Therefore the observable universe of patterns will appear to follow real probabilities even if the hypothetical "underlying" world involves complex ones.

Quantum Computing

I was amazed, in mid-2004, to receive a spam e-mail encouraging me to apply for a job doing research into quantum computing – and for a startup company, not a government lab. Twenty years ago that would have seemed ridiculous. Today, it's just reality.... Quantum computing is not yet commercial technology, but it's getting there. The first quantum computers will be relatively simple, specialized machines; but their successors will be more general and powerful, and quite possibly we'll see a rate of improvement just as impressive as we've seen in the traditional computing industry.

It seems clear, purely based on superior miniaturization, that quantum computers will be much more powerful than conventional ones. The capability of exploiting quantum nonlocality and related phenomena in a computing context is even more exciting. However, physics theory places some interesting limits on just how profoundly superiod quantum computers can become.

For a while, some theorists thought that quantum computers might be able to compute "uncomputable" things – things that ordinary digital computers just can't compute. This turns out not to be the case. But even so, there are many things they can compute *faster* than ordinary computers. And in computer science, very frequently, speed makes all the difference.

What does "uncomputable" mean? Something is uncomputable if it can't be represented in terms of a finite-sized computer program. For instance, the number Pi=3.1415926235.... is *not* uncomputable. Even though it goes on forever, and never repeats itself, there is a simple computer program that will generate it. True, this computer program can never generate *all* of Pi, because to do so it would have to run on literally *forever* – it can only generate each new digit at a finite speed. But still, there is a program with the property that, *if* you let it run forever, then it *would* generate all of Pi, and because of this the number Pi is not considered uncomputable. What's fascinating about Pi is that even though it goes on forever and doesn't repeat itself, in a sense it only contains a finite amount of information — because it can be compactly represented by the computer program that generates it.

But it turns out that, unlike Pi, almost all of the numbers on the number line *are* uncomputable. They have infinite information; they can't be stored in, or produced by, any digital computer program. Although they're the majority case, these are strange numbers, because it's impossible to ever give an example of one of them. For, if one could give an example of such a number, by the very act of giving the example one would be giving a finite exact description of that number – but how can one give a finite description of a number that has infinite information and hence by definition has no finite exact description?

Mathematicians have proved that these uncomputable numbers "exist" in an indirect way, by showing that the set of all numbers on the number line is a *bigger kind of infinity* than the set of all computers[44]. Because of the strangeness of these ideas, some mathematicians think that the idea of a continuous number line should be discarded, and that mathematics should concern itself

---

[44] In the lingo of set theory: The set of computers has cardinality aleph-zero; the set of real numbers has cardinality aleph-one.

only with computable numbers, numbers that have some finite description that can be embodied in a computer program.

Uncomputable numbers have a lot to do with randomness. The digits of Pi are random in a sense, but yet in another sense they're non-random, because you can predict them if you know the formula. The digits of an uncomputable number are random in a much stronger sense: there's no formula or computer program that will allow you to predict exactly what they're going to be.

Because of this relationship with randomness, some scientists thought that quantum theory and uncomputability might be connected. They thought that quantum randomness might somehow allow quantum computers to compute these mysterious, seemingly ineffable uncomputable numbers. This notion was particularly appealing because of the conceptual resonance between uncomputability and consciousness. Uncomputable numbers exist, but you can never grasp onto them – they're elusive in the same vague sense that consciousness is. Perhaps quantum randomness, uncomputability and consciousness are facets of the same underlying mystery?

But, while it may be that in some sense these things all reflect the same mystery, it is *not* true that quantum computers can compute uncomputable numbers. In the mid-1980's, David Deutsch proved mathematically that a quantum computer can't compute anything special, beyond what an ordinary computer can do. The mysterious uncomputable numbers, which some mathematicians don't believe in, are also uncomputable for quantum computers. Deutsch was not the first to study quantum computing – there was earlier work by legendary physicist Richard Feynman (2000), Paul Benioff of AT&T Bell Labs (Benioff, 1998), and Charles H. Bennett of IBM Research (1995), among others. But Deutsch's paper set the field of quantum computing on a significantly more solid footing, mathematically and conceptually.

But if quantum computers can't compute anything new beyond what ordinary digital computers can then what good are they? Well, one other thing Deutsch discovered was that, in some cases, quantum computers can solve problems much faster than ordinary computers. They'll come up with the same answer, but their indeterminate nature allows them, in a sense, to explore multiple pathways to an answer at once, hence arriving at the right answer faster. The trick is that they can't be *guaranteed* to get to the right answer faster. In the worst case scenario, they'll take as long as an ordinary computer. But *on average*, they'll be vastly faster.

To understand the power of this, think about contemporary systems for encrypting information, for secure transmission over the Internet. The encryption algorithms in use today are based on factorization – to crack one of them, you'd need to be able to divide a very large number into its component factors. But factoring large numbers is an intractable problem for ordinary computers today. On the other hand, it's known in theory how, with a quantum computer, one can factor large numbers rapidly, on average. When these devices are built, we'll have to find different ways of creating codes…and fortunately, quantum computing also provides some of these.

Ordinary computers are based on bits – elementary pieces of information, which always take one of the two values 0 or 1. All traditional computer programs internally represent information as long sequences of bits. A word processing file, a computer program itself, the Windows operating system – all are represented as sequences of 0's and 1's, where each 0 or 1 is represented physically by the absence or presence of electrical charge at a certain position in computer memory, the absence or presence of magnetic charge at a certain position on a hard drive, etc. Quantum computers are based instead on what are called "qubits." A qubit may be most simply considered as the spin state of a particle like an electron. An electron can have spin Up or spin Down, or, it can have a superposition of Up and Down spin – spin, say, half Up and half Down; or three quarters Up and one quarter Down. A qubit contains more information than a bit – but in a strange sense, not in the same sense in which two bits contain more information than a bit.

Now, quantum theory is not the ultimate theory of the universe. The big thorn in the side of modern physics is the apparent irreconcilability of quantum physics with Einstein's general-relativistic theory of gravitation. Quantum theory talks about indeterminate waves; general relativity talks about curved spacetime; and no one knows how to translate between the two languages with complete accuracy. Mathematician Roger Penrose and a few others have speculated that the ultimate unified theory of quantum gravitation will yield a new kind of computing – quantum gravity computing – that *will* allow the computation of traditionally uncomputable numbers. Of course this can't be ruled out. No one knows what a quantum gravity bit would look like, nor how they would interact. However, the vast majority of physicists and computer scientists are rightly skeptical of Penrose's conjecture… it doesn't take a terribly sensitive nose to detect a scent of wishful thinking here.

Quantum Cognition

The next and final question I'll explore in this chapter is: Let's suppose one created (via quantum computing) or discovered an intelligent system whose cognitive operations were fundamentally based on quantum operations. How would such a mind work?

The answer to this question has many aspects. For instance, we will see here that quantum minds may use fundamentally different methods for storing knowledge in long-term memory, and may also possess types of short-term memory that are not possible for classical minds, alongside more classical types of short-term memory.

In my AI research work, to be briefly reviewed in Chapter 15, I have found it useful to decompose cognitive operations into two primary categories: probabilistic inference and evolutionary learning. The Novamente AI system is fundamentally based on these two operations; and one may argue that human brain dynamics reflects these two basic properties as well, with Hebbian learning (Hebb, 1948) serving as an approximate probabilistic reasoning engine, and Edelman-style Neural Darwinism (1987) serving as a neural implementation of evolutionary learning. Here I will explore the possibilities of quantum cognition by exploring how quantum dynamics allows both probabilistic inference and evolutionary learning to be done in fundamentally different ways. These differences, I suggest, are a large part of what will make future quantum minds so very different from anything else yet conceived.

Let's start with probabilistic reasoning. Although humans aren't always good at making correct probability estimates, I believe that, nevertheless, probabilistic notions are essential to the operation of the human mind – and to the operation of any mind. This is one of the few aspects of mind, I believe, that is common to both huge and modest resources AI systems. Hutter's infinite-resources AIXI system utilizes notions from statistical decision theory; and similarly, I believe, any modest-resources mind is going to do a lot of things that are equivalent to making probability estimates. Novamente uses probability theory according to a particular formalism called Probabilistic Logic Networks.

But what happens to probabilistic reasoning when one introduces complex, quantum probabilities?

One may, perhaps, view the distinction between classical and quantum minds as the distinction between minds that reason using real-number probabilities only, and minds that reason using both real and complex number probabilities.

Of course, real number probabilities are a subset of complex number probabilities; but if one knows one is dealing with real number probabilities only, one can make a lot of simplifications, so that a mind specialized for reasoning about elementary particles using quantum probabilities wouldn't necessarily be good at reasoning about macroscopic systems like baseballs and budgerigars using real probabilities.

The human mind is clearly specialized for real probabilities. The current Novamente design is similarly specialized. But one of the many flexibilities associated with a digital software implementation is the ability to shift between different underlying reasoning logics. Creating a Novamente that reasoned using quantum probabilities rather than real ones would require only small changes to the source code. In fact, if a Novamente system were presented with a lot of sensory data from the quantum domain, it would quite likely modify its inference code to make use of quantum probabilities – a large change in conceptual perspective, but not in terms of software code.

But what does reasoning using complex probabilities mean? This is hard for the human mind to grasp, but some aspects may be somewhat readily understood. For instance, in ordinary probability theory, if one has

```
P(X or Y) = P(X)
```

(read this "the probability of 'either X or Y' equals the probability of X") then this implies that Y either never happens or is a subcase of Y, so it implies that either Y=X or P(Y) < P(X). For instance one might have

```
P((a person is between age 20 and 30) or (a person is between age 25 and 26)) =
P(a person is between age 20 and 30)
```

On the other hand, in quantum probability theory, one can have P(X or Y) = P(X) even if Y is neither the same as X nor in any case "less likely than" Y. For instance, if P(X) = 1 and P(Y) = i and X and Y are independent then one has

```
P(X or Y) = P(X) + P(Y) - P(X) P(Y) = 1 + i - 1*i = 1   = P(X)
```

This simple calculation illustrates the different meaning that "or" has in quantum probability theory. The whole concept of "alternatives" doesn't mean the same thing in quantum probability theory as in ordinary probability theory, because quantum or-ness embodies the possibility of "synergetic" quantum effects between the alternatives. In the simple calculation given above what's happening between X and Y is a kind of "destructive interference" where the additional contribution of Y to "X or Y" is being cancelled out.

However, once the results of quantum probabilistic inference are collapsed and turned into macroscopic measurements, they will inevitably result in computations that follow classical probability theory – the difference, as always, being that some computations can be done vastly more quickly via the utilization of quantum effects.

If my above speculations about the foundations of physics are correct, then complex dynamical systems are necessarily "classical" in nature because they involve irreversible, chaotic dynamics which can only be reasoned about using real probabilities. However, this doesn't mean that complex dynamical systems can't be designed to use dynamically simple subcomponents that make use of quantum-dynamical computation-acceleration magic in predictable ways. All it means is that, if my speculations are right, there may never be a purely quantum mind – because mind relies on emergent pattern, and emergent pattern relies on complex dynamics, which relies on irreversibility, which kills time-symmetry and ergo kills complex probabilities and quantum weirdness.   But even an impurely quantum mind, able to dip strategically into the quantum domain for complex-probability-superpowered calculations, would be a wonderfully powerful and amazingly *different* thing.

Evolutionary Quantum Computing

In examples like factoring and searching, one is coaxing a quantum computer to behave with the inexorable precision of a standard digital computer. Another approach, discussed in a paper I wrote in 1999, is *evolutionary* quantum computing. Evolutionary quantum computing is an extension to the quantum domain of a technique in conventional computer science called "evolutionary programming," in which one creates a computer program solving a given problem by simulating natural selection.

This approach tries to infuse quantum computing with some of the creative imprecision of living systems (which of course, at the molecular level, are complex quantum systems in their own right).   This is an exciting idea technologically, and is also philosophically resonant if one believes (as I do) that evolutionary learning is a critical aspect of cognition in general, due to its power for creativity.

In standard evolutionary programming (Koza, 1991), one solves a problem via creating a population of candidate programs, and evaluating how well each of them does at solving the problem. One takes the better programs from the population and combines them with each other, in the manner of the DNA crossover operations by which, in sexually reproducing species, a mother and a father combine to yield a child. The population of programs evolves over time, by the dynamic of "survival of the fittest programs," where fitness is defined by efficacy at solving the problem at hand. This is a very effective approach to a variety of computer science problems. Novamente currently uses a related but more sophisticated technique, an "Estimation of Distribution Algorithm" extending the Bayesian Optimization Algorithm (Pelikan, 2002), which develops the philosophy of evolutionary programming in a direction influenced by probability theory.

The beauty of evolutionary computing in the quantum domain is that it provides a partial way of getting around the decoherence problem that plagues modern quantum computing (which is that, in most cases, after one sets up a quantum computing system, its components tend to "decohere" and hence lose their wonderful quantum weirdness). In theory, one can evolve a bunch of quantum "programs" solving a target problem *without ever looking inside the programs to see how they work*. All one has to observe is how well they solve the target problem. Observing a quantum system is one thing that causes it to decohere, because when you observe it, its wave function gets smeared up with yours – the famous observer/observed interaction.   The ability to create quantum computer programs without observing them along the way – because, to create programs in an evolutionary manner, one only has to observe their behavior — may be very powerful, as quantum computing progresses.

If quantum theory does play a role in the brain, it is probably more in the evolutionary-quantum-computing style than in the manner of quantum algorithms for factoring and searching. Edelman (1987) and others have argued that the brain is an evolving system, with different "maps" of neural connectivity competing with each other to serve various cognitive, perceptive and active functions. If neural subnetworks in the brain are really molecular quantum computing systems, then the brain may be an evolving quantum computer. Who knows? At the present time, evolutionary quantum computing is still in the domain of theory — just like Schor's codebreaking quantum computers, and all other quantum computers except extremely simple examples.

But as I noted above, it seems most likely that, even if the human brain does use quantum effects on a macroscopic scale, it uses them in a not very powerful way. Let's face it; the brain isn't all that brilliant. A cognitive system with really flexible access to the quantum domain would be orders of magnitude cleverer than the human brain, and would also have a fundamental understanding of the microworld that we lack. It would reason fluently with complex as well as real probabilities, and would be able to search huge spaces of possibilities in parallel. Like any mind, it would still be nothing more than a set of patterns associated with an intelligent system – but the metapatterns structuring this set of patterns would be quite different than in our purely-classical-physics-based minds. I will return to this point now and then in later chapters, pointing out places where mind structures and dynamics may well come out differently for quantum versus classical cognitive systems.

Hugo de Garis (2003, 2003a) has written a series of papers exploring the notion of quantum evolutionary computing in detail, focusing on the notion of quantum neural networks. He has spelled out in detail how one could construct a quantum neural network and evolve it the quantum way, allowing different sets of inter-neuron weights to be experimented with in different possible universes. Based on the particular set of numerical parameters he explores, he concludes that the quantum evolutionary process would give a 100 times speedup over the classical evolutionary process. This may seem not much but, on the other hand, how much smarter would you be if you could think 100 times faster? And DeGaris's calculations, while fascinating, are relatively simplistic – there is no doubt that massively greater speedups could be obtained via more sophisticated architectures. Hugo and I gave presentations on this topic at an invitation-only workshop on Quantum Computing at MITRE Corporation in May 2006, and the leading quantum computing experts in attendance were enthused about our (admittedly speculative) ideas.

Quantum Associative Memory

Relatedly, Mitja Perus (2005) has explored the relevance of quantum dynamics for memory processes. He has taken a standard neural network model of memory, the Hopfield network, and modified it to make use of quantum dynamics.

A standard Hopfield network consists of a set of formal neurons (simplified mathematical models of brain cells) interconnected with each other. Each neuron may fire or not at any given point in time, and when it fires it passes charge to other neurons along its connections to them. The connections are weighted, and it's the weights of connections between the neurons that differentiate different neural networks from each other. These weights can be used to encode memories, and a single neural network can encode a lot of different memories this way – but what's interesting is that no one memory is encoded in any one neuron, nor even in any small group of neurons. Rather, each memory is encoded as a pattern of activity across the whole neural network, or a large portion thereof. Memory is "holistic"; adaptation of connection weights on the local level leads to intelligent memory behavior on the global level.

And when a Hopfield network is presented with a stimulus that only partially overlaps with something in its memory, the neurons will iteratively fire in such a way that causes the network to relax into a configuration corresponding to some pattern the network has previously remembers. If the network has remembered the images of 100 peoples' faces, and is then shown an image consisting of part of one of these faces, or a similar person's face, it will automatically iterate into a state of activity corresponding to the "best match" to its input.

Improvising on this standard approach, Perus describes a scheme for encoding information in a quantum Hopfield-like network. In the lingo of quantum theory, each memory is encoded in an eigenfunction, and retrieval is done using quantum holography. Reading out memory then involves collapse of the wave function.

Less technically, what this means is that, in the quantum Hopfield net, information is stored even more holistically than in an ordinary Hopfield net. Not only is each memory spread spatially over the whole network, it's spread across multiple parallel universes as well. Different memories don't just physically overlap, they're quantum-entangled. This gives even more effective performance than ordinary Hopfield nets – more memories stored with a given number of neurons. However, remembering a memory stored in the quantum Hopfield net involves causing a collapse of the wave function – the recollection of a memory involves the macroscopic registration of the memory.

The bottom line: just as evolutionary quantum computing allows faster evolutionary learning, quantum Hopfield nets allow more compact storage of information. And this greater time and space efficiency is not purchased via mere "tricks," it's purchased via distributing learning and memory across multiple parallel universes.

One hypothesis is that the human brain actually uses these sorts of mechanisms. But as I noted above, my suspicion is that this is either false, or true only in a very limited sense. I suspect that the subjective experience of a mind that heavily uses multiple-universe computation for learning and memory will be very difficult for us humans to relate to – even though the greater time and space efficiency that comes along with this different experience is easy for us to understand.

A New Kind of Mind

So what can we say about the potential nature of quantum minds? I've discussed quantum learning, memory and reasoning – and in each case we see two fundamental aspects:

- Much more efficient operation than is possible for classical systems;
- Distribution of learning, reasoning or memory across multiple universes.

But what will be the subjective experience of a quantum mind? Clearly its mind will run much faster than ours, which will make some difference to its experience. But this probably won't be the most critical difference.

A quantum mind will directly experience quantum coupling with its environment, which may mean that it doesn't feel as distinct from its environment as we do. And the different subcomponents within it will also experience nonlocal quantum coupling with each other – meaning that quantum experience may have an even greater sense of holistic unity than our own experience. It may be very difficult for a quantum mind to conceive of one part of its mind as separate from the others – because its reasoning, learning and memory relies on the interpenetration of the different parts of the mind. Then there may be a direct experience of the process of decoherentization, when the quantum interpenetration gives way to macroscopic definiteness, and the multiple universes collapse to a good approximation of just one.

While these speculations are interesting, there is no way we can really know if they're correct. The key point I want to make for the moment is the scientific possibility of the existence of quantum minds that possess a type of cognition completely different from our own. I see little credibility in claims that the human brain is a macroscopic quantum system in a strong sense, but I do think that macroscopic quantum cognitive systems are possible and will be both powerful and fascinating. Constructing such systems should be one of the major goals of 21'st century science.

Conclusion

I'm pretty sure that understanding quantum physics fully on an intuitive level is just impossible for any human being – because our mind/brains are tuned on an unconscious level for the macroscopic, non-quantum world. Fully and intuitively understanding quantum physics requires transcending humanity. That is: whoever succeeds in understanding quantum physics, will have necessarily succeeded in transcending many of the unconscious restrictions that characterize biopsychological "human nature." Whoever fully and intuitively understands quantum physics will not be understandable by humans! But, nevertheless, it is valuable to seek to rationally incorporate quantum theory within our human perspective.

The relation between the physics issued discussed here and patternist philosophy is a subtle one. On the one hand, the intricacies of physical theory do not fundamentally affect the principles of patternist philosophy. The view of the universe as a system of interlocking patterns remains sensible whether the dynamics underlying the shifting of patterns over time obeys real or complex probability theory. On the other hand, the relation of emergent patterns to their physical substrate may vary qualitatively to a large degree based on the particulars of physical law – a fact that leads to some interesting conclusions, particularly when it is viewed from a subjective as well as objective perspective. The conclusion that conscious awareness may be associated with quantum decoherence is not a trivial one and lives on the boundary between physics and philosophy. We may talk about patterns in the quantum domain, but according to this hypothesis, we cannot talk about patterns as having Firstness until/unless they are decohered. In quasi-Vedantic terms, Firstness does not touch quantum-maya in the same way as it does the other levels in the hierarchy (which is not surprising as it is the lowest level) – but perhaps some sorts of quantum computers may verge on a kind of quantum Firstness as they directly manipulate the process of decoherence in an explicit way that we do not.

Appendix: A Wild Speculation on Subjective Reality and Exotic Probabilities

In this Appendix I'll present a fairly wild speculation that takes its cue from the above discussion of quantum physics and applies some quantum-based ideas in a totally different context. I am sufficiently unsure of the value or meaningfulness of these ideas that I've placed them in this Appendix instead of in the body of this or another chapter of the book: yet the train of thought is sufficiently fascinating (if only as high-class science fiction) that I feel compelled to present it anyway!

What I will argue here is that there are fairly interesting analogies to be built as follows:

Analogy 1 [objective-reality-centric]:
Objective reality is to subjective reality, roughly as classical-physics reality is to quantum-physics reality
Rationale:
In both cases, the latter contain phenomena that are more "nebulous" than the measurable phenomena in the former, but are still useful for explaining phenomena in the former.

Analogy 2 [subjective-reality-centric]:
Objective reality is to subjective reality, roughly as quantum reality is to objective reality
Rationale:
Objective reality is built up from subjective reality, and contains things that are not subjectively real but are useful for explaining things that are subjectively real. Similarly, quantum reality is built up from objective reality, and contains things that are not objectively real but are useful for explaining things that are objectively real.

I will use these analogies to motivate a novel and interesting hypothesis regarding the nature of entities within subjective reality.

Please note, I am not claiming that subjective reality *is* quantum reality. That would be silly. I'm just making analogies. These analogies are conceptually evocative, and also seem to lead in some interesting mathematical and conceptual directions.

First I need to say something about the relationship between quantum physics and objective reality. This is a distinction that I fudged past in my earlier discussion of objective vs. subjective reality. Objective reality, as I discussed it there, is about things that we can imagine hypothetically observing if we were in a different position, a different situation etc. However, the reality portrayed by quantum physics is a bit different, because it consists of things that we could never, in principle, observe, no matter what. I don't think that it makes sense to consider in-principle-unmeasurable aspects of the hypothetical quantum universe as part of "objective reality," exactly. Rather, these things exist in a different domain – they "exist" in a different sense. They exist in the sense that postulating their existence is useful for explaining things that exist in objective, "classical" reality. This is similar to the sense in which postulating the existence of things in "objective reality" is useful for explaining things that exist in subjective reality. From a subjectivist perspective, objective reality is a useful hypothesis because it helps us explain observed patterns in our qualia, and quantum reality is a useful hypothesis because it assists objective reality in this job. This is Analogy 2 listed above.

Explaining quantum reality properly seems to involve qualities like "probability wave" that are meaningless in the classical world. These qualities can never be measured because as soon as they come into contact with a measuring instrument, they "vanish" — quantum laws only work for unmeasured entities! So we can, if we like, say that all these unmeasurable quantum properties are not "real." That's fine, but they're still very useful for explaining the results of measuring devices. Just as the ball that's rolled behind the chair is not "real" in the sense of immediate subjective reality, but is still useful for explaining the results of immediate subjective observations at other points in time.

Similarly, subjective reality seems to be reasonably well explicable using qualities like arity, solidity, centrality, intensity and historicity (and of course, other qualities that I haven't explicated here). These qualities can be measured in various ways (e.g. by asking people questions, or by doing brain measurements and correlating the results with peoples' verbal responses) but they also (so I suggest, and I realize this will be obvious to some people yet controversial to others) seem to have aspects that can never be measured — that are not fully captured via verbalization and are hence, so to speak, beyond the domain of measurement.

This is the analogy 1 listed above, and it leads up to a very interesting question: What is the right way to measure and manipulate uncertainty within subjective reality? In other words, if one has propositions about subjective reality, and wants to quantify their uncertainty, what's the right way to do it?

Cox's Theorem (Cox, 1961) shows nicely that "probability" is the right way to handle uncerta inty – but Saul Youssef showed that ordinary real-number probability is not the only kind of probability there is. One can construct probabilities obeying all the rest of Cox's axioms besides real-number-ness, by using probabilities drawn from any of three other algebras: the complex numbers, the quaternions or the octonions. Youssef showed that quantum theory can be derived from the assumption that one should use complex numbers rather than real numbers to measure uncertainty, together with some other simple assumptions.

Which leads up to the very interesting hypothesis that, maybe, subjective reality also calls for a non-real-number type of probability?

There is significant intuitive support for this, in the form of the intuition that subjective reality contains things that are "inexpressible in words" – i.e. that vanish when crisply expressed in language, in a similar sense to how quantum uncertainty vanishes when measured by a classical instrument. This kind of intuition is why Amit Goswami and others have posited that the human unconscious is a quantum system. Goswami ties in this subjectively experienced uncertainty of the human mind with the hypothesis that the human brain is a quantum system, but I don't make this same connection. I think it's possible the human brain is a "macroscopic quantum object" in important senses, but the notion I'm putting forth here is quite different from that. I'm suggesting that subjective experience has some of the same mathematical/conceptual properties as quantum reality, regardless of whether or not the physical system associated with that subjective experience relies on macroscopic quantum dynamics for its intelligent functionality.

Basically, the concept here is that in the subjective world, the dichotomy between X and not-X has a different sense than in objective reality. In objective reality (by which I mean "measurable reality", a la classical reality) we can't have two alternative possibilities both occur – at any particular point in time, the ball is either behind the chair or not, and the rock is either hard or soft. On the other hand, in quantum reality the electron can pass through both slit 1 and slit 2 (in the classic double-slit experiment). In subjective reality can for instance, one really be both happy and unhappy at the same time, in the sense of mathematical superposition?

There is a trivial sense in which one can be both happy and unhappy at the same time: the mind/brain is a large, complex system, and one part of it can be happy while the other part is unhappy. But this is not what I'm talking about; I'm proposing something more radical. The question I'm asking is whether it makes sense to think of subjective states as unresolved superpositions among multiple possibilities. If the answer is yes, then this means one wants to use non-real probabilities to reason about entities within subjective reality. This would be quite a major and exciting conclusion.

Note also that merely concluding one should use non-real probabilities to model subjective states doesn't resolve the issue of what kinds of non-real probabilities to use: complex, quaternionic, or octonionic. In the case of quantum reality the math only works out so as to agree with experiment if one uses complex probabilities. The situation with subjective reality is far less clear to me at the moment.

My intuition suggests a role here for octonions. Octonionic probabilities to measure the uncertainties of qualia! But I have nothing to substantiate this at present, except for vague intuitions. This is a train of thought I will definitely keep percolating in the back of my mind, waiting for a breakthrough.[45]

---

[45] If one wants to show a role for octonions here, the key is to come up with an argument why probabilities of subjective events would be nonassociative ... this would mean they have to be octonions rather than any of the other possible algebraic entities.

Chapter 9
Free Will

In this chapter, I'll present a novel, system-theoretic explanation of the human-psychological phenomenon of "free will," in terms of the dynamics and interactions of different parts of the brain. The ideas also have natural extensions to other kinds of minds such as AI systems. The theory presented integrates neuropsychological observations of Libet, Gazzaniga and other neuroscientists, but it has more generality and is more thoroughly linked with other aspects of philosophy of mind.

My belief is that, from a philosophical point of view, free will is actually a pretty simple phenomenon – in spite of all the confusion that's been attached to it. What has made free will seem so confusing has been mainly its intersection with consciousness, which is genuinely confusing due to its position at the bridge between the objective and subjective realms. Subtracting consciousness from free will, one obtains a subtle but not particularly paradoxical dynamical phenomenon resulting from the processing constraints on pragmatic intelligent systems. Of course, the cognitive dynamics underlying free will in brains or AI systems may display tremendous intricacy and complexity, but these don't necessarily add philosophical subtlety.

Virtual Multiverse Modeling and Free Will

At the core of my analysis of free will lies the concept of "virtual multiverse modeling." Suppose one has a world whose governing dynamic is difficult to predict. In the language of dynamical systems theory, this means we suppose that the world has a high Liapunov exponent (Devaney, 2003), which means that a small region of state space at time t is dynamically mapped into a much larger region of state space at size t+s, even if s is small. Then, an intelligent system (let's call it a "brain" for short — though it may be computational or biological – because we'll largely be discussing human brains in this chapter), in order to plan for the future, must create a *virtual multiverse* inside itself: i.e. at time t it must model several different future states for time t+s, since it doesn't know which future state will actually occur. It must create a virtual multiverse with branch-points regarding its own external actions, and its own internal events, as well as external events not directly caused by itself. This is what our brains do all the time.

The notion of a "multiverse" here is obviously motivated by quantum theory — however, the analysis of free will I am proposing here is not a "quantum theory of free will"; it is compatible with both quantum and classical physics. What I mean by a multiverse is a model of reality like the one explored by Borges in his famous tale *The Garden of Forking Paths* (Borges, 1999). Borges portrayed the world as consisting of pathways defining series of events, in which each pathway eventually reaches a decision-point at which it forks out into more than one future pathway. Borges' "paths" are the "branches" of the mathematical "tree structures" used to model multiverses; and his decision points are the nodes or "branch-points" of the trees. Actual reality is then considered as a single "universe" which is a single series of events defined by following one series of branching-choices through the mathematical tree. The many-universes interpretation of quantum physics posits that the multiverse is physically real, even though we as individuals only see one universe; and that an act of quantum measurement consists of a choice of direction at a branching point in the multiverse tree. On the other hand, what I am hypothesizing here is that we perceive a *psychologically real* multiverse – independently of whether there is a real physical multiverse or not – and that free will has something to do (details to follow shortly) with the choice of directions at branch-points in this psychological multiverse.

We know that the cognitive portions of brains do not directly experience the external universe; they only experience their own models of the external universe. This is demonstrated by many experiments regarding perceptual illusions, for example (see the very good discussion in Maturana and Varela, 1992). What this means is that, even if we should happen to live in a strictly deterministic universe, we subjectively live in a multiverse in which several different possible branches are subjectively real at any given time. But most of these branches are very short-lived: they exist only conjecturally while we wait for the next percepts which will tell us which of the branches is actualized.

Furthermore, brains largely experience themselves only via their models of themselves. Brains, being complex systems, are hard to predict even for themselves, and so one part of a brain often must use a virtual multiverse to model another part.

When a brain triggers a real-world action, this action occurs in the external universe, and then registers internally in the virtual multiverse which models the external universe. The brain is then aware of a process of "collapse" wherein the multiple branches of the virtual multiverse collapse to a single branch. Furthermore, this collapsing process occurs rapidly, within the same *subjectively experienced moment* as the actual event in the physical universe. Note that a subjectively experienced moment is not instantaneous from a physical-reality perspective.

Similarly, when a part of a brain carries out an action, and another part of the intelligent system is modeling this first part using a virtual multiverse, then the action in the first part corresponds with a collapse to a single branch in the virtual multiverse contained in the second part.

What I suggest, then, is that the special feeling of "free will" that we experience consists primarily of the subjectively-simultaneous consciousness of:

- An event occurring in the external universe,
- A collapse-to-a-single-branch occurring in the brain's internal virtual multiverse.

Or else the simultaneous consciousness of:

- An event occurring in one part of the brain,
- A collapse-to-a-single-branch occurring in the virtual multiverse used by another part of the brain to model the first part.

The subjective simultaneity is only present when the two things occur at almost the same physical time, which generally occurs only when the event in question is either internal, or else an external event that's directly triggered by the brain itself.

Benjamin Libet (2000) has done some critical experiments showing that, in many cases, the "decision" to carry out an action occurs *after* the neural signals directly triggering the action have already occurred. This observation fits in perfectly with the virtual multiverse theory. Note that this time interval is sufficiently short that the action and the decision occur within the same subjectively experienced moment. In fact, Libet's results, though often presented as counterintuitive, are explained naturally by the analysis I propose here – it's the *opposite* result, that perceived-virtual-multiverse-collapses occurred *after* the corresponding actions, which would be more problematic.

Daniel Dennett (2003) analyzes Libet's results by positing that free will is a distributed experience which occurs over an expanse of time (the experienced moment) and a number of different brain systems, and that there is nothing paradoxical about the part of this experience labeled "decision" occurring minutely before the part of this experience labeled "action trigger." I agree with Dennett's general observations – and with most of his comments about free will (unlike most of his comments about consciousness) – but I am aiming to achieve a greater level of precision in my analysis of the phenomenon.

For example, suppose I am trying to decide whether to kiss my beautiful neighbor. One part of my brain is involved in a dynamic that will actually determine whether I kiss her or not. Another part of my brain is modeling that first part, and doesn't know what's going to happen. A virtual multiverse occurs in this second part of the brain: one branch in which I kiss her, the other in which I don't. Finally, the first part comes to a conclusion; and the second part collapses its virtual multiverse model almost instantly thereafter.

The brain uses these virtual multiverse models to plan for multiple contingencies, so that it is prepared in advance, no matter what may happen. In the case that one part of the brain is modeling another part of the brain, sometimes the model produced by the second part may affect the actions taken by the first part. For instance, the part (call it B) modeling the action of kissing my neighbor may come to the conclusion that the branch in which I carry out the action is a bad one. This may affect the part (call it A) actually determining whether to carry out the kiss, causing the kiss not to occur. The dynamic in A which causes the kiss not to occur, is then reflected in B as a collapse in its virtual multiverse model of A.

Now, suppose that the timing of these two causal effects (from B to A and from A to B) is different. Suppose that the effect of B on A (of the model on the action) takes a while to happen (spanning several subjective moments), whereas the effect of A and B (of the action on the model) is nearly instantaneous (occurring within a single subjective moment). Then, another part of the brain, C, may record the fact that *a collapse to definiteness in B's virtual multiverse model of A, preceded an action in A.* On the other hand, the other direction of causality, in which the action in A caused a collapse in B's model of A, may be so fast that no other part of the brain notices that this was anything but simultaneous. In this case, various parts of the brain may gather the mistaken impression that virtual multiverse collapse causes actions; when in fact it's the other way around. This, I conjecture, is the origin of our mistaken impression that we make "decisions" that cause our actions.

The "illusion" of free will, therefore, consists largely of a mistaken impression gathered by some parts of the brain about the ordering of events in other parts of the brain. It is a simplifying pattern that the mind recognizes in itself – not an accurate model in detail, but a pattern that provides significant explanatory power based on a compact set of ideas. This simplifying pattern consists of confusion between two different roles played by virtual multiverse models:

- Assisting in the determination of actions (which happens sometimes, and with a significant time lag);
- Registering already-occurred actions (which happens more often, and almost instantaneously).

Because in the former, multiple-subjective-moment case, virtual multiverse collapse precedes action-determination, the brain mistakenly infers that in the latter, single-subjective-moment case, virtual multiverse collapse also precedes action-determination. But in fact, in the latter case virtual multiverse collapse follows action-determination.

However, it is not an illusion or confusion that virtual multiverse modeling has an impact on actions taken in the brain. This kind of modeling is clearly a very valuable part of brain dynamics, due to the complex and hard-to-predict nature of the brain and world. Virtual multiverse modeling is necessary due to *practical indeterminism* within and outside the brain, which exists whether or not *fundamental indeterminism* does. It is necessary because internal and external events are often *indeterministic from the subjective perspective of particular, useful parts of the brain.* Furthermore, and critically, *the brain as a whole is often indeterministic from its own subjective perspective.*

Confabulation

Another side of free will is the "confabulative" aspect emphasized by Michael Gazzaniga in discussing his famous split-brain experiments. These experiments demonstrate that, even when there is a clear external cause of a human taking some action, it is possible for the human to sincerely and thoroughly believe that the cause was some completely internal decision that they took. The left hemisphere of a split brain has no experience of stimuli delivered exclusively to the right hemisphere (e.g. through the left eye). However, the left hemisphere has such a strong motivation to create explanations that it will make up "free will stories" corresponding to behaviors initiated by the isolated right hemisphere. For example, in one experiment, a split brain subject's left eye received a command to stand. The person stood – and then, when asked why she stood up, she responded (using the language center of her left hemisphere) that she wanted a soda. In another experiment, when the left and right hemispheres were each asked to pick an appropriate picture to accord with an image flashed only to that hemisphere, the left selected a chicken to match the chicken claw in the picture it saw, while the right hemisphere correctly chose a shovel to remove the snow it saw. When asked why the person chose those images, he replied that the claw was for the chicken, and the shovel was to clean out the shed (Gazzaniga, 1989).

Confabulation means that, when a certain branch in the virtual multiverse has been chosen, the brain looks for reasons *why* it was chosen. If no immediate reasons are available, it will use inference to create reasons. Often these inferences will be accurate; sometimes they will be erroneous. Split brain surgery creates a situation in which erroneous inferences of this nature are much more common than usual. It happens that in humans this explanation-generating inference tends to take place in the left brain hemisphere; but the same post-facto explanation-generating dynamic may be expected to exist in nonhuman intelligences as well, regardless of whether their brains display any hemispheric dichotomy.

Confabulation adds a third aspect to virtual multiverse dynamics: not only do virtual multiverse inferences/simulations affect actions, and actions cause updating of virtual multiverse simulations; but also, reasoning about actions causes inferred stories to be attached to the memories of virtual-multiverse collapses.

Free Will and Consciousness

Next, I will tie this theory of free will into my prior discussion of the much subtler phenomenon of consciousness.

Some aspects of human consciousness can obviously be understood by thinking about the virtual multiverse models that parts of the brain construct, in order to model the brain as a whole. These virtual multiverse models are used to help guide the dynamics of the whole brain (on a slow time scale), and they are also continually updated to reflect the actual dynamics of the brain (on a faster time scale, occurring within a single subjective moment). So, the feeling of consciousness is in part the feeling of events in the whole brain being rapidly reflected in the changes in the virtual multiverse models maintained in parts of the brain ... and these changes then causing further virtual-multiverse-model changes which then feed back to change the state of the whole brain again ... etc. The conscious feeling of the flow of time is actually a feeling of continual ongoing branch-selection in the virtual multiverse model of the whole brain – the feeling of briefly-explored possible futures being left by the wayside as the actualized futures are registered in the model.

Dennett (1992) analyzed human consciousness as a serial computer running as a virtual machine on top of a parallel computer (the "parallel computer" being the unconscious, which comprises the majority of brain function). However, I don't think this is quite right. Rather, I think human consciousness has more to do with the feedback between virtual multiverse modeler software (embodied in various parts of the brain) and massively parallel software (the rest of the brain). The virtual universe modeler software is not exactly a serial computation process; it may well explore multiple branches in parallel.

These comments, I think, explain a significant amount about the particular nature of human consciousness. Clearly, neither these nor any other ramifications of the virtual-multiverse theory of free will will ever explicitly solve the "hard problem of consciousness" as addressed above — the relationship between subjective awareness ("qualia") and physical phenomena. However, it is not hard to see that the virtual multiverse model does fit in naturally with the philosophical solution to the hard problem given above.

Suppose one accepts, as a specific consequence of the solution to the hard problem given above, the postulate that *a quale occurs when a system comes to display a pattern that it did not display a moment before; and the more prominent patterns correspond to the more intense qualia.* Then, it follows from the present theory of free will that intense qualia will tend to be correlated with significant activity in the whole-brain virtual multiverse modeler. This provides an explanation for the oft-perceived correlation between consciousness and free will (free will also often being associated with significant activity in the whole-brain virtual multiverse modeler).

Predictions

What I have proposed here is a conceptual model of free will in terms of virtual multiverse modeling, but it also leads to some specific empirical predictions. For one thing, study of the human brain, as brain imaging improves, should allow us to localize the brain's multiverse modeling faculties (assuming these exist, as I hypothesize), and then to study whether the dynamics of interaction between these faculties and the rest of the brain are indeed as I have hypothesized.

Regarding artificial intelligence, the hypothesis made is that if an AI program is created with a virtual-multiverse-modeling faculty that is embedded into its overall dynamic process in a manner roughly similar to how this embedding occurs in the human mind/brain, then the AI will describe its decision-making experiences in roughly the same way that humans describe their experience of free will.

Chapter 10
Emotion

Much of my thinking on the topic of emotions has centered on the role emotions may (or may not) play in advanced artificial intelligences. Emotions play an extremely important role in human mental life – but it is not, on the face of it, clear whether this needs to be the case for minds-in-general (e.g. for AI's).

The vast majority of human emotional life is distinctly *human* in nature, clearly not portable to systems without humanlike bodies. Furthermore, many problems in human psychology and society are caused by emotions run amok in various ways – so in respects it might seem desirable to create emotion-free artificial intelligences.

On the other hand, it might also be that emotions represent a critical part of mental process, and human emotions are merely one particular manifestation of a more general phenomenon – which must be manifested in *some* way in any mind. This is the perspective I'll advocate here. I think the basic phenomenon of emotion is something that any mind must experience – and I will make a specific hypothesis regarding the grounding of this phenomenon in the dynamics of intelligent systems. Human emotions are then considered as an elaboration of the general "emotion" phenomenon in a peculiarly human way. There are a few universal emotions – including happiness, sadness and spiritual joy – which any intelligent system with finite computational resources is bound to experience, to an extent. And then there are many species-specific emotions, which in the case of humans include rage, joy and lust and other related feelings. On the other hand, the emotions of AI's may be very different from those of humans, and AI's need not be anywhere near as emotion-governed as humans.

What Is Emotion?

Let's begin with an idea drawn from the cognitive psychologist George Mandler (1975): Emotions have two aspects, which may be called *hot* versus *cold*, or "conscious-experiential-flavor" versus "neural/cognitive structure-and-dynamics" – or, using my preferred vocabulary, *subjective* versus *objective*.

From some conceptual perspectives, the relation between the qualia aspect and the objective, scientific aspect is problematic. But as you've seen in previous chapters, I follow a philosophy in which qualia and patterns are aligned – each pattern comes along with a quale, which is more or less intense according to the "prominence" of the pattern (the degree of simplification that the pattern provides in its ground). In this approach, the qualia and pattern aspects of emotion may be dealt with in a unified way.

So what is the general pattern of "emotion"? Dictionary definitions are not usually reliable for philosophical or scientific purposes, but in this case, a definition from dictionary.com is actually a reasonable place to start:

Emotion. *A mental state that arises spontaneously rather than through conscious effort and is often accompanied by physiological changes; a feeling:* the emotions of joy, sorrow, reverence, hate, and love.

One problem with this definition is its use of the mixed-up, excessively polysemous word "conscious." I will replace this with the term "free will" which, in the previous chapter, I sought to define in a general, physiologically and computationally grounded way. Thus I arrive at a definition of an emotion as:

Emotion. *A mental state that does not arise through free will, and that and is often accompanied by physiological changes*

"Free will," as I have explained it, is a complex sort of quale, consisting primarily of the registration of an (internal or external) action in an intelligent system's "virtual multiverse model," roughly simultaneously with the execution of that action. This generally goes along with the construction of causal models explaining what internal structures and dynamics caused the action. Sometimes, though, these two aspects are uncorrelated, giving the peculiar and fascinating feeling of "I don't know why I decided to do that."

Mental states that do not arise through free will, are mental states that are registered in the virtual multiverse model only considerably after they have occurred, thus giving a feeling of "having spontaneously arisen." This often goes along with the property of arising through such a large-scale and complicated – or opaque — process that detailed causal modeling is difficult. But sometimes, these two aspects are uncorrelated, and one can rationally reconstruct why some spontaneous mind-state occurred, in a reasonably confident way.

What causes mental states to register in the brain's virtual multiverse model in a delayed way? One cause might be that these mental states are ambiguous and difficult to understand, so that it takes the virtual multiverse modeler a long time to understand what's going on – to figure out which branch has actually been traversed. Another might be that the state is correlated with physical processes that inhibit the virtual multiverse modeler's normal "branch collapsing" activity – and that the branch-collapsing only proceeds a little later, once this inhibitory effect has diminished.

In the case of human emotions, the "accompaniment with physiological changes" mentioned in the above definition of emotion seems to be a key point. It seems that there's a time lag between *certain kinds of broadly-based physiological sensations in the human brain/body,* and *registration of these sensations in the human brain's virtual multiverse modelers.*

There are many reasons why this delay might occur. The phenomenon may be a combination of different factors, for instance:

- Since, in a state of strong emotion, the virtual multiverse modeling system is receiving constant powerful inputs from parts of the brain/body it has almost no control over, it doesn't bother to carry out detailed modeling (since this would be a waste of resources).
- Emotions bollix the virtual multiverse modeler because they are so profoundly indeterminate, and significant clarity about a given emotion-moment is often provided only based on the future of that emotion-moment. Each moment of an emotion helps us to interpret the previous moment as well as the following

moment. What I'm feeling right now is far clearer in the light of what I'll be feeling a moment from now. What this means is that, in the middle of an emotion, the virtual universe modeler doesn't know how to branch. It can branch in a very general way – "I'm now happy, not sad" – but its detailed branching-activity is in flux.[46]

And so, in regard to emotions, a flexibly superposed subjective multiverse is maintained, rather than a continually collapsed subjective universe that defines a single crisp path through the virtual multiverse. This helps explain both the beauty and the confusingness of emotions.

Regarding the second hypothesized factor, the obvious question is: Why do the broadly-based partly-physical sensations we humans call "emotions" have this strange relationship with time? This may be largely because they consist of various types of data coming in from various parts of the brain and body, with various time lags. A piece of sensation coming in from one part of the brain or body right now may have a different meaning depending on information about what's going on in some other part of the brain or body – but this information may not be there yet. When information gathering and integration regarding a "distributed action pattern" requires this kind of temporally-defused activity, then the tight connection between action and virtual-multiverse-model collapse that exists in other contexts doesn't exist anymore. Ergo, no feeling of "free will" – rather, a feeling of things happening in oneself, without a correlated "decision process." A strong emotion can make one feel "outside of time."

Furthermore, while it's easy to make a high-level story as to what made one sad or happy or feel some other emotion, it's not at all easy to make up a story regarding the details of an emotional experience. Usually, one just doesn't know – because so much of the details of the emotional experience have to do with physiological dynamics that are opaque to the analytical brain (unless the analytical brain makes a huge, massively-effort-consuming push to become aware of these normally unconscious processes, as is done in the course of the mental disciplines involved in some wisdom traditions).

So we have arrived at a more specific, technical, "mechanistic" and hypothetical definition of emotion:

Emotion. *A mental state marked by prominent internal temporal patterns that*
- *are not controllable to any large extent by the virtual multiverse modeling subsystem, or*
- *have the property that their state at each time is far more easily interpretable by integration of both past and future information,*
- *probably, though not necessarily, involve complex and broad physiological changes.*

What does this mean regarding the potential experiencing of emotions by nonhuman minds? Clearly, in any case where there's diverse and ambiguous information coming in from various hard-to-control parts of an intelligent system, one is not going to have the "usual" situation of virtual multiverse collapse. One is going to have a sensation of major patterns

---

[46] Of course, many physical-world situations could present this same kind of property: that the present, hard-to-determine state is reasonably-clarified only by reference to the future. However, these are not the physical situations in which we generally operate; they are not the ones that our systems are tuned for.

occurring inside one's own mind, but without any "free will" type "decision" process going along with it. This is, in the most abstract sense, "emotion." Emotions in this sense need not be correlated with broad physiological dynamical patterns, but it makes sense that they often will be.

Emotions in Humans, AI's and in General

Now we turn to the question of emotional typology. Humans experience a vast range of emotions. Will other types of minds experience completely different emotion-types, or is there some kind of general system-theoretic typology of emotions?

I think there will be a small amount of emotional commonality among various types of minds – certain very simple emotions have an abstract, mind-architecture-independent meaning. But the vast majority of human emotional nuance is tied to human physical embodiment and evolutionary history, and would not be emulated in an AI mind or a radically different biological species.

Any system that has a set of goals that remain constant over a period of time, can experience an emotion I call "abstract happiness," which is ***the emotion induced by an increasing amount of goal-achievement***. On the other hand, it can also experience "abstract sadness," i.e. ***the emotion induced by a decreasing amount of goal-achievement***. These emotions can become quite complex because organisms can have multiple goals, and at any one moment some may experience increasing achievement while others experience decreasing achievement.

Different flavors of happiness are then associated with different sorts of goals. For instance, there is the goal of increasing the amount of harmony (defined as, say, the amount of similarity with and the amount of emergent pattern produced together with) between the system and the rest of the universe. What I call "spiritual joy" is the feeling of increase in inner/outer harmony – i.e., the feeling of increasing achievement of the "inner/outer harmony" goal.

But why should increasing goal-achievement cause emotion in the sense I've defined it above? There are two aspects to this:

1. Factors tied to human evolutionary history;
2. Factors that are more based on information processing, and may apply beyond the human domain.

Due to the existence of these second factors, I suspect that happiness, sadness and spiritual joy are emotions with some universality. Due to the former factors, the specific flavor that these general emotions have in human beings is almost definitely peculiarly *human* in character.

In humans, achieving a goal like finding sex or finding a good place to sleep or killing prey or producing babies *naturally induces* broad and uncontrollable physiological changes. Achieving more abstract goals, in humans, tends to associatively bring forward patterns and processes associated with achieving these simpler primordial goals – thus activating broad patterns of physiological activity in ancient parts of the brain, and other parts of the body.

The evolutionary-history-bound nature of human emotions is well depicted in a snatch of dialogue from William Gibson's novel *Pattern Recognition* (2003, p. 69) – a discourse by an advertising executive on the importance of humans' odd cognitive architecture for his trade:

> *"It doesn't feel so much like a leap of faith as something I know in my heart."*
> *"The heart is a muscle," Bigend corrects. "You 'know' in your limbic brain. The seat of instinct. The mammalian brain. Deeper, wider, beyond logic. That is where*

advertising works, not in the upstart cortex. What we think of as 'mind' is only a sort of jumped-up gland, piggybacking on the reptilian brainstem and the older, mammalian mind, but our culture tricks us into recognizing it as all of consciousness. The mammalian spreads continent-wide beneath it, mute and muscular, attending its ancient agenda. And makes us buy things."

"... [A]ll truly viable advertising addresses that older, deeper mind, beyond language and logic."

What of specific human emotions like lust, rage and fear? Clearly these exist because we have specific physiological response systems for dealing with specific situations. Fear activates flight-related subsystems; rage activates battle-related subsystems; lust activates sex-related subsystems. Each of these body subsystems, when activated, floods the brain with intensive and diverse and hard-to-process stimuli, which are beyond the control of "free will" related processes. Many of the responses of these body subsystems are fast — too fast for virtual multiverse modeling to deal with. They're fast because primordially they had to be fast – you can't always stop to ponder before running, attacking or mating.

Clearly, a large portion of human emotion has to do with the virtual multiverse modeler's difficulties in modeling actions that come from the "older, deeper mammalian mind" and the yet more archaic reptilian brainstem. Yet, this kind of awkward fusion of old and new brains is not the sum total of emotion, human or otherwise. Let's return to the notion of abstract happiness as emotion which accompanies goal-achievement. When a human achieves a goal, the mammalian cortex responds in much the same way as it responds to the achievement of goals like finding food, getting sex, escaping from an enemy, or winning a fight. But the induction of these mammalian circuits is not the only reason for the virtual multiverse modeler to get confused into relative inactivity. There is also the fact that when a goal is achieved, not by a specific localized action, but by a complex coordinated activity pattern among many system components, this activity pattern may well have the property of being hard to model by the virtual multiverse modeler subsystems. So, peculiarities of human evolution aside, it seems some kinds of goal achievement are more likely to cause emotion than others, purely on information-processing grounds.

And what about the emotions of future AI systems? There's no doubt that, unless an AI system is given a mammal-like motivational system, its emotional makeup will vastly differ from that of humans. An AI system won't necessarily have strong emotions associated with battle, reproduction or flight. Conceivably it could have subsystems associated with these types of actions, but even so, it could be given a much greater ability to introspect into these subsystems than humans have in regard to their analogous subsystems.

Overall, my conclusion about AI emotions is that:

- AI systems clearly will have emotions.
- Their emotions will almost surely include, at least, happiness and sadness and spiritual joy.
- Generally AI systems will probably experience less intense emotions than humans, because they can have more robust virtual multiverse modeling components, which are not so easily bollixed up – so they'll less often have the experience of major non-free-will-related mental-state shifts.

- Experiencing less intense emotions does not imply experiencing less intense states of consciousness. Emotion is only one particular species of state-of-consciousness.
- The specific emotions AI systems will experience will probably be quite different from those of humans, and will quite possibly vary widely among different AI systems.
- If you put an AI in a human-like body with the same sorts of needs as primordial humans, it would probably develop every similar emotions to the human ones.

Later on I'll briefly discuss these issues in terms of the specific structures and dynamics of the Novamente AI system. The conclusion will be that there will likely be very major differences between Novamente psychology and human psychology as regards emotions: the strongest emotions of a Novamente system may be associated with the most complexly unpredictable cognitions it has — rather than, in humans, with phenomena that evoke the activities of powerful, primordial, opaque-to-cognition subsystems.

On the other hand, what can we say about emotions in the case of hybrid human-computer intelligence architecture like the "global brain mindplex" posited in Chapter 2 above? In this case, it seems, the main source of difficult-to-model unpredictability in the mindplex's mind will be the human component. Thus, the subjective experience of a global brain mindplex would likely be one of continually being swung around by strong emotions, corresponding to complex patterns of change in the human mass mind. Perhaps any disappointment future humans feel in losing some of their autonomy to the emergent metamind will be partially compensated by the knowledge that they're driving the metamind crazy!

Chapter 11
Evolution[47]

Hegel, in his *Logic*, proposed that the universe has two major aspects: Being and Becoming (the latter emerging as the synthesis of Being and Nothingness).   This rather obvious insight has a correlate in patternist philosophy; I often like to think of the universe as containing two central dynamics, which I call "autopoiesis" and "evolution."   Evolution is a well-known concept, though misconceptions regarding its nature are disappointingly commonplace even in the scientific literature.   Autopoiesis, on the other hand, is a more obscure term drawn from the system theory literature meaning, roughly: self-generation, self-creation, and self-perpetuation. Autopoiesis gives conservation and repetition of existing structures (i.e. maintenance of being); evolution gives the creation of new structures (i.e. becoming).

In this chapter and the next, I present a more "dynamical" view of the mind than I've given so far, based on the perspective of evolution and autopoiesis as the two primary "forces" in the mind. Many of the ideas given here are drawn from my previous books, but are recast here in a more compact and hopefully more elegant form; and there are also some significant new advances. Although some attempt will be made here to draw attention to other thinkers' related ideas, a fuller set of references to prior related ideas may be found in these earlier books.

Basic Principles of Evolution by Natural Selection

The reader who lacks a basic background in evolutionary biology is referred to any of the many excellent books on the subject, e.g. the writings on the subject by Daniel Dennett (1997), Richard Dawkins (1990) or Stephen Jay Gould (2002).   Here I'll give only a very quick and superficial run-through of the main currents of thinking in evolutionary theory, before launching into my own modifications, extensions and interrelations.

One way to understand evolution by natural selection is to contrast it with *artificial selection*. Artificial selection is what breeders do. They know that different plants and animals of the same species have widely varying characteristics, and they also know that a parent tends to be similar to its offspring. So they pick out the largest from each generation of tomatoes, or the fastest from each generation of horses, and ensure that these superior specimens reproduce as much as possible. The result is a consistently larger tomato, or a consistently faster horse.

Artificial selection is well understood, both in practice and in terms of mathematical theory. With artificial selection, however, the selection is merely a tool of the breeder, who sets the goal.   This is where natural selection differs.   If one wishes to use artificial selection as a model for natural evolution, one must answer the question: what is the goal and who sets it? The simple, brilliant answer arrived at simultaneously by Charles Darwin and Alfred Russel Wallace in the mid-1800s was: the environment sets the goal, which is species-level survival.

---

[47] The material in this chapter is partly drawn from my 1993 book *The Evolving Mind*

Much of the evolution of organisms may be understood, they observed, by assuming that organisms act so as to maximize the number of descendants (similar to themselves) they will produce. Of course, there is some fuzziness here: there are immediate descendants and far-distant descendants. But if one assumes that the goal of an organism is to maximize the number of descendants it has on various time scales, then one obtains a fairly good model of many aspects of biological evolution.

This is a fairly obvious statement, because classes of organisms with the tendency to produce a lot of descendants (similar to themselves) will tend to be the ones that flourish most over time, occupying more and more of the world. It is a tautology in the same sense that all mathematical statements are tautologous. It's a simple but nontrivial mathematical statement: If there are N objects falling into various classes, and each object may spawn zero or more new objects, then classes of objects that are more prolific at object-spawning will tend to dominate the population. This theorem follows almost immediately from the insight that the growth of groups within populations is exponential, an insight that Darwin derived partly from reading Malthus's pioneering work on population dynamics.

Darwin, in *The Origin of Species*, summarized his theory of natural selection as follows:

> *If under changing conditions of life organic beings present individual differences in almost every part of their structure, and this cannot be disputed; if there be, owing to their geometrical rate of increase, a severe struggle for life at some age, season or year, and this certainly cannot be disputed; then, considering the infinite complexity of the relations of all organic beings to each other and to their conditions of life, causing an infinite diversity in structure, constitution and habits, to be advantageous to them, it would be a most extraordinary fact if no variations had ever occurred useful to each being's own welfare, as so many variations have occurred useful to man. But if variations useful to any organic being ever do occur, assuredly individuals thus characterized will have the best chance of being preserved in the struggle for life; and from the strong principle of inheritance, they will tend to produce offspring similarly characterized. This principle of preservation, or the survival of the fittest, I have called Natural Selection.*

Herbert Spencer's phrase "survival of the fittest" captures one interpretation of Darwin's ideas admirably well: the idea of the struggle for existence, according to which various organisms battle it out for limited resources, and the best of the bunch survive. This interpretation of natural selection combined with Mendelian genetics, forms the theory we call "Neo-Darwinism" or "strict Darwinism" — an evolutionary research program which essentially dominated biology throughout the middle of this century. But it must be noted that the "struggle for resources" perspective is not implied by the general logic of evolution. There may be many other ways to achieve a large number of descendants in one's same category, besides competing for resources with other categories of organism. Struggling for resources is only one way for evolution to occur and not necessarily the most interesting or most prevalent way.

In the words of Stephen Jay Gould:

> *This strict version [of Darwinism] went well beyond a simple assertion that natural selection is a predominant mechanism of evolution.... it emphasized a program for research that almost dissolved the organism into an amalgam of parts, each made as perfect as possible by the slow but relentless force of natural selection. This "adaptationist program" downplayed the ancient truth that organisms are integrated*

entities with pathways of development constrained by inheritance - not pieces of putty that selective forces of environment can push in any adaptive direction. The strict version, with its emphasis on copious, minute, random variation molded with excruciating but persistent slowness by natural selection, also implies that all events of large-scale evolution (macroevolution) were the gradual, accumulated product of innumerable steps, each a minute adaptation to changing conditions within a local population. This "extrapolationist" theory denied any independence to macroevolution and interpreted all large-scale evolutionary events (origin of basic designs, long-term trends, patterns of extinction and faunal turnover) as slowly accumulated microevolution (the study of small-scale changes within species). Finally, proponents of the strict version sought the source of all change in adaptive struggles among individual organisms, thus denying direct causal status to other level sin the rich hierarchy of nature with its "individuals" both below the rung of organisms (genes, for example) and above (species, for example). The strict version, in short, emphasized gradual, adaptive change produced by natural selection acting exclusively on organisms. (1983, p. 13)

In its simplest form, strict Darwinism views an organism as a bundle of "traits." Random genetic variation causes traits to change, little by little, from one generation to the next. When a trait changes into a better trait, the organism possessing the improved trait will tend to be more successful in the struggle for existence, as compared to those organisms still possessing the old trait. Thus the better trait will tend to become more and more common. This point of view encourages the idea that every aspect of every organism is in some sense "optimally" constructed (Dupre, 1987).

The view which Gould urges, on the other hand, admits that it is impossible to decompose an organism into a vector of traits: that the qualities which characterize an organism are deeply interconnected, so that small changes induced by random genetic variation will often display themselves throughout the whole organism. An organism is produced by a vast and complex self-organizing process, which is triggered by its genome.

And Gould's view admits that "better" is not definable independently of evolutionary dynamics: the environment of an evolving organism consists partly of other evolving organisms, so that the organisms in an ecosystem may be "optimal" only in the weak sense of "fitting in relatively well with one another."

These two admissions form a large part of what one might call the "self-organizational" or "system-theoretic" theories of evolution — they interpret natural selection in terms of self-organization on the intra-organismic and inter-organismic levels. To put it a little differently, Gould and his ilk place a strong emphasis on the two phenomena of

- **Epigenesis:** that is, the creation of complex *phenotypic* forms from simple evolved *genotypic* forms;
- **Ecology:** that is, the strong influence of other evolving entities on the fitness of each individual evolving entity;

as well as the basic fact of differential reproduction based on fitness.

Unlike, say, evolutionism and creationism, or classical physics and relativistic physics, the primary relation between strict Darwinism and the self-organizational theory of evolution is not one of direct opposition. The difference between the two views is largely one of emphasis. Strict

Darwinism does not deny the existence of self-organization within organisms, nor does it deny the existence of complex ecological structures. And the self-organizational theory does not deny that, in some cases, evolution works by reinforcing traits that are just plain *better* than the alternatives in a way that does not depend sensitively on the details of local environments. However, the two views differ dramatically in their estimates of the *relative frequencies* of situations in which different phenomena are important. Strict Darwinism estimates that self-organization is only infrequently relevant to evolutionary analysis, and that trait-by-trait optimization is very often important. The self-organizational theory, on the other hand, gauges the relevance of trait-by-trait optimization to be small, and the frequency of significant self-organizing phenomena to be large. And the appropriate weighting of epigenesis and ecology is, I believe, critical to the application of evolutionary concepts to cognition.   For instance, by taking the right perspective on evolution, we can see very simply and clearly how emergent patterns of mental activity evolve by natural selection, as an automatic consequence of general principles of cognitive dynamics.

A very clear and simple exposition of the self-organizationist perspective on evolution was given by Robert Augros and George Stanciu in the 1980's, in their book *The New Biology* (1987).   Augros and Stanciu constructed what seems to me a fairly solid argument against the strict Darwinist approach in the evolution of species.   They describe evidence that organisms control their reproduction rates so as to avoid crowding their environment, meaning that competitive conditions don't arise that often.   They discuss the narrow niches that organisms often restrict themselves to, in order to avoid conflict with other species – for instance, a half-dozen species of bird can exist in the same tree by each colonizing a separate region of the tree.   And they point out there is no evidence that small gradual changes can accumulate to form big changes, such as the change from one species to another.   Rather, it seems, small changes in the genome must sometimes lead to huge changes in the organism developed from the genome – a consequence of subtle intra-organismic self-organization processes.

Note that all of these points are matters of relative frequency. Augros and Stanciu are saying that exponential population growth, competitive struggle and accretion of tiny changes into huge ones are *uncommon* occurrences. They are not saying that these things don't happen — all but the latter are positively known to occur, on occasion. Conversely, the strict Darwinists never said that these things happened *all the time* — only most of the time.

In short, the evidence suggests that in the ecosystem (as, I suggest, in other evolving systems like minds or societies), natural selection must be understood in the context of self-organization. The "fittest" organism is not always the one which beats out all its competitors in some sort of contest, but is more often the one which best makes a place for itself in the complex network of relations which its peers, together with the physical environment, create. There is a sort of contest to adapt most effectively, but it is a contest in which the playing field, and the rules of the game, vary a great deal from competitor to competitor (and are indeed created by a complex dynamic involving all of the competitors).

When the evolutionary process is viewed in the highly general way suggested by Gould and other systems-theory-oriented researchers, it can easily be generalized beyond the "evolution of species" context, and can be understood as nothing more or less than an analysis of the way the whole of a category of processes "reaches down" and affects each of the individual processes. The parts constitute the whole, but then the whole reaches down and guides the parts in their change over time.   This reach-down causes the elimination of some parts, the modification of others, and the creation of some new parts from old.   If one divides the parts of a system into categories, then categories containing parts that tend to produce a lot of new parts in that same category, will tend to proliferate – leading to a phenomenon of the "survival of the fittest categories," where fittest means "most profligate". In the standard evolutionary biology context, these "categories" are species and their members' organisms; in the evolving mindspace of an intelligent system, however, they may be abstract ideas and their members' concrete instantiations of these ideas. This very general and abstract view of the evolutionary process may be carried a bit further and connected with patternist philosophy, a step that I will take in a few pages, after a bit of preliminary exploration of related concepts.

Exploring the Concept of Fitness[48]

Much of my thinking about evolutionary biology has focused on coming to grips with the somewhat slippery notion of "fitness." This term "fitness" is used with many different meanings in evolutionary theory, and I won't discuss them all here. What I will do is to introduce a variation of the fitness concepts that I call "evolutionary fitness," which I define as "the tendency of a category of parts of a system to generate, in the future, the existence of a lot of system parts in the same category." A category of system parts that simply survives has more evolutionary fitness than one that dwindles; but the greatest evolutionary fitness is possessed by a category of system parts that, once it has some members present in a system, tends to increase its membership over time. One of the essential insights of evolutionary theory (a point that I'll elaborate below) may be formulated as: *Short-term evolutionary fitness generally predicts long-term evolutionary fitness.* That is, categories that persist themselves over the short-term future also tend to persist themselves over the longer-term future.

A simple example of the above principle is that, all else equal, a species that makes more babies now is likely to dominate more of the world in the future. But of course, this isn't all there is to it; there are a lot of other factors involved, which is why one makes abstract formulations rather than simply defining evolutionary theory in terms of rate of reproduction. A species that produced a huge number of babies but was overly inclined toward cannibalism, for instance, might have a very low evolutionary fitness except on a very short time scale. (I will return to this line of thinking a little later after I introduce the notion of "pattern sympathy.")

How does this very abstract notion of "evolutionary fitness" connect with the more concrete notion of fitness used by workaday evolutionary biologists? One of the big insights of evolutionary theory is that, in particular subsets of some systems (like ecosystems and, we'll see, minds and brains as well), it's possible to identify more specific "fitness functions" so that the evolutionary fitness of categories of system components, in a certain context, is proportional to the degree to which the components satisfy the fitness function. This is fascinating and important because it allows us to quantify in simple ways aspects of the evolutionary process that, without the identification of contextually appropriate "fitness functions," would remain comprehensible only in terms of very complex emergent ecosystem-level dynamics. For instance, it may be observed that in some contexts faster or stronger or smarter organisms are more likely to propagate themselves; or, in cognitive evolution, it may be observed that ideas embodying more intense patterns are more likely to survive and spread and prosper.

But these specific fitness functions like speed or compactness will never constitute a complete explanation of the evolutionary process. One of the insights of Gould and his ilk is that emergent pattern between organism and environment is incredibly critical for evolutionary fitness. Particular fitness functions correlating with overall evolutionary fitness are highly context-

---

[48] This section clarifies some issues that I discussed in a somewhat conceptually confused way in *The Evolving Mind*

dependent and time-variant, but the power of emergence as a fitness function is extremely general and pervasive.

This line of thinking leads to a perspective on evolution in which a central role is played by the concept of "emergence." It turns out that in many cases, a good way to estimate the evolutionary fitness of a category of entities is to see how much emergent pattern is generated between the entities in this category and the other entities surrounding them in the environment. In this sense, it is a decent guideline to interpret "fitness" in the sense of "most closely fit to the environment" — using the word "fit" as in the sentence "My coat fits well."

In *The Evolving Mind*, I suggested that one key indicator of the fitness of an entity in an environment is the "structural fitness," which I defined as the amount of emergent pattern generated between the entity and the environment it is embedded. Clearly this kind of fitness is not sufficient not make an entity "evolutionarily fit." But it is certainly a major *correlate* of evolutionary fitness, in the intuitive sense. To phrase it a little differently, what I hypothesized there is that: A population of entities evolves by natural selection if there is a reasonably strong correlation between the *structural fitness* of an entity and its *rate of reproduction*. Examples of this principle in the evolution of species are not hard to spot. Look at the parallel between a giraffe's long neck and the tall trees it eats from. The emergent pattern of tallness is not hard to spot!

This characterization of evolution has some interesting implications for the general relation between complexity and evolution. If Em(A,E-A) is large for every or most entities A in an population E, then this means that E as a whole has a lot of patterns in it. Because the emergent patterns between entities are patterns in the population as a whole. This gives us a very simple theorem: populations evolving by natural selection will necessarily be *structurally complex*. In other words, evolving systems tend to be highly patterned because they select their components based partly on having emergent patterns with neighboring components.

Anti-Selectionism

A related but different spin on evolutionary process is given by a train of thought I call "anti-selectionism." Probably the best example of this line of thinking is the excellent book *Evolution without Selection* by A. Lima de Faria (1990). More recently, complex systems science, with its emphasis on self-organization, has brought the anti-selectionist perspective a little more prominence. For instance, Stephen Wolfram, in his recent book *A New Kind of Science (2004),* presents a fairly strong variant of this viewpoint, though not quite as extreme as Lima de Faria's.

Lima de Faria's point is that most forms observed in organisms can be explained better in terms of self-organizing processes, than in terms of adaptation to environmental conditions. He gives on example after another of nearly identical patterns that have appeared independently in evolutionarily distinct species. Australian fauna provide outstanding examples of this. Tasmanian tigers looked a fair bit like African and Asian mid-sized predator mammals, but they were marsupials; their genetic material had relatively little in common with their near-lookalikes. And why do pandas look so much like bears, yet share more genetic material with raccoons? Why do brain corals twist around like brains, and why do the fractal branching patterns of our blood vessels look like the sprawling roots of plants? Not because the details of an animal's body are selected by evolution; rather, because the details of epigenesis are governed by complex dynamical systems that tend to have similar attractors.

In a similar vein, Wolfram takes on the frequent occurrence of the "golden ratio" $(-1 + \sqrt{5})/2$ in natural phenomena. Because the golden ratio occurs in mathematics as the solution to so many optimization problems, evolutionary biologists have assumed that its occurrence in the shapes of plants and animals is a result of evolution solving optimization problems, in the course of adapting organisms to their environments. But Wolfram shows that, in case after case, the golden ratio arises in simple self-organizing systems that model aspects of epigenetic development – without any selection at work at all.

Wolfram and Lima de Faria have excellent points, but they may be pushing the pendulum too far in the other direction. Traditional neo-Darwinist biology has stressed selection far too much, arguing that one after another particular trait must be optimized by evolution for some particular purpose. But yet, one does not want to underemphasize selection either: it is a powerful force for adapting the various parameters involved in the complex self-organizing epigenetic processes that build organisms from genomes.

From a patternist perspective, what these anti-selectionists are pointing out is that there are some highly intense patterns that recur again and again in different systems, and seem to have a value that goes beyond the particular characteristics of any one system they belong to. This is an interesting and nontrivial point. But from a general-evolutionist, patternist perspective, it is hardly surprising. For instance, if compactness of programs or speed of locomotion are important correlates of evolutionary fitness in various contexts, then it's hardly surprising that particular tricks for achieving program compactness or locomotion speed may recur again and again in various particular categories of entities. Anti-selectionism is meaningful only if one interprets it as "anti-gene-level selectionism"; that is, if one recognizes that selection on the gene level is important merely as an indirect means to selection on the level of emergent patterns within and among organisms.

Evolution and Continuous Pattern-Sympathy

So far, most of the discussion of evolution I've presented has been closely based on that I gave in *The Evolving Mind*. But, although the basic idea of evolution by natural selection is conceptually simple, there is a surprising amount of subtlety hidden beneath the surface. In the 12 years since I wrote *The Evolving Mind*, my thinking on the topic has advanced somewhat, and I've come up with what I think is a crisper way of formulating the relationship between evolution and patternist thinking. This new formulation involves the concepts of *pattern-sympathy* and *continuous pattern-sympathy*, which I will now define.

Let's say that a dynamical system displays "pattern-sympathy" to the extent that, when a pattern appears in the system, this pattern tends to continue into the future (with a probability greater than would have been the case if the pattern had not appeared in the system in the past). Some systems display generic pattern-sympathy; others display pattern-sympathy only with respect to patterns in some particular category (so that when a pattern in the right category appears in the system, this pattern tends to continue into the future). In Peircean terms, a pattern-sympathetic dynamical system is one that displays a temporal "tendency to take habits."

Next, let's say that a dynamical system displays "continuous pattern-sympathy" if, among patterns in the system, *continuance over the long-term* is reasonably well-predicted by *continuance in the short-term*. This is a little more than just saying that patterns tend to propagate themselves. It's saying that patterns tend to propagate themselves, and they tend to do so gradually over time. This simple notion, I suggest, is key to understanding the nature of evolution by natural selection.

To see why, begin by defining the "potential" of a certain gene at a certain time as the number of instances of the gene that will exist in the long-term future, divided by the number of instance of the gene that exist at the present time.   This may be roughly estimated by: the number of organisms containing that gene that will exist in the long-term future, divided by the number of organisms containing that gene now. It is then quite clear that one way for a gene to have a lot of potential is for organisms containing that gene to have a lot of offspring. This follows from the exponential mathematics of reproduction: more offspring now leads to lots more offspring later. As noted above, historically, Malthus highlighted the exponential mathematics of reproduction, and Darwin extrapolated the consequence of this: the set genes with long-term potential can be roughly estimated as the set of genes leading to short-term reproductive success. Or in the language I'm using here, genes – and sets of genes, and patterns among genes – are patterns that display continuous pattern-sympathy. The insight of natural selection theory is not that genetic patterns tend to persist over time (which is true but completely obvious), but rather that, due to the exponential nature of population growth, the genetic patterns that will persist over the long term can be predicted by looking at which genetic patterns proliferate over the short term. This distinction between time-scales is somewhat blurred by the common formulation of evolution as "survival of the fittest." A better formulation, in my view, would be "long-term survival of the short-term survivors," where the "survivors" in both the short and long term are not organisms but genes and patterns of combinations of genes.

But this only gets at part of the phenomenon of the origin and adaptation of species – because it leaves out ecology. What ecology says is that in some cases the "survivors" are not just genes or patterns of combinations of genes, but they may be purely phenotypic patterns (or combinations of genotypic and phenotypic patterns). Of course some phenotypic patterns may display continuous pattern-sympathy simply because they correspond with particular genetic pattern. But phenotypic patterns may also display continuous pattern-sympathy for another reason also: co-adaptation. When a genetic pattern persists and gives rise to a certain phenotypic pattern, this phenotypic pattern then forms part of the environment to which various organisms adapt, and it encourages the creation of other phenotypic patterns that "complement" it (in the sense that the original phenotypic pattern and the new ones give rise to significant emergent patterns). But if genetic pattern G has given rise to phenotypic pattern P, which has encouraged emergence of complementary phenotypic pattern Q, then even if genetic pattern G should disappear, there will be a tendency for a new genetic pattern H to appear, giving rise to a new phenotypic pattern P' that is also roughly complementary to Q.   For example, the existence of predators with certain properties may lead to the evolution of prey with certain properties – but then even if the original predators are extinct, the continued existence of these prey may lead to the evolution of new predators sharing properties with the original ones, in spite of genetic dissimilarity to the original ones. The relevant properties of the original predators are a phenotypic pattern that has persisted itself over time, without corresponding persistence of the underlying genetics, via co-adaptation. Co-adaptation dynamics leads to a tendency for phenotypic patterns to propagate themselves independently of their genetic basis – and this dynamic has the same exponential-growth dynamic that Malthus observed on the organism level and Darwin extrapolated to the gene level (though arguably with a considerably smaller exponent).   So, phenotypic patterns, independently of genotypic patterns, may display continuous pattern-sympathy as well.

The essence of evolution by natural selection, then, comes down to: *Long-term survival of the short-term surviving patterns, on both the genotypic and phenotypic levels.*

Evolution and Mind

The general question of how evolution relates to mind is a large and deep one, which I will briefly touch here and then explore more deeply in later chapters. What I will do here is to discuss two 20th century thinkers who explored this relationship interestingly – Karl Popper and Gerald Edelman. This combination of perspectives will hopefully get across a decent idea of the general intuitive match-up between thought and evolution that so many have perceived and expressed in different ways.

Let's begin with Karl Popper's "evolutionary epistemology." The basic flavor of this theory may be gleaned from the following passage in *Conjectures and Refutations* (Popper, 2002):

> *The method of trial and error is applied not only by Einstein but also, in a more dogmatic fashion, by the amoeba also. The difference lies not so much in the trials as in a critical and constructive attitude towards errors; errors which the scientist consciously and cautiously tries to uncover in order to refute his theories with searching arguments, including appeals to the most severe experimental tests which his theories and ingenuity permit him to design. (p.52)*

Popper himself never did much detailed work in this direction. However, Campbell (1974) explored Popper's point of view quite thoroughly. As he put it:

> *Evolutionary epistemologists contend, simply, that (exosomatic) scientific knowlege, as encoded in theories, grows and develops according to the same method as (and is, indeed, adaptationally continuous with) the embedded (endosomatic) incarnate knowledge shown ... in other organisms, including man. In the second case there is an increasing fit or adaptation between the organism and its environment... In the first case there is an increasing fit or adaptation between theory and fact...*
>
> *The highest creative thought, like animal adaptation, is the product of blind variation and selective retention ... or, to use Popper's phrase ... conjecture and refutation.*

At first, this might seem to be the strictest sort of strict neo-Darwinism. After all, the difference between the evolution of "ideas" in an amoeba and the evolution of ideas in a human being lies not in some vague ethic or "attitude" but in the fact that the human nervous system has a certain *structure*. This structure permits the results of various different trial-and-error experiments to *interact* in a complex and productive way. By continually emphasizing "blind variation and selective retention," the evolutionary epistemologists appear to be neglecting the role of structure in evolution.

However, evolutionary epistemology does not quite ignore structure entirely. In fact, Campbell himself has devised a sort of perceptual-motor hierarchy, consisting of ten levels. From the bottom up: nonmnemonic problem solving, vicarious locomotor devices, habit, instinct, visually supported thought, mnemonically supported thought, observational learning and imitation, language, cultural cumulation, and science. This hierarchy is endowed with a sort of pattern-oriented multilevel logic. As Plotkin (1982) put it:

> *If all knowledge processes are part of a nested hierarchical system, then information that has been laboriously gained by a blind variation ... at one ... level of the hierarchy may*

> *be fed upwards to some other...level where it can immediately operate as preset or*
> *predetermined criteria...*
> *[T]hese [are] **short-cut processes**.*

A *pattern,* according to the characterization I've given above, is precisely a "short-cut process." What Plotkin, following Campbell, is saying here is that higher levels of the hierarchy work partly by recognizing patterns in lower levels of the hierarchy. This is a crucial insight. But the evolutionary epistemologists do not pursue it very far, perhaps because of their emphasis on "blind variation." They give only the vaguest idea of how this hierarchy might operate. But still, this vision of mind as an evolving hierarchy of patterns gives a hint at a deep layer of truth. The really important idea here is that, in the context of cognitive evolution, there is a particular fitness function that correlates with evolutionary fitness – and this particular fitness function is "compactness", i.e. "being a shortcut process."

This idea has deep significance in the context of pattern philosophy. It suggests a quite general principle: that in evolving cognitive pattern-systems, evolutionary fitness of pattern-categories may be approximately predicted based on how much pattern-intensity is possessed by the patterns in the category.

This principle may be combined with the emergent-evolution principle mentioned above, which states that entities displaying more emergent pattern with their surrounding entities will often tend to have greater evolutionary fitness. What we arrive at is the idea that intense patterns tend to survive and flourish, as do patterns that spawn intense pattern when considered together with other patterns. A heuristic rule of "pattern intensity increase" comes to seem very natural as an overall descriptive rule for the evolution of patterns in complex cognitive systems. Of course, this kind of rule can never have universal applicability, though – the most one can hypothesize is that in intelligent systems this is generally the case, but naturally many exceptions will be found as well.

Combining this observation with the continuous-pattern-sympathy perspective, we arrive at the observation that in cognitive systems:

- Pattern intensity is an important component of pattern fitness.
- Those patterns that survive for a brief period are atypically likely to survive for a long period.

All in all, then, if one observes a pattern in an intelligent system that is both intense and has lasted a short while; the odds are relatively high that it will last a long while.

Brain Function as Evolution

Complementary to the above ideas about *mind* and evolution, another group of theorists has speculated about parallels between *brain function* and evolution. In our opinion, the most impressive of these efforts is Gerald Edelman's theory of neuronal group selection, or "Neural Darwinism." Edelman won a Nobel Prize for his work in immunology, which, like most modern immunology, was based on C. MacFarlane Burnet's theory of "clonal selection" (Burnet, 1962)), which states that antibody types in the mammalian immune system evolve by a form of natural selection. From his point of view, it was only natural to transfer the evolutionary idea from one mammalian body system (the immune system) to another (the brain).

The starting point of Neural Darwinism is the observation that neuronal dynamics may be analyzed in terms of the behavior of neuronal *groups*. The strongest evidence in favor of this conjecture is physiological: many of the neurons of the neocortex are organized in *clusters*, each one containing say 10,000 to 50,000 neurons each.

Once one has committed oneself to looking at groups, the next step is to ask how these groups are organized. A *map*, in Edelman's terminology, is a connected set of groups with the property that when one of the inter-group connections in the map is active, others will often tend to be active as well. Maps are not fixed over the life of an organism. They may be formed and destroyed in a very simple way: the connection between two neuronal groups may be "strengthened" by increasing the weights of the neurons connecting the one group with the other, and "weakened" by decreasing the weights of the neurons connecting the two groups.

This is the set-up, the context in which Edelman's theory works. The meat of the theory is the following hypothesis: the large-scale dynamics of the brain is dominated by the *natural selection of maps*. Those maps which are active when good results are obtained are strengthened; those maps which are active when bad results are obtained are weakened. And maps are continually mutated by the natural chaos of neural dynamics, thus providing new fodder for the selection process. By use of computer simulations, Edelman and his colleague Reeke have shown that formal neural networks obeying this rule can carry out fairly complicated acts of perception.

In general-evolution language, what is posited here is that organisms like humans contain chemical signals that signify organism-level success of various types, and that these signals serve as a "fitness function" correlating with evolutionary fitness of neuronal maps.

This thumbnail sketch, it must be emphasized, does not do justice to Edelman's ideas. In *Neural Darwinism* and his other related books and papers, Edelman presents neuronal group selection as a collection of precise biological hypotheses, and presents evidence in favor of a number of these hypotheses. However, I consider that the basic concept of neuronal group selection is largely independent of the biological particularities in terms of which Edelman has phrased it. I suspect that the mutation and selection of "transformations" or "maps" is a necessary component of the dynamics of *any* intelligent system.

As we will see later when we discuss AI, this business of maps is extremely important to Novamente. Novamente does not have simulated biological neurons and synapses, but it does have software structures called "nodes" and "links" that in some contexts play loosely similar roles. We sometimes think of Novamente nodes and links as being very roughly analogous to Edelman's neuronal clusters, and emergent inter-cluster links. And we have maps among Novamente nodes, just as Edelman has maps among his neuronal clusters. Maps are not the sole bearers of meaning in Novamente, but they are significant ones. A little later on we will be rigorously defining Novamente maps, and exploring their conceptual importance for AI cognition.

There is a very natural connection between Edelman-style brain evolution and the ideas about cognitive evolution presented earlier. Edelman proposes a fairly clear mechanism via which patterns that survive a while in the brain are differentially likely to survive a long time: this is basic Hebbian learning, which in Edelman's picture plays a role between neuronal groups. And, less directly, Edelman's perspective also provides a mechanism by which intense patterns will be differentially selected in the brain: because on the level of neural maps, pattern intensity corresponds to the combination of compactness and functionality. Among a number of roughly equally useful maps serving the same function, the more compact one will be more likely to survive over time, because it is less likely to be disrupted by other brain processes (such as other neural maps seeking to absorb its component neuronal groups into themselves). Edelman's neuroscience remains speculative, since so much remains unknown about human neural structure and dynamics; but it does provide a tentative and plausible connection between evolutionary neurodynamics and the more abstract sort of evolution that patternist philosophy posits to occur in the realm of mind-patterns.

Chapter 12
Autopoiesis

The next theme I will introduce is that of "autopoiesis" — the notion that minds are "autopoietic" in the sense that they can meaningfully be thought of as "constructing themselves." Simple as it is, this concept has engendered a great amount of confusion in the systems theory literature. A whole host of thinkers – Francisco Varela (1978), Humberto Maturana (Maturana and Varela, 1992), George Kampis (1991) and Robert Rosen (2002), for example – have put forth the "self-constructing mind" idea, but have tied it with notions of the uncomputability of mind, in ways that I think are incorrect and/or confusing. In *Chaotic Logic* I spent a lot of space trying to make sense of the relationship between self-construction and computability. I'll return to that issue a little later — but first I want to spend a few paragraphs elucidating the psychological significance of the "autopoiesis" concept.

The classic example of an autopoietic system is a proto-organism formed within a primordial soup of molecules. Much of the beauty of the collection of structures within an organism lies in the way it's configured so that the different parts of the organism all collaborate on each other's manufacture and maintenance. This is autopoiesis – self-production. It provides the framework within which organismic reproduction (quite different from the simpler pattern-replication one sees within e.g. crystals) occurs.

Human brains and Novamente AI systems are more rigidly structured than primordial cells, but this doesn't prevent them from being genuine autopoietic systems. Autopoiesis is about whether a system in a meaningful sense continually constructs itself – not about how rigidly or intricately structured the system is.

To see the value of the "autopoietic systems" framework for understanding the mind, let's take a simple heuristic example — purely for expository purposes, without any pretense of detailed realism. Let us think, in an abstract way, about the relation between a mental process that recognizes simple patterns (say lines), and a mental process that recognizes patterns *among* these simple patterns (say shapes). These shape recognizers may be understood as subservient to yet higher-level processes, say object recognizers.

If the shape recognizer has some idea what sort of shape to expect, then it must partially *reprogram* the line recognizer, to tell it what sort of lines to look for. But if the line recognizer perpetually receives instructions to look for lines that are not there, then it must partially *reprogram* the shape recognizer, to cause it to give more appropriate instructions. Assuming there is a certain amount of error innate in this process, one may achieve some interesting feedback dynamics. The collection of two processors may be naturally modeled as a self-generating system.

Next, consider that the mapping between line-recognizing processes and shape-recognizing processes is many-to-many. Each shape process makes use of many line-recognizing processes, and the typical line-recognizing process is connected to a few different shape-recognizing processes. A shape-recognizing process is involved in *creating* new line-recognizing processes; and a group of line-recognizing processes, by continually registering complaints, can cause the object-recognizing parents of the shape-recognizing processes to create new shape-recognizing processes.

What this means is that *the reprogramming* of processes by one another can be the causative agent behind the *creation* of new processes. So the collection of processes, as a whole, is a self-generating system. By acting on one another, the mental processes cause new mental processes to be created. And, due to the stochastic influence of errors as well as to the inherent chaos of complex dynamics, this process of creation is unpredictable. Certain processes are more *likely* to arise than others, but almost anything is *possible,* within the parameters imposed by the remainder of the network of processes that is the mind.

This example, as already emphasized, is merely a theoretical toy. The actual processes underlying shape and line recognition in the human brain are still a matter of debate. Vision processing in Novamente is currently relatively simplistic but in future will likely demonstrate this sort of complexity. More relevant examples will be given later on, as appropriate. But the basic concept should be clear.

Computability

Now I'll return to the somewhat technical – but philosophically essential – issue of the sense in which complex systems like minds may be said to be "computable." As I've noted above this is a controversial point in the systems theory community. Much like quale-aficionados in the domain of consciousness theory, autopoiesis-aficionados in the domain of systems theory believe that they have identified an aspect of complex systems like brain/minds that cannot be captured via traditional computational/empirical models. The "is autopoiesis uncomputable?" debate has not attracted as much attention as the "are qualia unphysical?" debate, but the two debates have the same basic character.

My illustrative example of autopoiesis in a (vision-enabled) AI system, given above, would probably be rejected by Varela and others as not being a true example of autopoiesis because the self-creation of the system occurs within a purely computational framework: the self-creation is a matter of a bunch of pieces of code reprogramming each other. But I think this is first-class autopoiesis even though it's computational – I don't think there is any other type of autopoiesis that's more genuine, or even fundamentally different. To elucidate this point will take a few pages, and I'll begin by explaining what "computability" means for the non-math/CS nerds in the audience.

"Computability" is a technical term but its morphology gets across the basic idea – something is computable if, in principle, it could be represented or produced by some computer. Recall that above, in our discussion of quantum computing, we described Deutsch's result that the class of things computable by quantum computers is the same as the class of things computable by ordinary, classical computers (although quantum computers may compute the solutions to some sorts of problems much faster on average). Formalizing the notion of computability requires some mathematics defining the notion of a computer in an abstract sense — basically as a machine that takes a finite set of rules and applies them to a memory store consisting of a large, potentially but not actually infinite set of rewriteable states. Something is computable if it can be accurately mathematically modeled by some system that is a computer in this very general sense – i.e. if it can be modeled in terms of finite sets of rules iteratively modifying some large finite list of entities with changeable states.

Computing theory contains a very nice idea called "universal computation," which says basically that any computer can simulate any other computer, though doing this simulation may slow it down a bit (see e.g. Davis, 1982). This notion of universality is why it makes sense to talk about "computability" plain and simple instead of always "computability on this or that computer." But of course, the one thing universality doesn't address is efficiency of computation: it may be possible in principle to simulate my brain on a PowerPC with a sufficiently large number of Zip disks for auxiliary memory, but it would be damn slow, and potentially the Big Crunch might come before I finished a single nontrivial thought.

As I noted earlier, computing theory provides one way of mathematically formalizing pattern theory: one can define X as a pattern in Y if X is a program for computing Y and X is simpler than Y. This definition begs the question "program on what computer?", and this is where universal computing comes to the rescue: since any computer can simulate any other, then if X and Y are sufficiently complex entities (more complex than the computers under consideration), whether X is a pattern in Y or not isn't likely to depend on which computer you use to define "program."

According to all known physics theories, the universe is a computable system: i.e., the universe may as well be conceived as one very large computer, albeit one with architecture very different from the computer I'm now using to type these words. Quantum theory introduces some strangeness here, but it turns out that one can model quantum systems in terms of special "quantum computers," and one can prove that quantum computers' only advantage over ordinary computers is speed. A quantum computer can't solve any problems or represent any knowledge beyond what an ordinary computer can do – but it can solve some problems far faster (Deutsch, 1985, 1997, 2000).

Not everyone agrees that the universe is a computable system, however. Some scientists and philosophers have argued that complex systems like brains and bodies are fundamentally uncomputable, and have tried to use abstract mathematics about uncomputable infinite structures to model these systems. And this has been tied in with autopoeiesis – dynamic self-creation – which is why these issues are being discussed in this chapter. It's been argued that complex living and thinking systems are autopoietic, and that autopoiesis is fundamentally an uncomputable thing.

My own perspective on these issues is — as often seems to happen — a bit idiosyncratic. I believe the insight that the mind and body are self-creating is correct and profound — but also that it is consistent with the computational perspective. In this section I will seek to clearly communicate the sense in which I think it makes sense to consider complex systems as both self-creating and computable.

Harking back to the earlier discussion of objectivity and subjectivity, I emphasize that just because I say a system is "computable" doesn't mean I think that computation is the only worthwhile perspective to take on that system. What I mean is that, insofar as a system can be studied and analyzed and predicted via scientific methods, involving measurement and related notions, then that system can be considered as computable. There are also nonmeasurable aspects of being, such as quantum reality and subjective reality – and I don't claim that, for instance, a computational (or any other physical) model of a human brain captures the subjectivity of that brain. The difference between myself and Varela, Rosen and kin is that they think there is some way of understanding complex systems that is fully scientific yet involves fundamentally noncomputable entities. While I can't rule this out, I remain skeptical, and I don't think this hypothesis is necessary in order to explain any our current scientific observations, nor does it provide a maximally simple explanation of our current scientific observations. Science is about measurements and any theory explaining measurements could be replaced with a functionally equivalent computational theory. Now you can say measurements don't capture everything, and I won't argue with you there, but then you're making a different sort of argument. Uncomputable mathematical models may well relate to objective reality in the same way that electrons and qualia do – as useful tools for understanding objective reality. But this is quite different from stating that certain objects in objective reality are "fundamentally uncomputable" due to their autopoietic nature – an assertion I do not support.

Because of the computability issue, my use of the term "autopoiesis" might be rejected by some systems theorists, and would possibly have been rejected by Varela, who coined it. However, as in many other places, I have avoided the temptation to make up a new word, or to use annoying subscripts (e.g. autopoiesis$_1$). I am referring to the same intuitive concept as Varela and others, even though the theoretical framework in which I ground it is different.

George Kampis and Component Systems

Next I'll briefly discuss the ideas of one particular systems theorist – George Kampis — who believes autopoietic systems to be uncomputable. I think Kampis is wrong in his thoughts on autopoiesis and computability; but unlike many others, at least he's wrong in an interesting way.

Kampis, in his excellent book *Self-Modifying Systems in Biology and Cognitive Science* (1991), presents a conceptual model of a self-constructed, autopoietic system called a "component system." A component-system is, quite simply, a collection of components, each of which can act on other components to produce new components. Discussing component-systems seems to be a good way to get to the bottom of these issues regarding autopoiesis and computation – although the discussion in this section will get a bit technical, so I'll have to beg tolerance from non-techie readers….

Kampis defines an abstract component-system via the following properties:

a) There is a finite set of non-dividable and permanent building blocks, drawn from a given pool.
b) There is an open-ended variety of the different types of admissible components, built up from the building blocks according to some composition rule (which may be explicit or implicit).
c) The components of the system are assembled and disassembled by the processes of the system such that every admissible component is also realizable. (p.199)

For illustrative purposes, Kampis suggests that the reader visualize the "non-dividable building blocks" as LEGO blocks, and the "admissible constructions" as different possible structures buildable out of LEGO blocks. One must merely imagine that each LEGO structure contains some appropriate means for acting on other LEGO structures to produce new LEGO structures.

Harking back to the discussion above, his main biological example of a component system is a "molecular soup" full of organic molecules acting on one another to form new molecules. Psychologically, on the other hand, one is supposed to think of ideas acting on each other to produce new ideas.

Having introduced Kampis's terminology, I can now describe the central thesis of *Self-Modifying Systems*, which is one I do not fully agree with: That *biological and psychological systems, being component-systems, are fundamentally uncomputable.*

This thesis combines two distinct claims:

- **Claim 1**: Formal component-systems display uncomputable behavior.
- **Claim 2**: Formal component-systems are good models for biological and psychological systems.

The first claim is a semi-mathematical result, which Kampis calls his "Main Theorem." I call it semi-mathematical because Kampis doesn't really define his terms or present his arguments with the full rigor of a mathematical proof. The second claim, on the other hand, is obviously a *scientific* hypothesis.

In my book *Chaotic Logic* I argued against these claims in a systematic way. The crux of the discussion there goes as follows:

- It is argued that formal component-systems are actually not uncomputable in any pragmatically meaningful sense.
- "Self-generating" systems are defined, roughly, as computable analogue of Kampis's component-systems.
- It is argued that self-generating systems are a good model of biological and (more relevantly) cognitive systems.

Why do I believe formal component-systems can pragmatically be considered computable? In brief, Kampis's argument for uncomputability is basically that the set of possible "admissible components" generable by a component-system is uncountable, hence cannot be generated by any computer program. The interesting thing, however, is that for any particular component-system in nature, it is not possible to demonstrate this posited uncountability. Any particular component-system is going to demonstrate only a finite number of components in any finite period of time. This is why I find Kampis's statement unimpressive.

If one posits any particular finite observer, then with respect to that observer, a whole lot of component-systems are going to appear uncomputably complex. But one can't generalize from this to say that component-systems are uncomputably complex in an objective, observer-independent sense.

Because of my differences with Kampis over the computability issue, in *Chaotic Logic* I coined the new term, "self-generating system," that basically refers to what Kampis calls a component-system, but with the difference that self-generating systems are explicitly ***computable***. But I haven't used that term very often in subsequent writings, having fallen into the habit of using Varela's more poetically evocative term "autopoiesis" instead!

In *Chaotic Logic* and *From Complexity to Creativity*, I presented a fairly thorough conceptual model of cognitive systems, which is explicitly based on the self-generating systems approach. That is, the mind is modeled as a set of components that are all acting on each other and transforming each other. Each component is considered as a pattern recognized by other components, and also as an agent recognizing patterns in other components. The depth of the "psynet model of mind" presented there lies in the hypotheses made about the particular types of agents and patterns involved: but the overall framework is very Kampis-like, except for the fact that the components involved are explicitly intended to be reflect processes implementable in computer software, or quantum-computable systems like brains.

In a similar vein, in Chapter 15 I will discuss my Novamente AI system, which is inspired by the psynet model, and is a collection of components called Atoms and MindAgents, which act on each other in specific ways. Again: a computable component-system, aka a self-generating system. Not all Novamente components act on each other at each time cycle; rather, probabilistic choices are made, based on certain properties of the components. Atoms and MindAgents can be removed from the system ("annihilated"), and new ones can be created. Of course, there is a huge amount of subtlety in the makeup of the Atoms and MindAgents, and the particular actions undertaken by them; this is where ideas from neural nets, logic, evolutionary programming, etc., come in. Not just *any* self-generating system can be a mind; only self-generating systems with the right sort of mix of components can qualify. But all this necessary detail should not distract one from the underlying truth that the self-generating systems framework is the natural computing paradigm with which to understand intelligence.

Creativity and Subjectivity

These rather technical points lead into some more exciting ideas, regarding the nature of creativity in minds and biological systems. Kampis complains that computable systems are bad models of creativity, because computer programs can never create anything beyond what has been put into them — a very old argument. This is true in the same sense that mathematical theorems are never truly original creations — they are all contained in the basic axioms of mathematics. Yet still, one may say that from the perspective of any particular observer X, a computer program Y may appear to output things that go amazingly beyond what was put into Y. But then some other observer Z, who understands Y better, may then see how these things were predictable in terms of Y.

This leads us to a basic principle of systems theory, which received its definitive modern form in the work of Ross Ashby (1954): *the essence of creativity is the interplay between rules and randomness.* Or, another way to put this is: Creativity is the interplay between the known and the unknown. From the point of view of any observer, some aspects of the world are known and some are unknown. So, systems recognized in the world and observed to be characterized by certain properties – certain "rules" – can't really be assumed to be fully characterized by these properties. There may always be unknown aspects. And these unknown aspects may result in behavior that appears truly novel, truly and fantastically creative, relative to the particular observer.

From a pure objective-reality perspective, this may seem like a lame sort of creativity: it's just creativity relative to some observer, not true and pure and total creativity! But from a subjectivist perspective, wherein the objective world is just a particular (though particularly powerful) subjective construct, this sort of creativity is quite fundamental. Because from a subjectivist perspective, the existence of a particular observer is a given, anyway: the observer is just as fundamental as the objective reality observed.

The Cognitive Equation

The "self-generating systems" framework for modeling autopoiesis is very general – a point which leads up to the question: what distinguishes mental self-generating systems from self-generating systems in general?

The most basic distinguishing factor, I suggest, is that the former is largely filled with components that are concerned with *recognizing patterns*. In *Chaotic Logic* I proposed that a mind is a **self-generating network of components that are largely concerned with recognizing patterns in each other**.

In that book, published in 1994, I gave an idealized formalization of this kind of self-generating dynamic was given there, and somewhat grandiosely named it "The Cognitive Equation." My work since then designing practical AI systems has enhanced my understanding of these issues a bit, and I now believe that this "cognitive equation" should be presented in a slightly different, though conceptually very similar, form. So, here I'll give you the Cognitive Equation, in both the 20th and 21st century styles.

First, the Old Cognitive Equation. Basically, this formalization involves a population of components (like in a Kampis component-system, described above) which includes an anti-relationship, so that some components may have anti-compnents, and when a component and anti-component meet they annihilate each other. This mechanism of anti-components may seem a little awkward, but it's basically a way of mathematically allowing components to be removed from a system. In Novamente, in practice, this is carried out by a software object called a Forgetting MindAgent – but from the point of view of mathematical and conceptual formalization, there is no problem with conceiving of a Forgetting MindAgent as creating anti-components corresponding to the components it is about to cause the system to forget.

The Old Cognitive Equation is a dynamical equation which involves an iterated series of steps, defined as follows:

1. All components in the population act on the entire current population of components, with a probability p determined by some function.
2. All components whose anti-components have been created are eliminated.
3. All anti-components are eliminated.
4. The set of patterns in the remaining set of components (or some appropriate subset thereof) is recognized.
5. These patterns are embodied as components, and become the new population.
6. Return to Step 1.

The basic idea here is that, at each time step, the existing population of components act on each other, and then is replaced en masse by the set of patterns in the population. The big question of course, is how one gets from step 4 to step 5 – how are patterns represented as components? There are many ways this can be accomplished, if one restricts attention to appropriate subsets of the space of all patterns in the population.

(Although I've called this an "equation," I haven't presented it here in a symbolic format with an = sign. In *Chaotic Logic* I did so, but in fact the mathematics didn't add much, and I think the "pseudocode" form given here is actually clearer.)

In *Chaotic Logic* I proposed that this neat little iterative process could be considered the "essential dynamic equation of mind." Now I consider this a bit of an overstatement, and I would prefer to phrase the statement a little differently. I would hypothesize instead that this is not a dynamic equation that is ever precisely implemented, but rather an *idealized dynamic*, which is *approximated* by real systems.[49]

Now I would prefer to formulate the basic mental dynamic in a more pragmatic form, as follows (the **New Improved Cognitive Equation**):

---

[49] Of course, this is really what I meant back then, but at that stage I was prone to overly extreme formulations, even more so than I am now!

1. All components in the population act on the entire current population of components, with a probability p determined by some function.
2. Many of these components act by recognizing patterns in other components, or in sets of other components, and embodying these patterns in newly created components.
3. Some of these components act by keeping records of what components have existed, or acted, in the recent past.
4. All components whose anti-components have been created are eliminated.
5. All anti-components are eliminated.
6. Return to Step 1.

The difference between these two formulations of the "cognitive equation" should be clear. For one thing, I've now pushed the "pattern recognition" step back into the component-activity part of the process, recognizing that in real systems there is no all-seeing guru recognizing patterns in the system and placing them into the system as components. For another thing, I've admitted that there may be some things going on in the system besides explicit pattern recognition. Real systems are messy, and real minds have some aspects that are critical to their mental function, and others that help "bridge" their mental functioning with the underlying hardware ... and others that are just there. The uniquely mental aspect of a mind's dynamic activity, however, consists in precisely the dynamic captured in the idealized "cognitive equation" – the passage, over time, from a system of components, to a system whose components are *patterns* in the previous system of components.

Finally, by in step 3 incorporating temporal record-keeping, I've made an explicit place for dynamical pattern recognition – recognition in the way a system changes over time. The *Chaotic Logic* formulation permitted this but did not emphasize it. I've found that this kind of pattern recognition is particularly important for some aspects of intelligence, such as procedure learning.

No intelligent system can explicitly embody all patterns in itself, as components of itself. But we hypothesize that intelligent minds *try* to do this, about as hard as they can. This is what makes them intelligent minds; this is what allows them to develop complex structures, to fill themselves up with patterns. Recognizing patterns among components representing perceptions and actions leads to cognition; recognizing patterns among components recognizing ideas and memories leads to advanced cognition and sophisticated goal-achieving behavior. The patterns thus recognized become new ideas, grist for the mill of pattern-recognizing components. Recognizing patterns in the behavior of pattern-recognizing components, allows a system to improve its pattern-recognition ability.

Mind Attractors

The "cognitive equation" I've given above is not all that useful of an equation, really, compared to equations in hard sciences like physics. I don't know how to solve it in any practical situation, and nor do I know how to approximate it on a computer in an interesting way, except via building a complex AI system like Novamente, which involves a lot of other ideas beyond the Cognitive Equation as well. The main value of the Cognitive Equation at present would seem to be purely conceptual value: it is a formulation that seems to drive thought in interesting directions. For instance, it leads one to the notion of a "mind attractor," which has a wide variety of useful and stimulating ramifications.

The cognitive equation, like many other nonlinear iterations, can lead to all manner of complex dynamics. Among other things, it can lead to self-perpetuating subsystems: i.e. sets of components that produce one another, thus perpetuating each other over a long period of time. I like to call such self-perpetuating subsystems "mind attractors."

A couple technical points arising regarding these so-called "attractors" are worthy of attention. Firstly, a mind attractor is not necessarily a mathematical attractor of the cognitive equation as a dynamical system (see Devaney, 1989 for a formal definition of "attractor") – because it may involve only a small subset of the component population at any given time. More precisely, a mind attractor is a special kind of "invariant subspace" of the set of component populations: the subspace in which a particular set of components exists.

Secondly, a mind attractor generally will not truly stick around forever. Rather, mind attractors are often more similar to what Mikhail Zak (1991) has called "terminal attractors." They stick around for a while and then quite possibly vanish.

In the Novamente AI design, we call mind attractors by the special name "maps." Essentially, a Novamente map is a set of Atoms with the property that, once a few elements of the map become active, it generally occurs that all the elements of the map will become active during a certain interval of time. Maps are patterns in Novamente, just as mind attractors are, generally speaking, patterns in a self-generating system.

As a simple toy example of an attractor of the cognitive equation, consider two entities f and g, defined informally by:

```
f(x) = the result of executing the command "Repeat x two times"
g(x) = the result of executing the command "Repeat x three
times"
```

Then, when f acts on g, one obtains the "compound"

```
f(g) = the result of executing the command "Repeat x three times"
the result of executing the command "Repeat x three times"
```

And when g acts on f, one obtains the "compound"

```
g(f) = the result of executing the command "Repeat x two times"
the result of executing the command "Repeat x two times" the result of
executing the command "Repeat x two times"
```

Now, obviously the pair {f,g} is a pattern in {f(g),g(f)}, since it is easier to store f and g, and then apply f to g to get f(g) and apply g to f to get g(f), than it is to store {f(g), g(f)} directly. So f and g, in a sense, perpetuate one another according to the cognitive dynamic! The set {f,g} is a pattern in the products generated from the set {f,g}; it is a ***fixed point attractor*** of the cognitive equation.

Of course, a fixed-point mind attractor is merely the simplest kind of attractor that a self-generating pattern dynamic can have. There may also be limit cycles and strange attractors. In fact, it is to be expected that self-generating pattern dynamical systems possess chaotic attractors, as well as more orderly strange attractors. Unfortunately, there is at the present time no mathematical theory of direct use in exploring the properties of this kind of dynamical system. Empirical exploration, such as we are doing with Novamente, seems the only feasible path.

Physical Attractors and Mind Attractors

How do these thoughts relate to more conventional dynamical systems theory ideas? The brain or Novamente, like other extremely complex systems, are unpredictable on the level of detail but roughly predictable on the level of structure. This means that the dynamics of the physical variables of such a system display a strange attractor with a complex structure of "wings" or "compartments." Each compartment represents a certain collection of states which give rise to the same, or similar, patterns. Structural predictability means that each compartment has wider doorways to some compartments than to others.

The complex compartment-structure of the strange attractor of the physical dynamics of an intelligent system determines the macroscopic dynamics of the system. There would seem to be no way of determining this compartment-structure based on standard numerical dynamical systems theory. Therefore one must "leap up a level" and look at the dynamics of *mental processes*, represented in the brain by interacting, inter-creating neural maps, and in Novamente by interacting, inter-creating C++ objects in RAM. The dynamics of these processes, it is suggested, possess their *own* strange attractors called "mind attractors" or "maps."

Each state of the network of mental processes represents a large number of possible underlying physical states. Therefore mind-level attractors take the form of *coarser structures*, superimposed on physical-level attractors. If physical-level attractors are drawn in ball-point pen, mind-level attractors are drawn in magic marker. On the physical level, a mind attractor represents a whole complex of compartments. But only the most densely connected regions of the compartment-network of the physical-level attractor can correspond to mind attractors.

The Dual Network

Further, there may be some overall structure to the set of mind-level attractors. I have posited one such structure, which I call the "dual network" — indicating, as the name suggests, is a network of pattern/processes that is simultaneously structured in two ways.

The first kind of structure is *hierarchical*. In a hierarchy, simple structures build up to form more complex structures, which build up to form yet more complex structures, and so forth; and the more complex structures explicitly or implicitly govern the formation of their component structures. In a mind attractor context, this means attractors within attractors within attractors.

The second kind of structure is *heterarchical*: different structures connect to those other structures which are *related* to them by a sufficient number of pattern/processes. This means attractors that overlap with or mutually stimulate other attractors.

One can see this duality easily enough in the simple domain of word definitions: there is a hierarchical ontology of definitions as one can see in a dictionary schema like WordNet[50] (Feldbaum, 1998; an online dictionary with a hierarchical ontology that places daschund within the dog category, dog within the animal category, and so forth), but there is also a sprawling network of interdefinitions, as one would find by linking each word to the other words that occur in its definition. The hierarchical and heterarchical aspects of WordNet or any other ontology-incorporating dictionary interpenetrate and reinforce each other. This example lacks the dynamical aspect of the dual network posited to be part of intelligent minds, but it possesses the structural aspect. An AI system containing pattern recognition and concept generation

---

[50] http://wordnet.princeton.edu/perl/webwn

components corresponding to the different words in WordNet would possess a full-fledged psychological dual network.

Psychologically speaking, the hierarchical network may be identified with command-structured perception/control as well as with hierarchical ontologies of declarative knowledge. On the other hand, the heterarchical network may be identified with associatively structured memory, and with the more free-flowing control processes that lead to spontaneous creativity. To the extent that the mind attractors associated with memory, action and perception exist within this kind of overarching framework, the dual network itself may be considered a kind of very abstract mind attractor. And the existence of this sort of coordinated network tells us something about the coordination of memory and perception. The perceptual-motor hierarchy and the structurally associative memory are functionally somewhat separate but they also overlap and this is critical to their ability to deal with subtle situations where memory and perception/action blur. It fits in with the modern notion that memory stores processes rather than facts or images – processes that may generate facts or images in an appropriate way depending on the context (Rosenfield, 1989).

The overlapping of hierarchy and heterarchy works most elegantly if the heterarchical network is fractal in nature – so that the clusters in the fractal heterarchy correspond to hierarchical groupings in the control hierarchy. This may in fact be one of the reasons why we tend to view our concept-space in terms of hierarchical ontology. In the mind, knowledge is active rather than passive, and groupings of concepts are conveniently associated with processes for learning new concepts and reasoning about old ones (using the given grouping as a context and bias). But the control of learning and reasoning processes is effectively done using hierarchical structures, and so it is natural to divide up concepts hierarchically as well due to the correspondence between concepts (treated as contexts) and learning/inference control.

Mind Systems

A related notion to "mind attractor" is that of a "mind system." In this book I have touched upon many different entities that may profitably be considered psychological *systems*: for example, linguistic systems, belief systems, minds and realities. All of these systems, we suggest, are strange mind attractors. Elements of mind, language, belief and reality exist in a condition of constant chaotic fluctuation. The cognitive equation gives the overarching structure within which this creative chaos occurs.

As we shall see as the following chapters progress, the assertion that each of these systems is an attractor for the cognitive equation has many interesting consequences. It implies that, as Whorf and Saussure claimed, languages are *semantically closed,* or very nearly so. It implies that belief systems are *self-supporting* — although the nature of this self-support may vary depending on the "rationality" of the belief systems. It implies that perception, thought, action and emotion form an *unbroken unity*, each one contributing to the creation of the others. And it tells us that the relation between mind and reality is one of *intersubjectivity*: minds create a reality by sharing an appropriate type of belief system, and then they live in the reality which they create. This is an augmentation, not a contradiction, to the ideas on subjective and objective reality presented earlier.

How all these systems work together to help create minds is a complicated story, but it is a story that takes place within the framework of mind as a self-generating pattern dynamic, a collection of components continually recreating each other based on patterns recognized in each other.

Focused Consciousness as a Builder of Mind Attractors

Now I come to one of the subtlest points in mind dynamics: the relationship between autopoiesis and awareness. At first blush the two might seem largely unrelated; yet in fact they are bound together very closely. The portion of the mind concerned with focused attention – the "focus of consciousness" is both:

1. The focus of the most intense awareness in the mind (the most intense qualia);
2. The part of the mind most intensely involved in creating mind attractors: autopoietic subsystems of the mind.

I will bolster this point largely by reference to the neuroscience literature, and then talk a bit about implications for AI.

Because of the philosophical debate concerning the concept of "consciousness," for decades, neuropsychologists shunned the term – a situation that has (thankfully) begun to clear up during the last 15 years or so. But even so, there has been a great deal of excellent work on the neural foundations of conscious experience, usually under the more neutral term "attention".

In particular, two recent discoveries in the neuropsychology of attention stand out above all others. These discoveries correspond to two observations made by Charles S. Peirce in the 1800's.

Peirce wrote: "The brain shows no central cell. The unity of consciousness is therefore not of physiological origin..." Now neuropsychologists have shown that, indeed, conscious processes are distributed throughout the brain, not located in any single nexus.

And Peirce proclaimed that "the unity of logical consistency ... and also the unity of the individual object ... [lie] not in the operations of the intellect, but in the *quale-* consciousness upon which the intellect operates." Now neuropsychologists have shown that one major role of consciousness in perception and cognition is precisely that of *grouping*, of *forming wholes*.

Writing in the late 1990's, the neuroscientists Rizzolati and Gallese (1998) named the issue nicely by positing two basic ways of approaching the problem of attentiveness. The conventional approach rests on two substantial claims:

> *That in the brain there is a selective attention center or circuit independent of sensory and motor circuits; and*
> *That this circuit controls the brain as a whole. (p. 240)*

In its most basic, stripped- down form this first claim implies that there are some brain regions exclusively devoted to attention. But there are also more refined interpretations: "It may be argued ... that in various cerebral areas attentional neurons can be present, intermixed with others having sensory or motor functions. These attentional neurons may have connections among them and form in this way an attentional circuit" (p.241).

This view of attention alludes to what Dennett (1991) calls the "Cartesian Theater." It holds that there is some particular *place* at which all the information from the senses and the memory comes together into one coherent picture, and from which all commands to the motor centers ultimately emanate. Even if there is not a unique spatial location, there is at least a single unified system that acts *as if* it were all in one place.

In their paper, Rizzolati and Gallassi contrast this conventional view with their own "premotor" theory of attention, of which they say:

> *First, it claims that ... attention is a vertical modular function present in several independent circuits and not a supramodal function controlling the whole brain. Second, it maintains that attention is a consequence of activation of premotor neurons, which in turn facilitates the sensory cells functionally related to them.*

In this view, consciousness is an emergent phenomenon involving perceptual and action-oriented brain subsystems, and of course also involving cognition and memory as well. It is a phenomenon that can emerge in many different places in the brain. This idea is consistent with more recent thinking on the topic, for instance Gerald Edelman's theory of consciousness and the brain, as discussed above, and other recent neurobiological research. Edelman, building on his Neural Darwinist theory of brain function, essentially proposes that consciousness consists of a *feedback loop* from the perceptual regions of the brain to the "higher" cognitive regions (Edelman, 1990).

But what does neuropsychology tell us about the *role* of this distributed consciousness? It tells us, to put it in a formula, that consciousness serves *to group disparate features into coherent wholes.* This conclusion has been reached by many different researchers working under many different theoretical presuppositions. There is no longer any reasonable doubt that, as Umilta (1988) put it some years ago, "the formation of a given percept is dependent on a specific distribution of focal attention." Patterns of attention shape patterns of perception – which in turn shape patterns of attention, both directly and by means of emergent dynamics involving action and cognition.

Visual perception, for example seems to proceed in two stages.. First is the stage of elementary feature recognition, in which simple visual properties like color and shape are recognized by individual neural assemblies. Next is the stage of *feature integration,* in which consciousness focuses on a certain location and *unifies the different features* present at that location. If attention is not focused on a certain location, the features sensed there may combine on their own, leading to the perception of illusory objects.

In *From Complexity to Creativity,* I referred to this sort of theory of attention as a theory of Perceptual-Cognitive-Active (PCA) Loops. Of course, there are all sorts of complex linkages involved here, not just a loop, but the overall structure of attention can be modeled as a cycling of information between perception, cognition and action, forming disparate impressions into coherent wholes. In the simplest case the interaction between the three parts of the loop is as follows: Perception molds the sensory impressions together, cognition gives memory-based feedback on which possible holistic arrangements are most probable, and action guides the sensory organs to pick up more information according to the patterns embodied in the evolving holistic arrangement.

This sort of process may have originated evolutionarily in the low-level perception/action domain, but now, of course, our attention is often applied to entirely abstract mental content. The fact that, in the human brain, perception and action oriented processes are often reused for abstract attention-focusing, is indicated by the frequent use of mental imagery in abstract thought. However, not all abstract thought involves mental imagery in any direct sense; sometimes the processes involved really are purely abstract. Even in the most abstract cases of attention, however, one still has:

- A collection of elements that are important in some context ("perceptual features" or something else, e.g. abstract mental content);
- The ability to focus in on various elements in the collection to a greater or lesser extent, and generally to shift some focus onto other related elements as well (this is the role of "action");
- The ability to create coherent wholes by linking ("binding") together elements in various ways (this is the role of cognition in attention).

Now, a *maximally coherent* whole is not usually desirable, because thought, perception and memory require that ideas possess some degree of flexibility. The individual features of a percept should be detectable to some degree, otherwise how could the percept be related to other similar ones? The trick is to stop the coherence- making process just in time – not too early, not too late. And "just in time" must be determined on a context-dependent basis; there is not necessarily a unique optimal level of coherence for all situations.

Of course, the really critical step in this process is Step 3, which does the actual "coherentization" of the inputs to the perceptual-cognitive-active loop. How this is done will obviously depend upon the details of the mental system in question; Novamente and the human brain use different mechanisms. One way or another though, what we propose is that the cognitive end of the attentional loop *creates mind attractors* out of its inputs, by in some way strengthening appropriate connections between elements in its input set

The reflexive nature of awareness comes in when patterns involving the current state of the self are fed into the PCA loop as data. Coherent systems are then formed, binding parts of the current self together, and possibly binding them with external stimuli or actions. After all, relationships involving the state of the self ("I am thinking about X at this time") are perfectly valid inputs into the PCA loop, just as much so as anything else. It so happens that, when an intelligent system is thinking about X, and X has a lot of importance in the system, then relationships such as "I am thinking about X at this time" are fairly likely to have a high importance too, via the mechanism of activation spreading. Whether these reflexive relationships are of any use or not, is context-dependent.

Chapter 13
Science[51]

This book is a work of philosophy, but in my everyday work-life I am mainly a scientist, and in nearly every chapter here I have played around the boundary of philosophy and science. In this chapter I'll address the issue of science quite explicitly, giving a patternist explanation of science itself. This could be viewed as a digression from the book's main theme of the philosophy of mind, but I don't think it really is – for many reasons, including the fact that science is a cognitive phenomenon which resembles and reflects many other cognitive phenomena. Also, a deep understanding of the nature of science is necessary for appreciating the relationship between philosophy and science in the book's other chapters, as well as some of the specific scientific hypotheses made therein.

The philosophy of science that I present here owes much to prior philosophers of science, but also has some original aspects. I have given it the somewhat awkward name "sociological/computational probabilism," or SCP for short. My main goal in formulating the SCP approach has been to develop a philosophical perspective that does justice to both the *relativism and sociological embeddedness* of science, and the *objectivity and rationality* of science. This is somewhat similar to my goals in the first part of Chapter 5 where I sought a view of mind synthesizing the subjectivist and objectivist perspectives. I have also been motivated by an interest in clarifying the relationship between probability theory and the validation of scientific theories.

The SCP philosophy has its roots in several different places, including patternism as well as:

- Contemporary mathematical learning theory, particularly probability theory and the theory of algorithmic information (Chaitin, 1988);
- Imre Lakatos' and Paul Feyerabend's conflicting but related philosophies of science (Lakatos, Feyerabend and Mottolini, 1999).

It draws ideas from these sources, but adheres precisely to none of them.

Two main issues are addressed here (though the SCP framework may be used to address other issues as well). First: What is a workable and justifiable way to compare two scientific theories, or two scientific approaches? And secondly and more briefly: what modes of thinking,

---

[51] The ideas given here resemble some that I presented in *Chaotic Logic*, which also discusses Lakatos, the history of science, and the dynamics of human belief systems; but the treatment here introduces a number of important new ideas.

what types of cognitive dynamics, should occur and be reinforced inside the mind of the scientist, in order for science to progress effectively?

I would like to emphasize the relative modesty of my objectives here. Feyerabend (1975) has argued that the only universal rule characterizing high-quality scientific progress is "anything goes." Any other rule that anyone proposes, he suggests, is going to have some exception somewhere. I think this is quite probably true. On the other hand, this kind of slipperiness is not unique to the philosophy of science; it also holds true in some of the sciences, e.g. biology and linguistics. The goal in these "semi-exact" sciences is to find principles that are – to borrow a phrase from computer science — "probably approximately correct." That is, one wants conclusions that come very close to being true in nearly all cases – but one accepts that various sorts of exceptions will exist. My ambition in this chapter is to outline a philosophy of science that is simple and clear, and that provides a "probably approximately correct" model of the human endeavor we call science.

What Is This Thing Called Science?

What is this thing called science? Feyerabend correctly points out that science is a sociological phenomenon – one among many types of activity carried out by human beings. My approach to defining of science takes this rather obvious observation as an explicit starting-point. This is different from many approaches to defining science that are found in the philosophy-of-science literature, which are more abstract and impersonal in nature.

Begin with the assumption that there is a community of human beings who each has their own notion of reality. I then posit the existence of something called the Master Observation Set (MOS): a massive compendium of all scientific datasets to be considered in the formulation of scientific theories by individuals in this community. Currently the MOS is a hypothetical entity, but the increasing habit in the scientific community of posting observational and experimental data on the Internet is rapidly making it a palpable reality.

The nature of the data in the MOS will be explored a little later on, in the course of our discussion of computational probabilism, where we will think of the MOS as consisting of datasets which contain a combination of "raw data" and "metadata." But let us leave this issue for later.

The MOS may be considered as a "fuzzy set," where the degree of membership of a dataset in the MOS is defined as extent of the belief in the accuracy of the dataset possessed by the average person in the community. The key point is that each of these humans in the community basically agrees that the data records and metadata in the Master Observation Set reflect reasonably accurate observations of their own realities. There may be disputes over the accuracy of particular observations or datasets, but so long as these don't involve the bulk of observations in the MOS, things are still okay.

Given this background, I can now define what I mean by a "scientific theory." Namely, a scientific theory as a set of procedures with inputs and outputs specified as follows:

> The inputs are drawn from the set of subsets of the MOS
> The output is a prediction about some subset of the MOS

The relation between the output subset B and the input subset A may be defined by some predicate. For instance, A could be data regarding the state of a population of yeast cells in a Petri dish at a point in time T and a temperature of 30 degrees Celsius; B could be data regarding the state of the same population of yeast cells one minute later, assuming the temperature in the meantime has been gradually decreased to 0 degrees Celsius.

Different scientific theories provide procedures dealing with different sorts of subsets A and predicting things about different sorts of subsets B. A scientific "Theory of Everything" would try to make the best possible prediction about any subset B based on any subset A. No single, coherent scientific Theory of Everthing currently exists; but the body of modern science as a whole may be considered as an attempt at such.

In less technical terms, what I am saying is that, to be scientific, ultimately a theory must boil down to a set of prescriptions for predicting some almost-universally-accepted-as-real data from some other almost-universally-accepted-as-real data. Most of the time this data is quantitative, but it certainly doesn't have to be. It just happens that, among humans, quantitative data seems to have an easier time fitting into the category of "almost universally accepted as real." Data represented in terms of words tends to be slipperier and less likely to achieve the standard of near-universal perceived validity.

As already emphasized, I have intentionally defined science in terms of a social community, rather than in terms of an objective reality. This is in line with both Feyerabend and Lakatos's emphasis on science as a social activity. The problem with defining science in terms of an objective reality is obvious: the definition of objective reality is deeply bound up with particular scientific theories. Do we mean classical objective reality, quantum objective reality or something else? Whatever else science is, we know it is a social activity, and can define it as such without being preferential to any particular scientific theory about the nature of the universe, or any particular research program.

The main topic of this chapter is how to judge the quality of a scientific theory or a scientific approach. However, in pursuing this topic we will also implicitly touch a related subject, which is how to judge the quality of approaches to understanding the universe in general – be they scientific or not. Feyerabend criticized Lakatos's philosophy of science for not explaining why science is superior to witchcraft or theology. This seemed problematic to Feyerabend not because he believed in the superiority of science – in fact he did not – but because Lakatos did. Lakatos would have liked his philosophy of science to explain the superiority of rationality to other modes of social and individual cognition. On this topic we conclude that, in a sense, what Lakatos desired is partially achievable. One can apply some of the same approaches used to compare scientific theories, to compare nonscientific approaches. Furthermore, the sociological definition of science allows us to consider the sense in which some theories traditionally viewed as nonscientific may be considered as scientific in some important senses after all.

An interesting example of a theory that is not traditionally considered scientific is the Buddhist philosophy of consciousness, alluded to in Chapter 3 above. This philosophy is extremely refined, delineating numerous types of consciousness and their interrelationship; it has much of the rigor and detail that is typically associated with science. Arguably, it has more rigor and detail than some social sciences do today, and more than biology did before the molecular-biology revolution. However, it fails the above definition of science, unless – and this is an important unless — one restricts the community of humans involved to those who have achieved a certain level of "spiritual advancement" in the Buddhist religion. It makes many statements about phenomena that are not in the overall human Master Observation Set, but are rather specifically thought to be observable only by spiritually advanced human beings. This is a theory that is scientific with the respect to the community of the Buddhistically advanced, who can test and check its statements, and unscientific with respect to the rest of us.

Now, one may argue that this isn't so different from a theory making statements about phenomena regarding underground caves on the dark side of the moon, where most humans will never go.   Are theories about caves on the dark side of the moon scientific with respect to humanity as a whole, or only with respect to those humans who actually visit the dark side of the moon and crawl through the caves and see what's there?   The key question is whether the non-moonwalking humans back on Earth agree to admit the data on the interior of the caves on the dark side of the moon into the Master Observation Set.   If they don't agree to admit the data – perhaps they believe there's a government conspiracy to fake this data to hide the aliens who live in the caves up there – then theories about this data don't qualify as scientific with regard to the community of humans, but only with regard to the community of moonwalking humans.   So the difference between this case and the Buddhist psychology case is, basically, that nearly all humans are willing to accept the data collected by astronauts observing the caves on the dark side of the moon; whereas most humans are not willing to accept the data collected by Buddhist monks in the course of their meditations.

So Buddhist psychology is scientific, but only with respect to a limited community.   This means the ideas developed here may all be applied to it, so long as one restricts all consideration to the relevant community.   But the actual situation is a little stronger than that.   Some of the approaches we will discuss for evaluating scientific theories and approaches may also be used to evaluate things like Buddhist psychology and witchcraft, even in the context of communities where these things fail the definition of science given here.

A Naïve Computational/Probabilist Approach

Having defined what science is, we may now approach the question of defining what makes one scientific theory better than another.   I'll start by presenting a point of view that I call "naïve computational probabilism."   This is a philosophy of science, motivated by recent research in algorithmic information theory, which is mathematically sophisticated and elegant but – for reasons to be explored in the following section — historically and psychologically naïve.

The basic idea is that a scientific theory should be judged on three criteria:

- What is its scope of applicability?
- How simple is it?
- How accurate are its predictions, on average?

Scope, simplicity, predictivity.   The ideal is an extremely simple theory with extremely accurate predictions and broad scope.   In practice one must strike a balance, sacrificing some simplicity and scope in order to get more explanatory value.

The accuracy of predictions of a scientific theory T may be formalized in many different ways.   For instance, one may begin by defining a mathematical metric (distance measure) on the Master Observation Set, and then proceed by averaging — over all pairs (A,B) where A and B are subsets of the MOS that lie in the scope of the theory – the distance between T's predictions of B based on A and the actual state of B.

The simplicity of a theory T may be formalized by assuming that all theories are expressed in some particular formal language, designed for interpretation by some abstract machine M.   One may then look at the length of the statement of the theory T and the amount of time T takes to make predictions about subsets of the MOS that lie in its scope. These two criteria of space and time complexity may then be averaged in various ways.

This approach to assessing the quality of scientific theories is rooted in algorithmic information theory and statistical decision theory. It connects nicely with recent developments in theoretical computer science. For instance, if one combines it with Marcus Hutter's mathematical formalization of the notion of intelligence, one arrives at the conclusion that intelligence may be achieved by learning good scientific theories.

The downside of this approach, on the other hand, is that its practical applicability is limited by the relative lack of formality of real scientific theories and real scientific practice. I will review the nature and consequences of this informality in an abstracted form, in the following section. I believe that there is some deep truth at the kernel of this "naïve computational/probabilist approach," in spite of its practical shortcomings – and after reviewing the various problems that one confronts in applying these abstract ideas to actual science, I will return to similar themes again, in the context of a neo-Lakatosian analysis of research programs, and finally in the context of a "pragmatic unified approach" to the philosophy of science.

The Pragmatic Failures of Probabilism

Two of Lakatos's philosophy-of-science concepts strike me as particularly important and useful. The first is his highly savvy critique of "probabilism" – the perspective that competing scientific theories may be compared against each other by assessing which one is more probable based on the given evidence. The second is his notion of "progression versus regression" as a way of comparing scientific research programs, even when these research programs use very different languages and present incommensurable views of reality.

Let us begin with Lakatos's issues with probability theory as applied to the validation of scientific theories. After brushing aside some fallacious arguments made by earlier philosophers to the effect that science is nonprobabilistic in principle, Lakatos (1999) poses the very cogent objection that, although a probabilist perspective on science may be correct in principle, it is utterly infeasible in practice. The problem is that to properly apply probability theory to a scientific theory, that theory must be completely formalized. And, as Feyerabend has argued in detail, even relatively simple and standard scientific theories like Newtonian mechanics are far from completely formalized when applied in practice. (In fact, even pure mathematics is far from completely formalized when applied in practice; the Mizar[52] project has illustrated this in a fascinating way by carrying out a complete formalization of a reasonably large subset of modern mathematics.) Using Newtonian mechanics in reality requires a substantial phase of "setting up the problem," in which informal reasoning is used to match aspects of real-world phenomena to variables and equations in the theory. And this informal "setting up the problem" phase is carried out quite differently for Newtonian mechanics than for quantum mechanics, for example – it truly is "part of the theory."

In principle, it may seem, no such theory-dependent setting-up phase should be necessary. It should be possible to define a set of "empirical observations" in a theory-independent way, and then assess various theories in terms of their predictions on this set of empirical observations. Feyerabend, however, argues that this is not possible because observation is so theory-dependent that different theories may differ on what is being observed in a given real-world situation. This is

---

[52] www.mizar.org

an excellent point, worth exploring a little – although my conclusion will be that it doesn't hold up completely.

Within the context of contemporary science, there is fairly good agreement on what constitutes an observation. General relativity, classical mechanics and quantum field theory are "incommensurable" theories in the sense that they provide mutually contradictory conceptual and formal perspectives on the world, and there is no clear and universal way of taking a phenomenon as expressed in the language of one of these theories, and translating it in the language of the others. But even so, it seems the task of defining a theory-independent set of laboratory observations for comparing these theories would not lead to any significant conceptual problems (though it would be a Herculean endeavor, to be sure). But there are subtle conceptual issues here – of which Feyerabend was well aware — which arise when one analyzes the situation carefully.

Let's suppose that – as envisioned above — we created a Master Observation Set, consisting of the outputs of all scientific laboratory instruments of all kinds that have ever been used. We could then perform predictive experiments, of the form: Feed a given scientific theory one portion of the Master Observation Set, and see how well it predicts some other related portion. The same predictive experiments could be performed for different scientific theories, thus providing a common basis of comparison. The expected prediction error of each theory could be tabulated, thus making the probabilist foundation of science a reality.

Applied to fundamental physics theories, what would we learn from this exercise? We would find that classical mechanics fails to correctly predict a lot of observations involving microscopic phenomena and some involving macrosopic phenomena (involving gravity, light and electromagnetism, for example). It would also be found mathematically and computationally intractable in very many cases. General relativity would be found to fail on some microscopic phenomena, and to be mathematically and computationally intractable for a huge number of microscopic and macroscopic phenomena. Quantum field theory would be found erroneous for some gravitational phenomena, and would be found mathematically and computationally intractable for anything except microscopic systems or very simple macroscopic ones.

Certainly, some issues of theory-dependence of observation would arise in constructing the Master Observation Set, but it seems to me that they would not be so terribly severe. For example, it's true that (as Feyerabend notes) general relativists tend to focus on those anomalies in Newtonian mechanics that general relativity can address, ignoring the other anomalies. However, this is theory-dependence of focus; the general relativists would not deny the admission of these other Newtonian anomalies into the Master Observation Set. Similarly, some modern physicists who question the current estimate of the mass of the top quark suggest that the Fermilab scientists who have produced this estimate have erred in deciding which observations in their laboratory constitute "quark observation events" and which constitute "noise." They contend that the classification of events versus noise has been done with a specific view towards obtaining a specific value for the mass of the top quark by studying these events. This may perhaps be the case – but even if it is, the problem would be avoided by adding all the raw output of the Fermilab equipment to the Master Observation Set.

Another problem that must be dealt with is the fact that the application of a scientific theory to a scientific dataset within the Master Observation set is not fully automated. Each dataset comes with some "metadata" telling what the dataset measures. To take a single representative example, in the case of gene expression data there is a standard format called

MIAME[53]. According to the MIAME approach, a dataset giving the expression levels of genes observed in a tissue sample from some organism consists of two parts: a table of numbers indicating genes' names and their quantitative expression levels, and some metadata information telling the conditions under which the data was gathered: the type of organisms, its age, the temperature of the room, the type of microarrayer used to gather the data, etc. In general, before a dataset can be analyzed according to some scientific theory, some work must be done, using the metadata to case the data in the language of the scientific theory. Feyerabend argues, in effect, that this metadata-interpretation process is informal and is a critical part of scientific theories. He suggests that because it is informal, this metadata-interpretation may potentially be done differently by different scientists operating within the same theoretical tradition.

It seems to me that the metadata problem is an important one, but is not fatal to the probabilistic program. If a universal metadata language is agreed upon by adherents to various theories, then one can salvage the probabilistic program by one of two strategies:

- Define a scientific theory as a kind of activity carried out by its human adherents in order to predict one set of data from another. The notion of probabilistic validation is perfectly useful in this case, just as much as if a theory is considered as purely formal without any human component.
- Define a scientific theory to consist of a theory as traditionally conceived, plus a set of formal mappings that translate metadata (expressed in the universal metadata language) into theoretical expressions.

Either one of these expansions of the ordinary definition of scientific theory would seem to work, as a way of salvaging the probabilist program.

The only problem remaining, then, is the main one that Lakatos complained about. He argued that, even if this kind of probabilistic approach to comparing scientific theories is theoretically possible, it is pragmatically absurd.

This is an interesting point to reconsider in the context of modern technology. With more and more scientific datasets placed online for public scrutiny – indeed, many high-quality journals now make this a precondition for publication – my proposed Master Observation Set isn't nearly so farfetched now as it would have seemed in Lakatos's day.

But clearly, the formation of the MOS as a database is not the major problem here. The bigger issue is one of combinatorial explosion and computation time. There is no feasible way to survey all possible subsets A and B of the MOS, and ask how well a given theory predicts B from A. The only way to work around this seems to be to define meaningful samplings of the set of all subset-pairs of the MOS. For instance, imitating crossvalidation in statistics, one can divide the MOS into N equally-sized subsets, and then iteratively leave each subset out, seeing how well a theory can predict the left-out subset from the rest. Once one starts doing things like this, a subtle kind of theory-dependence can creep in, because different ways of subdividing the MOS may lead different theories to perform better! However, this is not such a severe problem; and it seems that, in the future, some form of crossvalidation on an online MOS may well be feasible – and may provide a good way of assessing scientific theories.

The statistical issue of "overfitting" also arises here: ideally, one would like theories to make predictions of subsets of the MOS that were not used in the creation of the theory. Lakatos observes that this is a major criterion often used in the assessment of scientific theories.

---

[53] http://www.mged.org/Workgroups/MIAME/miame_checklist.html

Frequently but not invariably, this takes the form of a theory predicting observations that were not known at the time of the theory's creation.

Altogether, my conclusion is that a probabilistic approach to comparing scientific theories is not only possible-in-principle – it may become possible-in-practice once existing technologies and data-publication practices advance a little further.

However, the possibility of a probabilistic approach is one issue – and the issue of whether science actually works this way, in practice, even in an approximate sense, is quite another. Lakatos's claim is that, even though probabilism is possible in practice, it is so computationally intractable that it essentially bears no relationship to the pragmatics of scientific-theory assessment. Unfortunately, I think this is almost the case.

The Prevalence of Biased Probabilistic Inference Heuristics

What are the most important ways in which real-world scientific-theory-validation differs from the idealized view of science as cross-validation testing on the MOS that I have presented above? One big difference, as noted above, is that the translation from metadata to theoretical formulations is nearly always left informal in practice. However, there is another difference that is even more essential. Because of the often large effort involved in applying scientific theories to observations to obtain predictions about other observations, a significant amount of merit tends to be given to "predicted predictions" on the MOS.

That is, suppose a scientific theory is determined – through detailed calculations or computer simulations — to make prediction P for the values of observation-set B, based on the values of observation-set A. Then, this knowledge will be used to extrapolate inferentially, and guess the predictions P' made for the values of other observation-sets B', based on other observation-sets A'. These predictions P' will then be compared to reality, even though they haven't been derived in exquisite detail like the predictions P. This inferential extrapolation is typically carried out, not using generic probabilistic inference, but using heuristic inference methods supplied by the theory itself. In fact this is a major part of any decent scientific theory – the tools it provides allowing the scientist to analyze a new situation without taking a detailed-data-analysis approach.

In reality, because we can't do cross-validation or detailed out-of-sample testing on the MOS, we're always looking at approximations – which wouldn't be so troubling, if it weren't for the fact that the approximations are computed using approximation techniques that are part of the theories being tested!   This is the sort of reason that science begins to look like a kind of opportunistic anarchism, as in Feyerabend's philosophy. This the reason why Lakatos is basically right about probabilism – until we can really construct the MOS and deploy vast amounts of computational power to assessing competing theories using crossvalidation, we had better use some method besides probability theory to compare scientific theories.

Research Programs

These observations about the practical limitations of probability theory lead us directly into Lakatos's theory of "research programs." Since each scientific research program provides its own heuristic short-cuts to true probabilistic validation, and since truly accurate crossvalidation or out-of-sample validation on the MOS is infeasible, some other way of assessing the quality of a research program is required – and Lakatos actually had some concrete suggestions in this direction. Note the language I'm using here – we're now talking about assessing the quality of a *research program*, not an individual theory. This is essential.

For example, we're now talking about assessing quantum mechanics, not quantum mechanics' explanation of the helium atom. Of course, one can often compare two theories of some particular phenomenon – say, competing explanations of the behavior of liquid helium. But this comparison nearly always takes place within a common research program (quantum electrodynamics, in the case of liquid helium). Sometimes, a cross-research-program analysis of some particular phenomenon is possible, but this can't be counted on. Generally, if one has two incommensurable research programs – two research programs that "speak different languages" in a fundamental sense, like quantum versus Newtonian mechanics, or Freudian versus behavioral psychology – then one faces major problems in doing specific-example-based comparisons. There are difficulties of formalization and validation such as mentioned above … and then one also faces problems of the theory-driven selection of illustrative examples.

But how can one assess the quality of a research program except by assessing the ability of its component theories to make predictions about the MOS? Lakatos suggests a distinction between "progressive" and "regressive" research programs. In essence, a research program is progressive if it is creatively and adaptively growing in response to new information.

I like to formalize this as follows: A research program is progressive if, when confronted with a significant amount of new data added to the MOS, it can generally predict this data either:

- Without modification, or else
- With modifications that are relatively simple compared to the complexity of the new data added.

A regressive research program, on the other hand, deals with qualitatively novel datasets by means of modifications that are equal or greater in complexity to the new data itself, or else by tactically decreasing its scope to avoid the problems encountered.

Of course, a research program may be progressive at some points in its history and regressive at other points. Predicting the future progressiveness or regression of a research program is itself a difficult problem of probabilistic inference! So it may seem that, from a practical perspective, we have done nothing but push the problem of probabilistic inference up from the level of scientific theory accuracy to the level of research program progressiveness. There is some justice to this criticism – however, I think the correct response is that, in some ways, judging research programs progressiveness is an easier problem.

The idea that judging research programs may in a sense be easier than creating scientific theories, is related to various results from modern computer science. For example, in Chapter 2 I mentioned a computational technique called the Bayesian Optimization Algorithm (Pelikan, 2002), which solves optimization problems by maintaining a population of candidate solutions. Each candidate in the population is evaluated as to its "fitness" – its efficacy at solving the optimization problem – and a probabilistic model is then constructed, embodying patterns that are common to many of the fit candidates and not many of the unfit candidates. These patterns are then used to generate new candidates, which are more likely than random candidates to be fit, because they embody patterns known to characterize fit candidates. The problem with this algorithm is that finding a good probabilistic model of the population is not so easy – in a sense, one has replaced one's original optimization problem with a new one: finding an optimal model of the fit elements in the population. The beauty of the algorithm, however, is that this new, higher-level optimization problem is actually considerably easier. It is easier to recognize patterns characterizing reasonably-good solutions – once one has some reasonably-good solutions to study – than to find good solutions in the first place. This is analogous to the situation in the evolution of science, as one sees by mapping BOA "candidates" into "scientific research programs" and BOA "models" to "heuristics indicating scientific research program quality." It's easier to find patterns characterizing scientific research program quality – as Lakatos has done, based on many examples of both high and low quality research programs – than to find quality scientific research programs directly.

A good example of a research program shrinking scope in response to challenging new data can be found within my own primary research area, artificial intelligence. The contemporary field of AI is dominated by a perspective that I call "narrow AI," which holds that the problem of engineering and describing software programs with general intelligence can be addressed via the process of engineering and describing software programs carrying out highly specific tasks. AI researcher Danny Hillis expressed part of the philosophy underlying this perspective with his assertion that intelligence consists of "a lot of little things,"[54] cobbled together in a fairly simple way to yield overall coordinated functionality. On the other hand, there is an alternate research program that I call AGI (Artificial General Intelligence), which holds that the essence of intelligence is holistic, and hence that the correct theories about how general intelligence works will not have much to do with software programs narrowly tailored to highly specific tasks. In the early days of narrow AI (the 1960's and 1970's), researchers sought to construct narrow AI programs carrying out highly ambitious tasks, but their theories failed to accurately predict the behavior of these programs – specifically, the programs worked far less intelligently than the theories suggested they would. The result has been, not to discard the narrow-AI approach, but simply to stop trying to make ambitious software programs – thus, in practice, narrowing the scope of the narrow-AI research program to "the explanation of the behavior of software programs carrying out specialized tasks." For instance, in the domain of automated theorem proving (a subspecialty of AI), few researchers seek to create truly autonomous theorem-proving software anymore; the scope has narrowed to the study and creation of software that proves theorems in a semi-automated way, with significant human aid.

This narrowing of scope of the narrow-AI research program seems to me indicative of regressiveness. However, narrow-AI advocates would argue that this regression is only temporary – that after the problems within the newly narrower scope have been solved, the narrow-AI research program will be ready to broaden its focus again. After enough "little things" have been fully understood, then perhaps the project of piecing together little things to make a big thing – an artificial general intelligence – will seem as simple as the original narrow-AI theorists thought it would be. On the other hand, the AGI research program has many fewer practical achievements to its name than its narrow-AI competitor, on a conceptual level it has been comparably progressive, at least in the last few years — as AGI research is explaining more and more cognitive phenomena each year, and continually leading to more and more interesting software systems. This example illustrates the difficulty of applying these philosophy-of-science ideas to make normative judgments about practical situations. Predicting whether a regressive research program will become progressive again, or whether an early-stage, speculative progressive research program will indeed flourish and blossom – these are not easy problems (though I have my own strong opinion regarding the narrow-AI/AGI case!).

Ptolemaic astronomy, with its epicycles, is the classic example of a regressive research program generating modifications whose complexity is qualitatively "too large" relative to the

---

[54] Personal communication

new data being incorporated. Feyerabend argues against Ptolemaic astronomy as an example of overcomplexity, on the grounds that the Copernican perspective added other kinds of complexity that were even more complex than the Ptolemaic epicycles. Feyerabend has a good point regarding the relativity of simplicity-assessment – however, I ultimately think he's incorrect, and that, relative to overall human judgment, Ptolemaic astronomy is just plain more complicated.

This example illustrates a major problem with my formalization of the progressive/regressive distinction: it's dependent upon the measure of simplicity one defines. This may seem to be a hopelessly subjective issue. However, the subjectivity may be at least partially gotten under control by an appeal to algorithmic information theory, as was done above in the context of the naïve computational/probabilist approach. If one defines a computational model (using, say, a kind of abstract computer such as a Turing machine) of what a "theory" is – say, if one defines a theory as a computer program running on some particular abstract machine – then one can define simplicity based on algorithmic information theory or its variants. A simple entity is one that runs fast and takes up little memory. A modification to a scientific theory is simple if it requires little code and doesn't make calculations with the theory humongously slower. There is some arbitrariness in the weighting of space versus time complexity here, and in practice scientists seem to weight space complexity much more – i.e., brief theories are considered high-quality, whereas no one minds much if a theory entails horrifically complex calculations even to explain phenomena that intuitively appear quite simple.

Without this kind of appeal to an "objective" measure of simplicity, it seems, the theory of research programs is doomed. Thus in our definition of science, it would probably be wise to insert a clause stating that the individuals involved, as well as agreeing on the probable reality of most of the Master Observation Set, should roughly agree on what's simple and what's complicated. Given this addition to the definition of "science," we may say that scientific theories come along with a method for validation by definition.

It's worth noting that, in this approach, a research program cannot be validated in the abstract: it has to be validated as a methodology for action adopted by a particular set of intelligent agents (primarily humans, at the moment). In order for the theory to get modified to deal with new situations, someone has got to do the modifying.

A Pragmatic Synthesis

Returning to the naïve computational/probabilist approach defined earlier, we may now ask: What have we managed to salvage? In essence, we have discarded the probabilistic-accuracy criterion as being too difficult to assess in practice, and replaced it with a shift up from the level of theories to the level of research programs; but we have not managed to get rid of the computational-simplicity aspect. We are still talking about simplicity, not in the context of simplicity of theories, but in the context of simplicity of modifications to research programs.

What is the conclusion about the comparison of different scientific theories and approaches? It seems that reality dictates a mixed approach.

In the case of theories existing within the same research program, or research programs that are not too incommensurable, the naïve computational/probabilist approach is closer to reality. We really do compare theories based on which ones make more accurate predictions within their pertinent domains.

On the other hand, when research programs are too strongly incommensurable, the probabilistic approach is bollixed in practice by issues relating to different interpretations of the same data, and different heuristics for estimating predictivity in various contexts. One must resort to a cruder, more high-level approach: the comparison of the quality of the research programs in general. However, in order to compare the progressiveness of two research programs, one requires a standard of "simplicity" that spans theories within both of the programs.

Furthermore, the same approach used to compare incommensurable scientific research programs may be used to compare nonscientific belief systems. The probabilist aspect of the ideas presented above is peculiar to scientific theories – but the notion of progressive vs. regressive is not tied to probabilism. One may assess and compare the progressiveness of two different religions, for example. The question of who will perceive this is valuable, however, is another issue.

The Lakatosian notion of progressiveness embodies the scientific ideal of "progress" – which, as Feyerabend points out, is not a universal idea. Aristotelian science, for example, was more attuned to the ideal of stability – of finding a collection of ideas that would explain the universe adequately, once and for all. The belief systems of precivilized tribes also tend to be oriented toward stability rather than progress.

And so, Feyerabend is correct that there is no "objective" way to compare scientific theories or research programs. Different research programs breed different intuitive notions of simplicity, and hence there is subjectivity in the calculation of which research program requires the more complex modification to deal with a new dataset.   There is also a value judgment implicit in the assessment that "progressiveness" and the ability to gracefully incorporate new information is a good ting.

The dependence on the value judgement that "progress is good" doesn't worry me much. But the dependence on a subjective measure of simplicity is more troubling. If the probabilistic assessment of theory quality is impossible due to computational intractability, and the assessment of research program progressiveness depends on a subjective simplicity measure — then can science progress? Must we abandon, with Feyerabend, the vision of science as a progression through a series of better and better research programs?

I think not. The key lies in the definition of science as a human enterprise – and more generally, an enterprise carried out by particular communities of particular minds.

Human Nature

When David Hume treated the problem of induction (in *A Treatise on Human Understanding*), he noted a familiar infinite regress. Suppose we predict the future from the past using some predictive methodology – then how do we know this predictive methodology is workable? We know this because in the past the predictive methodology seemed to work. But, how do we know the predictive methodology will continue to work in the future? Well, we have to use some predictive methodology to govern this prediction. Hume pointed out that humans do not suffer from this regress except in pathological cases, because our "human nature" provides a way out.   Essentially, he argued, we are hard-wired to use certain base-level predictive methodologies, and we do so because our brains tell us to, independently of any abstract reasoning we may carry out.

Combining Hume with Darwin, we see that – if one is willing to use science to help resolve issues in the philosophy of science, a classic Hofstadterian "strange loop" (Hofstadter, 1979) — this pushes the problem of determining the hard-wired predictive methodology out of the domain of human psychology and into the domain of evolution by natural selection.   And how did natural selection come up with this hard-wired predictive methodology for us? The same way that natural selection comes up with everything else: by chance-guided, selection-focused physical self-organization. We reason inductively the way we do because we can't help it; chance-guided, selection-focused self-organization made us this way and further self-organizing dynamics has reified it.

Similarly, the solution to the problem of the progress of science is provided by the fact that we humans have an innate sense of simplicity. Our sense of simplicity is guided by the beliefs and theories that we hold, and the scientific research programs or other organized traditions that we work within – but only in pathological cases is a scientist's sense of simplicity *completely dominated* by the research program in which he works. There is a universal human sense of simplicity, about which our individual senses of simplicity cluster – and it is this that allows us, eventually, to arrive at virtual consensus on which research programs are progressive and which are regressive. Individual, belief-guided variation in the sense of simplicity allows some people to carry on a long time with research programs that others believe are clearly regressive – but eventually, if a research program's ongoing data-driven modifications get too perversely complicated according to the innate human sense of simplicity, even the true believers' intuitions rebel and the research program is recognized as regressive.

One can see this happening in social science today, with the Marxist research program. Marxist true believers have a deeply Marx-influenced sense of simplicity, so that there are many modifications to Marxism that appear to them reasonably simple and elegant, although other humans tend to view them as overcomplicated and unconvincing "special pleading." However, as the modifications get more and more perversely complex according to the innate human sense of simplicity, more and more Marxists are impelled to abandon the faith and move on to different ways of looking at the world. Clearly, the probabilistic aspect has also played a role here: Marxism, on the face of it, would seem to have made many wrong predictions about major world situations. However, the Marxist theoretical framework is extremely fertile, and it always has "good" explanations for why its predictions weren't wrong after all. Marxism provides many powerful examples of the ability of research programs to influence human perceptions of the world. What has caused Marxism to gradually dwindle in support, in my view, is not just its failure to predict, but at least as critically, its failure to come up with humanly simple modifications to allow it to explain newly observed data. For instance, Marxist theorists have come up with many explanations to cover phenomena such as the emergence of the middle class and the democratizing nature of computer technology – but these explanations seem to involve a lot of "special pleading" and lack the elegance and power of Marx's original analysis of the capitalist class versus the proletariat.

We see from this analysis the profound truth of Feyerabend's assertion that science is first and foremost a human enterprise. To nonhuman intelligences, with a different innate sense of simplicity (no Humean "human nature"!), the human Master Observation Set might well lead to an entirely different set of hypotheses. These intelligences might look at our theories and find them interesting but a bit perversely complicated; and we might think the same thing about their own theories of the same data. Without the common ground of human intuition-about-simplicity, we'd have no way to resolve the dispute except using the massively computationally intensive probabilist approach – cross-validation and out-of-sample testing on the MOS.

Cognitive Analogues and Enablers of the Scientific Enterprise

In this final section, I will shift focus somewhat: away from the broad currents of scientific change, and into the mind of the individual scientist. I will somewhat cursorily explore the question: What sorts of thought processes are maximally conducive to scientific progress as discussed above? This is interesting from the point of view of theoretical psychology, and also from the point of view of artificial intelligence. One of the long-standing goals of AI research is to produce artificial scientists. It stands to reason that this quest should be informed by a proper understanding of the nature of the scientific enterprise.

Informed by Lakatos, Feyerabend and others, and using the language of probability theory and computation theory, we have reformulated the philosophy of science in a novel way. What does this tell us about cognition, on the individual level? As it turns out, all the conclusions drawn above may be ported from the "scientific community" level to the "individual mind" level, if one merely takes the step of replacing the Master Observation Set with an Individual Observation Set – the set of observations that an individual human mind has either made itself, or has learned through symbolic means (e.g. language) and believed.

In place of scientific theories, we have predictive procedures in the individual mind: procedures that take some information in the IOS and make predictions about other information in the IOS. Again, the quality of a procedure may be measured probabilistically. And again, for computational reasons, much of this quality-assessment is done via inference rather than direct evaluation on observations in the IOS. And again, much of this inference is heuristically guided by theories held by the mind.

The psychological analogue of a research program is a belief system, a tightly interconnected network of ideas, including a set of related and interlocking procedures for predicting parts of the IOS based on other parts. As noted above, on the societal level research programs are merely special kinds of human belief systems, and their progressiveness may be studied in about the same way as one would do for other belief systems (such as religions).

The theory-dependence of observation is extremely marked in human beings. It penetrates down to the lower levels of the perceptual cortex. We have a remarkable capacity to see what we want to see, and remember what we want to be the case.   A famous and very low-level example of this is the blind spot that we would each see in front of our nose, if our brains didn't perform interpolation so as to convince us that our perceived world, at a given instant, doesn't have a hole in it (Maturana and Varela, 1992).

And the theory-dependence of human observation and memory does, in fact, make it difficult for us to objectively compare belief systems. We humans are very good at not seeing things that don't match our beliefs. There are many examples of this in the history of science – scientists ignoring "anomalous" data points for example – but there are even more examples in other domains of human life, such as politics.

Human belief systems have a strong propensity to persist over time, because of their capability to bias observation and memory in their favor, and their ability to conceive inference control heuristics that bias probability estimates in their favor. In *Chaotic Logic* I described this propensity as a kind of "psychological immune system."

Counterbalancing this tendency of belief systems to persist, however, is the human mind/brain's quest for progress. We humans intrinsically value both novelty and simplicity. And these two values, as we have seen, lead to the possibility of a Lakatosian assessment of the progressiveness of research programs and other belief systems. We want our thought-systems to grow to encompass newly observed situations, and we desire simplicity of understanding, according to our individual variations on our innate human senses of novelty and simplicity. These two desires cause the human mind – in many cases – to allocate attention away from belief-systems that have become regressive in the Lakatosian sense. And of course, this dynamic on the *individual cognitive level* is what is responsible for the same dynamic on the sociological level, as is seen in the history of science.

This analysis makes clear one way in which science differs from most other human belief systems. The human mind has a tendency to grow *persistent* belief-systems, but it also – via its innate quest for novelty and simplicity – has a tendency to seek *progressive* belief systems. What is unusual about science is the way it values novelty and conceptual simplicity over stability and persistence. This seems to be associated with the fact that science seeks to overcome the individual bias present in the middle of the Vedantic hierarchy via appeal to the bottom rather than the top of the hierarchy. Novelty has to do with temporality, which is key to the physical realm and largely irrelevant in the upper spiritual realms.

This leads to the rather obvious conclusion that individuals who value stability and persistence may be good scientists, but they will be unlikely to create new research programs. The progress from one research program to the next is always made by a human being who, psychologically, values novelty and conceptual simplicity over stability. (Of course, however, human psychology is complex, and the same individually may value novelty in one domain and stability in another – at this juncture one may introduce the notion of "subpersonalities" (see Chapter 18 below) and reformulate the previous statement to say that new research programs are created by individuals with powerful subpersonalities that value novelty and conceptual simplicity over stability.) And this means that science is much more likely to progress in cultures that reinforce these values, than in more "steady-state" oriented societies.

Finally, what does this tell us about the engineering of scientifically accomplished AI systems (the topic to which we'll turn in Chapter 14)? The main lesson is that a purely analytical, probability-calculation-based approach will probably never be workable. It will run into the same problems that characterize naïve computational probabilism in the philosophy of science. All the needed probabilities can never be calculated, so heuristics must be used – but these heuristics inevitably wind up being theory-driven, and hence they do not provide an objective way of comparing theories. Probabilistic theory assessment must be augmented by a more system-theoretic sort of theory assessment, such as is provided by the notion of progressiveness. And the notion of simplicity is key here. If an AI program finds different things simple than humans do, it is going to find different scientific theories preferable. This may be a good thing, in some contexts; we may wish AI scientists to derive theories no humans would ever think of. On the other hand, if its innate notion of simplicity is too different from ours, an AI may find itself unable to learn effectively from human scientific knowledge – it may be able to absorb the abstract theoretical statements, but not the intuitions that surround them and guide their use.

Chapter 14
Language

Sometime in the period 2000-2002, a European philosophical magazine asked me to write 300 words about language and its general significance. I can't remember what the publication was or why they were interested, but as I recall they were gathering similar blurbs from a variety of individuals they deemed interesting. I saved the text I wrote for them, anyhow – it goes as follows:

Language is above all a system for reflecting on itself. It is a system of distorting mirrors, each symbol, grammatical rule or meaning reflecting parts of the outside world and other parts of language in its own peculiar way. And as it reflects itself, it slowly changes itself, evolving not only new words and grammars but fundamentally new forms.

The archaic root of language was mental abstraction, which led the mind to create symbols signifying abstract patterns that don't actually exist in the world, but only in its model of the world. Symbolism brought proto-language, as observed among apes and one-year-old humans. And language emerged as the reflection of symbolism on itself – the evolution of complex webs of symbols standing for symbols. Language creates an inner cosmos capable of competing with the outer, hence severing the mind from the world.

Then language was projected on the outside world, in the form of languages for manipulating physical objects: engineering, machinery, physics. And when the mind itself came to emulate the machine, becoming a factory for the systematic production of complex linguistic forms — we had the birth of advanced reason, as we see in mathematics, science and philosophy. Then, computer programming, which synthesizes machinery and advanced reasoning in an intriguing way.

But the evolution isn't finished. Over the next two centuries, computer programming languages will interact with human languages in ever more complex ways, resulting, for example, in living texts — documents that search for other related documents on the Net, rewrite themselves, battle and mate with each other.

In fact, we can't fully envision where the self-reflection of language will lead it next, because our tools of thought are themselves limited by language's current condition! We must merely watch, and feel, the expansion of language around us and within us.

Reading this text in hindsight, only major conceptual point I feel I left out there was the role of language in binding together individual minds into an emergent group mind – the social "kinds of mind" mentioned in Chapter 2 above are built out of language to an even greater extent than our individual minds are.

In the rest of this chapter I'll explore these sorts of points in a slightly more elaborated way – mostly focusing on abstract philosophy, but at the end, then, connecting the philosophical points to more concrete issues in applied computational linguistics (one interesting test of how well you understand something being: whether you can implement it in a computer program!).

Benjamin Lee Whorf

As an entrée into the discussion of language, I'll introduce the work of Benjamin Lee Whorf, the radical ethnolinguist[55], who wrote a number of beautiful and influential essays on American Indian languages and the philosophy of language and mind. Whorf's claim to fame is the theory of "linguistic relativity," which states that people from different cultures think differently because their languages are structured differently.

The theory of linguistic relativity is often called the "Whorf Hypothesis"; or else the "Sapir-Whorf hypothesis," a more accurate name which honors both Whorf and his mentor, Edward Sapir. It is, to say the least, a highly controversial idea. However, research done in the last two decades verifies that it is accurate, at least in certain contexts. For example, the sociologist Alfred Bloom (1981) has demonstrated that Chinese tend not to think in terms of counterfactuals — they tend not to construct ideas of the form " If x were true, then..." unless they believe that x, in fact, has a reasonable chance of being true. This means that most Chinese will have to struggle to answer a question like, "If the government were to outlaw fingernails, how would you react?" They will be far more likely than Westerners to answer: "But the government hasn't and won't!" This phenomenon cannot be unrelated to the fact that the Chinese language lacks a convenient method for marking counterfactuals.

John Lucy's (1992) recent work on Yucatecan language has drawn yet more precise and rigorous Whorfian conclusions, showing that the ways Mayans describe pictures differently from Americans is specifically related to the differences in ways their languages describe things. For instance, in cases where English requires the speaker to specify whether a quantity is singular versus plural (e.g. dog vs. dogs), but Yucatec does not ("dog" is treated like "water" in their language, as a continuous quantity), English speakers are more likely to remember whether they saw one or many dogs in a picture.

Whorf's hypothesis of linguistic relativity has received a lot of press. But much less attention has been paid to Whorf's philosophy of language, thought and reality, which is what led him to linguistic relativity in the first place. Actually, in his general philosophy, Whorf was a sort of precursor of the current trend of crossbreeding modern science and mystical religion. The urge for scientific/spiritual synthesis seems to have been the main psychological motive underlying all his work.

His first attempt to combine science and spirituality was a rather awkward hybridization of Episcopalianism with general relativity theory. But gradually, over the course of years, Whorf's scientific focus shifted to ethnolinguistics, and his religious beliefs drifted toward Theosophy. And in the last essay of his short life — "Language, Mind and Reality," published in *Theosophist* magazine in 1942 — he achieved the synthesis he had been searching out for so long. He proclaimed that the forms of a person's thought are controlled by inexorable laws of pattern of which he is unconscious. These patterns are the un-perceived intricate systematizations of his own language — shown readily enough by a candid comparison and contrast with other languages,

---

[55] It's interesting to note that Whorf never earned his living as a linguist — his natural inclination was toward physical science, and throughout his life he remained employed as a chemical engineer.

especially those of a different linguistic family.... And every language is a vast pattern-system, different from others, in which are culturally ordained the forms and categories by which the personality not only communicates, but also analyzes nature, notices or neglects types of relationship and phenomena, and builds the house of his consciousness.

This is the science half of the Whorfian synthesis: the empirically testable hypothesis that the deepest patterns of thought are mostly also patterns of language. The reader may have noticed how nicely Whorf's phrasing fits in with patternist philosophy. "[I]nexorable laws of pattern" control the mind — this is precisely the point of view I've repeatedly championed above. Whorf is saying that the most complex and subtle patterns in the mind are precisely linguistic ones. And this is almost indisputable — so long as one, following Whorf, takes a suitably general view of what constitutes a "language."

And, in his *Theosophist* essay, Whorf immediately followed up this scientific hypothesis with a remarkably lucid and beautiful statement of his spiritual views. This is such a tremendous passage that I can't resist an extended quotation:

> *This doctrine is new to Western science, but it stands on unimpeachable evidence. Moreover, it is known, or something like it is known, to the philosophies of India and to modern Theosophy. This is masked by the fact that the philosophical Sanskrit terms do not supple the exact equivalent of my term "language" in the broad sense of the linguistic order. The linguistic order embraces all symbolism, all symbolic processes, all processes of reference and of logic. Terms like Nama refer rather to subgrades of this order — the lexical level, the phonetic level. The nearest equivalent is probably Manas, to which our vague word "mind" hardly does justice. Manas in a broad sense is a major hierarchical gradein the world-structure.*
>
> *It is said that in the plane of Manas there are two great levels, called the Rupa and Arupa levels. The lower is the realm of "name and form," Nama and Rupa. Here "form" means organization in space ("our" three-dimensional space). This is far from being coextensive with pattern in a universal sense. And Nama, name, is not language or the linguistic order, but only one level in it, the level of the process of "lexation" or of giving words (names) to parts of the whole manifold of experience, parts which are thereby made to stand out in a semi-fictitious isolation. Thus a word like "sky," which in English can be treated like "board" (the sky, a sky, skies, some skies, piece of sky, etc.) leads us to think of a mere optical apparition in ways appropriate only to relatively isolated solid bodies. "Hill" and "swamp" persuade us to regard local variations in altitude or soil composition of the ground as distinct THINGS almost like tables and chairs. Each language performs this artificial chopping up of the continuous spread and flow of existence in a different way. Words and speech are not the same thing. As we shall see, the patterns of sentence structure that guide words are more important than the words.*
>
> *Thus the level of Rupa and Nama — shape-segmentation and vocabulary — is part of the linguistic order, but a somewhat rudimentary and not self-sufficient part. It depends upon a higher level of organization, the level at which its COMBINATORY SCHEME appears. This is the Arupa level — the pattern world par excellence. Arupa, "formless," does not mean without linguistic form or organization, but without reference to spatial, visual shape, marking out in space, which as we saw with "hill" and "swamp" is an important feature of reference on the lexical level. Arupa is a realm of patterns that can be "actualized" in space and time in the materials of lower planes, but are*

themselves indifferent to space and time. Such patterns are not like the meanings of words, but they are somewhat like the way meaning appears in sentences. They are not like individual sentences, but like SCHEMES of sentences and designs of sentence structure. Our personal conscious "minds" can understand these patterns in a limited way by use of mathematical or grammatical formulas.

Whorf's articulation here is closer to poetry than to academic prose — it represents not merely the recital of an hypothesis, but the outpouring of a vision — a vision that is at once personal and universal, and at once scientific, philosophical and spiritual.

What Whorf is saying here, to put it in very prosaic language, is that it is the lower levels of the mental hierarchy which deal with physical space (*Rupa*) and the recognition and categorization of objects (*Nama*). These processes involve relatively simple patterns. The highest levels of the mental hierarchy, on the other hand, deal with patterns that are much more abstract, that speak of relations between relations between relations — this is *Arupa*. And ordinary consciousness resides at an intermediate level on this hierarchy — not quite as high up as the loftiest reaches of abstract pattern, but high enough to get some idea of what *Arupa* is all about.

In Vedantic terms, what we have here is an interpretation of *anandamaya*, the Realm of Bliss, as a universe of patterns and abstract forms, shifting and fluctuating, giving the world its underlying structure. The specific forms of language are seen as instantiations of these general, spacetime-transcending (Arupa) forms.

Whorf defines language very generally, as something that "embraces all symbolism, all symbolic processes, all processes of reference and of logic." But what is the recognition of a pattern if not a process of reference and a symbolic process? A pattern is a representation as something simpler, and this implies both reference and symbolism. So, according to Whorf's very broad idea of language, linguistic order encompasses any kind of systematic patterned order. What Whorf is experiencing here, and expressing in his Theosophical terminology, is precisely the filtering down of forms from the Realm of Bliss to the Mind, through the medium of Intuition.

Rather than just repeating the spiritual insights of others, however, Whorf makes a unique and powerful contribution here, by virtue of his linguistic knowledge and interests. He teaches us that language is a large part of what holds consensus reality together. If many of us tend to understand things the same way, this is because we all use the same basic language of thought — where by a "language of thought" I mean, not a collection of words, but merely a systematic syntax for manipulating mental symbols, i.e. a collection of high-order patterns relating other high-order patterns. Whorf's central assertion is that it is language itself that holds reality together — that most of what filters down from *anandamaya* to *manomaya* via *vignanamaya* is specifically linguistic structure.

Unsurprisingly enough, while modern academics have by and large dismissed Whorf as an eccentric extremist, they have adopted many of his ideas, draped in less dramatic phrasings. Listen, for example, to Searle (1983):

> *I am not saying that language creates reality. Far from it. Rather I am saying that **what counts as** reality — what counts as a glass of water or a book or a table, what counts as the same glass or a different book or two tables — is a matter of the linguistic categories that we impose on the world.... And furthermore, when we experience the world, we experience it **through** categories that help shape the experiences themselves. The world doesn't come to us already sliced up into objects and experiences; what counts as an object is already a function of our system or representation, and how we perceive*

the world in our experiences is influenced by that system of representation. The mistake is to suppose that the application of language to the world consists of attaching labels to objects that are, so to speak, self identifying. On my view, the world divides the way we divide it.... Our concept of reality is a matter of our linguistic categories.

Searle's emphasis on "categories" is reminiscent of Lakoff's (1987) *Women, Fire and Dangerous Things*, the title of which refers to an aboriginal language thatgroups women, fire and dangerous things together under one categorical name. It also reminds of Hilary Putnam's formal-semantic theorem, to the effect that

'Objects' do not exist independently of conceptual schemes. **We** *cut up the world into objects when we introduce one or another scheme of description.*

It has become acceptable in modern academic philosophical and anthropological circles to admit that language guides our *categorization* of the world. If Whorf were still around, how would he react to this? I suspect he would observe that categorization is just the simplest kind of patternment: that language does guide the way we group things together, but it also guides our perceptions and cognitions in subtler ways. And Whorf might also be a bit amused to find the claim that "our concept of reality is a matter of our linguistic categories" in the same essay as the statement that "I am not saying language creates reality. Far from it." It would seem that contemporary thinkers like Searle find Whorfian ideas useful, but they want to avoid controversy by marking a sharp distinction between "our concept of reality" and "reality." Surely Whorf would identify this distinction as a manifestation of Indo-European-language-induced thought-patterns. Lucy's work seems to have stimulated a partial rehabilitation of Whorf's ideas within academia, and it seems that cognitive science comes closer to Whorf's vision each decade.

The Psychology of Language Production

Whorf's "mystical" idea of linguistic forms passing down through Intuition into the mind is closely reminiscent of contemporary and classical models of sentence production. It brings us back to the first detailed psychological model of language production, conceived by the German neurologist Arnold Pick. Pick (1935) gave six stages constituting a "path from thought to speech":

1. Thought formulation, in which an undifferentiated thought is divided into a sequence of topics or "thought pattern," which is a "preparation for a predicative arrangement ... of actions and objects;"
2. Pattern of accentuation or emphasis;
3. Sentence pattern;
4. Word-finding, in which the main content words are found;
5. Grammatization — adjustments based on syntactic roles of content words, and insertion of function words;
6. Transmission of information to the motor apparatus.

This sequential model, suitably elaborated, explains much of the data on speech production, especially regarding aphasia and paraphasia. And it can also be read between the lines of many more recent models of speech production; the details have been modified, but the basic idea is the same.

Pick's stages of sentence production connect naturally with the better-known ideas of Chomsky's transformational grammar. In Chomsky's view, language consists of "deep structures," representing the inner structure and meaning of sentences, and "surface structures," which are the sentences we actually speak, write and hear. Transformation rules map deep structures into surface structures. Chomsky's "deep structure" corresponds to the emphasis pattern and sentence pattern of Pick's steps 3 and 4; whereas Chomsky's "surface structure" is the result of Pick's step 5. Modern computational linguistics (such as the dependency grammar to be briefly reviewed below) rejects much of the specific apparatus that Chomsky attached to the deep vs. surface structure distinction, but retains the philosophical essence of the distinction.

Taking a broader view, it is plain that there is a substantial amount of overlap between Pick's steps 1-5. The "two" processes of idea generation and sentence production are not really disjoint. In formulating an idea we go part way toward producing a sentence, and in producing a sentence we do some work on formulating the underlying idea. In fact, it is quite possible that the process of producing sentences is inseparable from the process of formulating thoughts.

In *From Complexity to Creativity* I gave a novel mathematical model of sentence production, incorporating these ideas. In the present context, however, the point is: Where do the "deep structure" patterns, the emphasis and sentence patterns, come from? Conventional psychological wisdom has it that they come from the cognitive areas of the mind. Whorf's encompassing vision simply extend this idea: they say that the most abstract semantic/syntactic sentence patterns actually come from a "higher" region, a region of cultural universals, a collective unconscious, an abstract realm of forms.

Highly creative writers are able to dip very deeply into the higher realms, to pull out abstract sentence-patterns and concept-structures that others cannot reach. But what they are doing is just a more ambitious version of what we all do when we formulate sentences. It is just a question of how far up in the universal dual network you reach, to grab your abstract forms. Eventually, when one goes far enough up, one is grabbing patterns of such subtlety and complexity that they extend far beyond the domain of consciousness, and present themselves to the mind as a priori emergences — Intuitive forms, ringed around with fluctuating patterns from *anandamaya*.

Nietzsche, Language and Consciousness

Dancing through Whorf's ecstatic visions of linguistic reality is the relationship of language and that mysterious entity, consciousness. In 1882 Nietzsche hypothesized that language is what distinguishes the conscious from the unconscious:

> ... *Man, like every living being, thinks continually without knowing it; the thinking that rises to **consciousness** is only the smallest part of all this — the most superficial and worst part — for only this conscious thinking **takes the form of words, which is to say signs of communication**, and this fact uncovers the origin of consciousness.*
>
> *In brief, the development of language and the development of consciousness (**not** of reason but merely of the way reason enters consciousness) go hand in hand...The emergence of our sense impressions into our own consciousness, the ability to fix them*

*and, as it were, exhibit them externally, increased proportionately with the need to communicate them to **others** by means of signs...*

[C]onsequently, given the best will in the world to understand ourselves as individually as possible, "to know ourselves," each of us will always succeed in becoming conscious only of what is not individual but "average"...

*This is the essence of phenomenalism and perspectivism as **I** understand them: Owing to the nature of **animal consciousness**, the world of which we can become conscious is only a surface- and sign-world, a world that is made common... (The Gay Science)*

Nietzsche interpreted the high degree of consciousness which we humans display as a socio-cultural phenomenon, an exaggeration of animal consciousness which evolved together with language — which evolved, in short, as a meme. But his view of the utility of consciousness was not quite so rosy as Dennett's. According to Nietzsche, only conscious thinking is forced into the straightjacket of language, and for this precise reason conscious thinking is much less fertile than unconscious thinking. Language is for social interaction, therefore that which can be put in the form of language is precisely that which is common rather than that which is individual, unusual, unique.

Yet one cannot conclude that Nietzsche felt linguistic, conscious thought to be unimportant or useless. His attitude was much more complex than that. In a draft of a preface for his never-written treatise *The Will To Power*, he wrote "This is a book for thinking, nothing else." But in the notes for that very book, he wrote of thinking:

Language depends on the most naive prejudices....

***We cease to think when we refuse to do so under the constraint of language**; we barely reach the doubt that sees this limitation as a limitation.*

***Rational thought is interpretation according to a scheme that we cannot throw off**.*

This is about as Whorfian a statement as one could ever hope to find. Nietzsche valued linguistic, conscious, rational thought immensely — for much of his life it was his only solace from physical suffering. But he did not trust it, he did not see it as objective; he refused to treat it as a religion.

Whorf's work focused on the differences in world-view implied by differences in linguistic structure. Nietzsche, on the other hand, saw certain very simple, very essential elements in common to all languages, and perceived that they played an essential role in the construction of the concept of an internal and an external world.

For instance, Whorf wrote of the way English, but not Hopi, refers to lightening as an object. Nietszche saw this objectification of non-objects — crucial in the construction of the external world — not as a peculiar feature of some languages, but rather as a consequence of the one central objectification involved in isolating the "self," the inner actor, as distinct from everything else.

Our bad habit of taking a mnemonic, an abbreviative formula, to be an entity, finally as a cause, e.g., to say of lightening "it flashes." Or the little word "I."

[H]itherto one believed, as ordinary people do, that in "I think" there was something of immediate certainty, and that this "I" was the given cause of thought, from which by

analogy we understood all other causal relationships. However habitual and indispensible this fiction may have become by now — that in itself proves nothing against its imaginary origin: a belief can be a condition of life and nonetheless be false.

The self, the "I", is understood as the basis of the linguistic concept of *subject*, of *actor*. Thus the construction of a self, and the construction of an external world, are perceived as closely related, as emanating from the same fundamental principles. The concept of *subject*, in Nietszche's view, is a prime example of the subtle inter-connection of language and thought. Our language assigns imaginary subjects to actions, and we correspondingly assign imaginary subjects to actions in our conscious and near-conscious thinking; we construct an external world based largely on subjects. And we postulate an imaginary entity called *I*, and attribute to this subject a host of actions that are actually due to the independent and interactive behavior of a number of different subsystems.

These "imaginary" subjects may be understood as the result of an overextended analogy. First, events are correlated with other temporally prior events — e.g. smoke is correlated with fire. Then, it is observed that in many cases it is useful, and hence satisfying, to explain a *large number* of different events in terms of one temporally prior entity. General concepts like"weather," "hatred," "patriotism," and so forth arise, each one out of the desire to explain a certain collection of effects with one entity. These concepts refer to definite collections of specific phenomena; they are simply tools for thinking and remembering.

But then what happens is that, when something cannot be explained in detail, a general concept is adduced as an "explanation." This is not always a mistake: given limited resources, a mind cannot explain everything in detail. It must learn to recognize which things can be explained in terms of well known ideas, and can be ignored until the pressing need to analyze them arises, and which things are anomalous, requiring special attention so that trouble will not occur when the need to analyze them arises. But it is a mistake sometimes: a general concept is adduced as an explanation for a phenomenon to which it simply does not apply. Thus "it flashes" for lightening.

"It bit me" is meaningful, it is a general explanation which could easily be backed up by a detailed explanation. But "it flashes" is not: this is a general explanation which is really unrelated to any detailed explanation. The only possible related detailed explanation would be of the form "this and that combination of atomspheric phenomena flashes" — but that is severely stretching the concept of it, and in any case it is not the sort of explanation that would come naturally to the mind of a non-meteorologist. "I did it" is problematic for the same reason "it flashes" is no good. It is not just shorthand for some detailed explanation ready at hand, it is an empty abstraction.

P.T. Geach, in *Mental Acts* (1957), has made this point in a particularly eloquent way:

The word 'I', spoken by P.T.G., serves to draw people's attention to P.T.G.; and if it is not at once clear who is speaking, there is a genuine question 'Who said that?' or 'Who is "I"?' Now, consider Descartes brooding ... saying 'I'm getting into an awful muddle — but then who is this "I" who is getting into a muddle?' When 'I'm getting into a muddle' is a soliloquy, 'I' certainly does not serve to direct Descartes' attention to Descartes, or to show that it is Descartes, none other, who is getting into a muddle. We are not to argue, though, that since 'I' does not refer to the man Rene Descartes it has some other, more intangible thing to refer to. Rather, in this context the word 'I' is idle,superfluous, it is used only because Descartes is habituated to the use of 'I' in expressing his thoughts and feelings to other people.

According to Whorf, this reification of the subject does not happen in Hopi and other non-Indo-European languages. But on this point I must side with Nietzsche. The grammatical manifestation of reification may vary from language to language, but I very strongly suspect that every language postulates some form of imaginary acting entity. This, unlike use of counterfactuals, emphasis on flux versus stasis, and other linguistically varying phenomena, is absolutely essential to the concept of language. It is an instinctive application of analogical reasoning to the act of naming on which all communication is based, and no culture can escape from it. Humans cannot help but attach a certain amount of concrete reality to the symbols that they use. We can, as Nietzsche suggested, fight this tendency, but this is a battle that no one can ever completely win.

An interesting spin-off of this analysis of imaginary subjects is the theory that free will is an emotion inspired by language. Nietzsche's analyzed free will as

> *the expression for the complex state of delight of the person exercising volition, who commands and at the same time identifies himself with the executor of the order — who, as such, enjoys also the triumph over obstacles, but thinks within himself that it was really his will itself that overcame them. In this way the person exercising volition adds the feelings of delight of his successful executive instruments, the useful 'underwills' or undersouls — indeed, our body is but a social structure composed of many souls — to his feelings of delight as commander.* **L'effet c'est moi***: what happens here is what happens in every well-constructed and happy commonwealth; namely, the governing class identifies itself with the successes of the commonwealth.*

The feeling of free will, according to Nietszche, involves 1) the feeling that there is indeed an entity called a "self", and 2) the assignation to this "self" of "responsibility" for one's acts.

My own view, as recounted earlier, is that free will is a bit more complex than this – but conceptually, I think Nietzsche's perspective is right on. In my earlier discussion of free will I emphasized the cognitive structures underlying free will – but I think it is quite possible that these structures evolved together with the evolution of language. The description of ourselves as "I" may well have come along with the sophisticated virtual multiverse modeler: both language and complex virtual multiverse modeling seem to rely on a similar sort of formal sophistication.

This idea of coevolution of language and will is somewhat reminiscent of Julian Jaynes' theory in his book *The Origin of Consciousness in the Breakdown of the Bicameral Mind* (2000), but there are significant differences. Jaynes argues that in the time of Homer there was no such thing as free will – people conceived themselves as doing what the gods told them rather than what they decided – and also no such thing as consciousness. His evidence, Whorfianly enough, is that in Homer the characters carry out no internal dialogues and make no decisions; the gods always tell them what to do. I do think that this narrative and linguistic choice is revealing: probably the nature of will and self-awareness was very different back then. But the virtual multiverse modeler and self-models of humans were clearly very sophisticated at that stage of history, to enable such enterprises as the writing of books. What had not yet happened, apparently, was the widespread inclusion of the self-model within the self-model. That is: it may be the case that in the time of Homer people habitually modeled themselves, but didn't model their modeling of themselves. This would match up with the description of one's activities as controlled by the gods – but the actuality of one's activities being controlled by extensive self- and virtual-multiverse- modeling activity. But the flexibility of language allowed people to discuss each other's models of each other, and hence individuals eventually began to internalize others' models of themselves, and build their own self-model-including self-models – and modern consciousness emerged![56]

Contemporary Computational Linguistics and Its Limitations

All this discussion of "cosmic linguistics" may seem only loosely related to the science of linguistics as currently practiced – but the connection is closer than it might seem. I've spent a fair bit of time on linguistics issues in the past few years, in the context of creating computational linguistics systems – software systems that attempt to parse human language and map it into knowledge that can be reasoned on by AI systems, or used to guide search engines to do more intelligent search, etc. While the business of creating such systems is quite distinct from Whorf's ecstatic visions, the fundamental underlying ideas are not so different.

In fact, the *failures* of modern computational linguistic systems connect quite directly to Whorf's and Nietzsche's visions. No one has yet managed to make a software system that can "understand" human language in any reasonable sense, and the reasons for this are directly tied to the way language goes beyond its own domain and pervades everything else in the mind and world.

To give a little linguistic meat to the discussion, I'll talk a bit about computational linguistics – in particular, computational "language understanding"; the attempt to make software systems that can interpret natural language. The weaknesses of all software systems in this domain (including the ones I've personally been involved in!) lead directly back to the more philosophical views on language discussed above in the context of Whorf and Nietzsche.

There are many approaches to computational linguistics out there; for the purpose of this discussion I'll stick with one that is called the "dependency grammar" approach. From a philosophical perspective all of the common approaches are essentially identical, though from an applied-linguistics perspective there are of course many differences.

For sake of simplicity, I'm going to restrict attention to text processing here, ignoring the complexities of processing speech. Speech processing is important and interesting and involves emotional and social subtleties of deep philosophical significance, but, the points I want to make here can be made perfectly well without getting into it.

The approach to text understanding that I've taken in my computational linguistics work breaks the comprehension process down into a series of transformations roughly like this:

```
Text → Tokenizer → Morpheme Analyzer →

Parser → Semantic Mapper → Semantic Sructures
```

I'll now step through these stages one by one, and we'll see where things get subtle and cosmic.

Tokenization refers to the simple process of breaking a string of characters down into an array of words. This is simple enough in most cases, e.g. mapping "This is a string of characters" into (this, is, a, string, of, characters), but there can be plenty of complexities at this stage due to complex uses of punctuation.

---

[56] Of course, this kind of psycho-archeological hypothesis is fairly wild speculation; but my point is that, though Jaynes's exact speculation doesn't seem psychologically plausible to me, there is a variant of it that I think might be.

Morpheme analysis has to do with breaking words into stems and modifiers. For instance, "killer" is broken into "kill+er", "ran" is broken into "run+past", and so forth. This again is a fairly straightforward process, handled in English via a fairly small list of rules coupled with a long list of exceptions.

So far, these early-stage language processing tasks are carried out fairly well by relatively simple computational processes. The next stage, parsing, is where things begin to get complex. This is the stage that most of linguistics theory and practice have focused on, so far. There are many different grammatical theories, each of which contains a large database of linguistic rules, coupled with an appropriate parsing algorithm. The grammatical theory I've used the most in my own practical computational linguistics work is one called "dependency grammar," which works by associating words with "connectors" of various types. Parsing consists of matching up connectors from one word with connectors from another word. This is different from the typical "phrase structure grammar" approach that one learns in school as a child. Phrase structure grammar words by explicitly grouping the words in a sentence hierarchically into phrases, clauses, and so forth. Each phrase or clause has a "head" that controls its linguistic behavior. In dependency grammar, on the other hand, phrasal and clausal structure is left to emerge form the patterns of linkage: linkage between words is taken as fundamental and phrasal/clausal structure is taken as epiphenomenal. This enables simple explanations of many linguistic phenomena that can be explained in phrase structure grammar theory only via quite complex mechanisms. For instance, the grammar of questions in traditional phrase-structure grammar as developed by Chomsky becomes very complex, involving various kinds of "wh- movement"; but in dependency grammar questions are handled much more simply by the introduction of appropriate connectors.

To illustrate the nature of dependency grammar more concretely, let's consider a simple example:

```
The cat chased a snake
```

The specific dependency grammar I've worked with most, the "link grammar" (Sleator and Temperley, 1991), associates connectors with these words as follows. Different kinds of connectors are denoted by letters or pairs of letters like S or SX. Then if a word W_1 has the connector S+, this means that the word can have an S link coming out to the right side. If a word W_2 has the connector S-, this means that the word can have an S link coming out to the left side. In this case, if W_1 occurs to the left of W_2 in a sentence, then the two words can be joined together with an S link. So, the connectors of the words in our example sentence are:

Words	Formula
a, the	D+
snake, cat	D- & (O- or S+)
Chased	S- & O+

If you remember high school grammar, you will note that the connector types correspond loosely to basic "parts of speech", e.g. D for determiner, O for object, S for subject, etc.

Linking up the connectors with each other appropriately, we find the link grammar parse structure for this sentence is:

```
        +----------------Xp----------------+
        +------Wd----+           +-----Os---+      |
        |      +-Ds-+---Ss--+    +-Ds-+     |
        |      |    |       |    |    |     |
        LEFT-WALL the cat.n chased.v a snake.n .
```

The cognate of this parse in standard phrase structure grammar is:

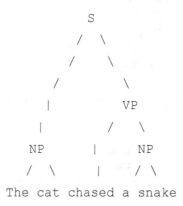

In this case, because the sentence is so simple, either approach is equally effective.

What parsing does, in general terms, is to specify the syntactic relationships between the words in a sentence. In the above example this was very simple, in other situations it's trickier, of course. Consider the classic sentence "I saw the man with the telescope." This leads to the two parses:

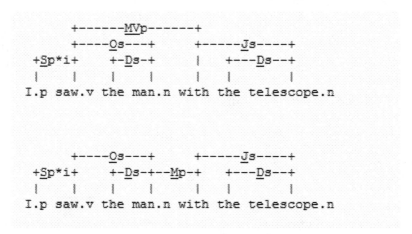

In the first parse, you're looking through the telescope and using the telescope as a medium to see the man. In the second parse, you're seeing the man (by what medium is not specified) and observing that the man is with a telescope. Selecting between these two parses depends on semantic interpretation, not syntax alone. (And then there's the alternate interpretation in which I'm trying to use the telescope as a saw!) This kind of ambiguity is manageable for simple sentences like this, but when you try more complex sentences, the multiple ambiguities interact with each other, resulting in a combinatorial explosion of partial parses. For instance, in applying computational linguistics to biological research articles one runs into sentences like

```
"Fragments from N-ras exons I and II containing the codons of
interest were amplified by polymerase chain reaction and analyzed
for the presence of point mutations by three different technical
approaches, including specific oligonucleotide hybridization,
direct sequencing, and single-strand conformation polymorphism
analysis."
```

This is not really such impenetrable prose – any human biologist can understand it – but no existing parser can handle this sort of text adequately. A large number of parses are produced – dozens to hundreds depending on the particular system – and then there is no good way to find the correct one(s) among the morass. Instead, current computational linguistics systems take an information extraction approach, in which the goal is to parse bits and pieces of the sentence that hopefully contain the most relevant chunks of knowledge.

After parsing is done, the next step in the computational language parsing pipeline is "semantic mapping" – the transformation of syntactic relationships into semantic ones. In principle a computer system could learn these rules but in practical computational linguistics systems, these days, the rules are put in by hand. A simple example rule is the one that translates an Ss link (as used in the above parse) into a "subject" relationship, i.e.

```
Input: A -- Ss -- B
Output: _subj(A,B)
```

In this case the mapping is a trivial relabeling, but the general need for this mapping layer is clear if one looks at the parse for "The snake was chased by the cat." Here we have

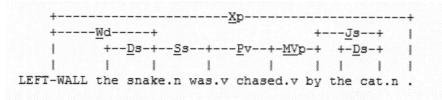

and one needs the mapping rule

```
Input: was -- Pv -- A -- MVp-- by -- Js -- B
Output: _subj(A,B)
```

This rule deals with the "passive voice" construction. Such mapping rules provide a layer of normalization, i.e. they map syntactically different utterances into semantically identical forms, where appropriate.

But unfortunately, simple rules like this don't go quite far enough. As an example, consider the following two outputs, from the RelEx language understanding system that some colleagues and I have been developing for the last few years:

```
Sentence 1: Osama believes that <statement>
_subj(believe, Osama)
that(believe, <statement>)

Sentence 2: Osama believes <statement>
_subj(believe, Osama)
_obj(believe, <statement>)
```

These two sentences basically mean the same thing, and yet, even after the mapping from syntactic relations into semantic relations, they're still represented differently by the language processing software. This is bad!

What we want is to get a single relationship

```
believe(Osama, <statement>)
```

out of both versions of the sentence (I've used the standard notation of predicate logic in representing the relationship above; of course, any number of equivalent or roughly equivalent knowledge representations could be used instead.)

This kind of representation can be produced easily enough from the second variant above via a rule

```
Input:
_subj(V,X)
_obj(V,Y)

Output:
V(X,Y)
```

But the first variant is problematic.

One way or another, to deal with this situation, the language processing system has got to know that

```
that(believe, <statement>) =
_obj(believe, <statement>)
```

i.e. that the object of a belief is what the subject is believing in. Rules like this can be explicitly encoded into a language processing system, or they can be learned by the system. Such learning may take place in a couple different ways:

- Through statistical language analysis, i.e. by feeding the system a massive corpus of texts, and having it pick up regularities. From many sentences of the form "X believes Y" and "X believes that Y" the system may be able to learn that the two expressions mean the same thing.
- Through embodied learning, or other sorts of correlation between linguistic and nonlinguistic information. From many cases where the sentences "X believes Y" and "X believes that Y" are both used to describe the same situation, the system may be able to learn that the two expressions both mean the same thing.

The statistical approach is very much in the spirit of the mainstream of computational linguistics today – which is only natural since the Internet has given us a massive amount of text to work with. The embodied-learning approach is more like how humans learn language, and seems to me the most promising approach, but is not being pursued nearly as avidly due to its greater complexity from a computer implementation approach, and because it requires explicit effort to interact with and teach the learning system. I'll say more about this in Chapter 15 when I discuss the Novamente AI project, which is taking precisely this sort of embodied-learning approach to language learning.

So, the limitations of modern computational linguistics systems as regards language comprehension are twofold. (Note that I haven't even touched on topics like language generation or conversational pragmatics – that would take us too far afield, and the general conceptual point I want to make can be adequately communicated even within the limitations of the comprehension domain.) First, parsing of complex sentences tends to be infeasible, because there are too many syntactically legal parses, and the selection of the semantically correct one tends to require world-knowledge which current computational linguistic systems don't have. Second, even where parsing can be done correctly, mapping of correct syntactic parse structures into semantic structures is difficult, because it requires a lot of rules that humans learn via experience, and that are entirely separate from rules governing grammatical legality.

And this brings us back to the broader, less technical ideas of Whorf and Nietzsche regarding the general nature of the language phenomenon. What modern linguistic science (computational or not) tries to do is to isolate particular aspects of linguistic patterning (e.g. syntax, semantic mapping, morphology) and deal with them in isolation. But this approach reveals severe limitations. Language is more holistic than that. Syntax parsing problems that look horrendously complex in isolation, appear far simpler when considered in a broader context of linguistic patterns existing on multiple levels. The human mind deals with language in a holistic way – babies learn language via learning phonology, morphology, syntax, semantics and reasoning all together (though with different emphases at different stages in the learning process). And this is because language is really a way of structuring both ideas and sounds/character-sequences and social interactions.

The phenomenon that Whorf pointed out so extensively and articulately, that the syntax of written and spoken language helps structure thought, is just part of the overall web of language with its multiple interpenetrating aspects and levels. Syntactic structures impact semantic structures, semantic structures impact syntactic structures; and there is also interplay with phonological and morphological structures. The exchange of spoken and written words between human minds occurs in a context of a broader sharing of trans-individual patterns. As we reviewed in Chapter 2 above, human minds are not really individual – they consist largely of patterns that emerge among the combination and interaction of multiple human minds – and much of this emergence is mediated by the holistic phenomenon of language, spanning all the levels from phonology through shared semantic understanding. Dissecting the various aspects of this emergent-mind-enabling linguistic patterment, as is done in modern analytical and computational linguistics, is very important work – but it is instructive how very difficult it is to understand any aspect of the linguistic whole in isolation. From an AI perspective, this leads up to points that will be made in Chapter 15 below, regarding the difficulty of achieving any aspect of humanlike intelligence in isolation, apart from the holistic dynamics of mind in which each aspect reinforces and builds on the others.

The need to compress our ideas into compact sentences that can be uttered or written in a practical way imposes the requirement that language contains some sort of illusions. There is just no way to compress complex information into compact form without some loss of information. However, this doesn't mean that the kind of systematic illusions that we see in human thought and language are necessary. But even so, the various illusions that characterize the human condition – self, free will, the specialness of individual human consciousness and all that – are all represented in our language, as Nietzsche observed in his discussion of the psychological significance of that wonderful, dangerous little word "I." The direction of causation is difficult to isolate here. The language exists as it does because of our psychology – we shaped our language based on the shapes of our thoughts. On the other hand, when we teach a baby to think, we teach it using language, and the syntax of the language a baby learns affects their forming mind along with and synergetically with the semantics.

Chapter 15
Toward Artificial Minds

The field of Artificial Intelligence is a flourishing subdomain of Computer Science, with numerous journals and conferences each year, and plentiful new PhD's entering the field. And, while AI is not big business on the level of word processors, computer games or operating systems, a number of profitable AI software companies exist, as well as successful AI-oriented divisions within larger companies. Yet what is meant by "AI" these days bears scant resemblance to the grand visions that accompanied the founding of the field in the middle of the previous century. AI these days nearly always means "narrow AI" – the creation of software programs that carry out highly specific functionalities that are typically considered "intelligent" when humans carry them out. Because of this I have adopted the terminology "Artificial General Intelligence" (AGI) to refer to the pursuit of software systems that display a wide variety of intelligent functionalities, including a reasonably deep understanding of themselves and others, the ability to learn how to solve problems in areas they've never encountered before, the ability to create new ideas in a variety of domains, and the ability to communicate richly in language. AI research flourishes; and AGI research, by and large, languishes.

In this chapter I summarize a set of ideas that I have developed during the last two decades, which has led me to what I believe is a novel and productive way of thinking about artificial intelligence, and also to a specific design for an AGI system: the Novamente AGI design[57]. As this is not a treatise on technical AI, the discussion will necessarily be somewhat sketchy. However, at the end I will get concrete and describe some of the specific learning experiments we are now doing with the Novamente system, aimed at having it learn the sorts of things that a human infant learns when interacting with the world, and considered as the first steps in a coherent educational program with the end goal of general intelligence at the human level and beyond.

Patternist Philosophy and AI

In the big picture, the relationship between patternist philosophy and AI is a relatively simple one. Patternist philosophy states that mind is made of pattern, and that intelligent systems are involved in recognizing and creating patterns in their environments in the service of achieving more or less complex goals. It follows from this that artificial intelligence almost surely must be

---

[57] The Novamente project is ongoing and the current software implementation of the Novamente AI design is somewhere between 30% and 60% complete depending on how you measure it. What is discussed here is mainly the AGI design rather than the state of the current implementation; however I will occasionally insert comments regarding what has currently been implemented and tested and what has not.

possible, since brains are merely physical mechanisms for giving rise to complex dynamic pattern systems, and other physical mechanisms with similar patternist potential must almost surely be constructible. Free will and emotion are analyzed in the patternist view as the subjective consequences of particular organizational patterns in intelligent systems. Subjectivity and objectivity are viewed as different perspectives on the universe – i.e. different ways of organizing patterns – and there is no reason why these two different modes of pattern organization may not be associated with computer programs or robots or intelligent gas clouds on Titan as well as human brains. Consciousness is connected with physical reality in several senses, for instance it seems that quantum decoherence is closely related to the experience of observation that lies at the heard of consciousness – but there is no evidence for anything special about human brains in the context of consciousness; we are not the only complex, self-organizing, decohered systems out there.

The patternist perspective gives a particular structure to the tasks of analyzing and synthesizing intelligent systems. About any would-be intelligent system, we are led to ask questions such as:

- How are patterns represented in the system? That is, how does the underlying infrastructure of the system give rise to the displaying of a particular pattern in the system's behavior?
- What kinds of patterns are most compactly represented within the system?
- What kinds of patterns are most simply learned?
- What learning processes are utilized for recognizing patterns?
- What mechanisms are used to give the system the ability to introspect (so that it can recognize patterns in itself?)

Now, these same sorts of questions could be asked if one substituted the word "pattern" with other words like "knowledge" or "information." However, I have found that asking these questions in the context of pattern leads to more productive answers, because the concept of pattern ties in very nicely with the details of various existing formalisms and algorithms for knowledge representation, learning and so forth.

The subtlety of AI in practice comes down to a distinction drawn back in Chapter 2: huge-resources versus limited-resources minds. If an incomprehensibly massive computer processor and memory store were made available, then one could program really simple algorithms for pattern recognition and extremely flexible algorithms for controlling actions, and just let pattern-learning roll, leading to arbitrarily powerful intelligence. This idea has been worked out in great detail by some theorists (e.g. Hutter and Schmidhuber, mentioned above), but it doesn't seem to lead anywhere pragmatically interesting. The hard part is achieving enough recognition of the right kinds of patterns (involving what's in the world, involving how to act in the world, and involving how one's own mind works) given the severely limited processing and memory resources that are actually available in the real world. Even if 1000 or a million times as much processing power and memory were available as what we have in contemporary computers, the same issue would still be there – the human brain is a kind of hodge-podge of specialized subsystems, each one specialized for recognizing and enacting certain kinds of patterns; and any actual AI system is going to have this same character. The art and science of AI then comes down to figuring out various useful tricks for recognizing and enacting various useful kinds of patterns – and then getting all these tricks to work together within a framework that also supports more generic pattern recognition and enaction in those cases where all the specialized tricks fail. This is not a very easy problem at all – we don't really understand how the brain does all this, and contemporary mathematics and computer science give us only very limited guidance. This is the problem on which I have focused the largest part of my time during the last decade, in fact.

In parallel with this book, I've written a (considerably longer and more technical) book focused exclusively on my approach to AI (*Engineering General Intelligence*); the ideal way to read this chapter would be to read it once before reading that other book, and once after. But the chapter isn't written with EGI as a prerequisite; my hope is that in spite of the skipping-over of nearly all the AI details, my main philosophical and conceptual points will get across.[58]

Originally my hope was to make this a chapter on "what the patternist philosophy of mind tells us about the structures and dynamics of AI's in general" – but after some work in this direction, my conclusion was that the answer is "not all that much." The question is too general, and the variety of possible AI's is too great. Rather, the right question to ask is: "Given _____ [fill in the blank with something intelligent] approach to AI, what does patternist philosophy tell one about how to fill in the details of particular AI systems following this approach, and how to interpret the structures and behaviors of these AI systems?" Used in this way, patternist philosophy can be extremely useful for AI, and this has wound up as the focus of the chapter. I briefly review here the essential ideas of my Novamente AI architecture and explain how patternist philosophy elaborates the way I believe they will lead to digital mind when fully implemented, tuned and taught.

Of course, it would be possible to also write parallel chapters discussing what patternist philosophy has to offer for other approaches to AI besides the SMEPH/Novamente approach – biological-modeling-based neural network AI, logical theorem-proving-based AI, and so forth. Such chapters would be interesting but I have not taken the time to write them. The focus of this book is not AI, anyhow. SMEPH/Novamente-based AI is treated here both for its particular interest to me (since I believe it has a higher likelihood of near-term workability than other contemporary AI approaches), and as an example of how AI approaches may be treated within the patternist perspective. I urge others to explore the relationship between patternist philosophy and alternate AI approaches.

Although understanding my own mind and the nature of possible AI minds were probably my main motivations for entering the domain of philosophy of mind, what I have finally concluded is that philosophy only takes you so far toward understanding AI, neuroscience, or any other particular instantiation of the general principles of intelligence. Naively I once thought philosophy could bring you further – but mind-philosophy leaves off and gives way to mind-science at a somewhat more abstract point than I had hoped, mostly because of the powerful role that resource limitations play in any practical intelligent system. Philosophically, mind is a relatively simple thing, once you strip away all the foolish misconceptions that have accumulated via over-reductionist science, self-deluding folk psychology, and other factors. Yet, making intelligent minds in actual reality is not a simple thing at all, because it takes a lot of tricks to get reasonably general intelligence to emerge from physical substrates that are severely limited in memory and speed. The subtlety is not so much in the general nature of what emerges (mind), but in the particular structures and dynamics that occur in particular types of minds, and most of all in the tricks via which the emergence occurs from physical mind-substrates in spite of resource limitations. And philosophy is not the right tool with which to explore these tricks – that is a job

---

[58] And even though I skip over most AI details, some sections of this chapter will still get annoyingly technical for some readers, much as in the "Quantum Reality and Mind" chapter above.

198 The Hidden Pattern

for neuroscience, computer science, and perhaps eventually a rigorous transdisciplinary "complexity science."

Or, to put it another way: An AI based on a wrong philosophy of mind is never going to achieve any kind of significant intelligence. An AI based on a correct philosophy of mind has a chance – but there are still many other things that must be gotten right in order to create a genuine AI system, and these things have to do with mathematics, education and software design, not philosophy.

Intelligence, General and Narrow

The distinction between general AI and narrow AI bears a bit more elaboration. Intelligence, as I've defined it in Chapter 4 above, is clearly a "fuzzy" concept – if intelligence is the ability to achieve complex goals in complex environments under limited resources, then an awful lot of things not naturally considered intelligent may be viewed as intelligent to a limited degree. For instance, a thermostat achieves a goal with a nonzero degree of complexity in environments of nonzero complexity, using quite limited resources – so according to my definition a thermostat is intelligent to a certain degree. Similarly, a program that plays chess matches my definition of intelligence, to a certain degree. What thermostats and chess programs lack, however is an appreciable scope of generality in their intelligence, as compared to the relatively broad scope of human intelligence. Chess is complex in a certain sense, but it doesn't have the breadth of diverse patterns that everyday life does. To achieve a level of complexity-of-goals comparable to human intelligence, an artificial system would need to display a "general intelligence" comparable to that of humans – emulating the way in which humans can deal with goals defined by a variety of qualitatively different contexts. Theoretically, one could imagine a game with the same richness of context and diversity of pattern that human life has, so that the analogue of a chess-playing program, in the context of this game, would have a level of general intelligence similar to that of a human. But there are no games in any human culture that have this property.

Most of the academic and commercial AI field, today, is concerned with what I call "narrow AI" – meaning with programs that have the rough level of intelligence of a chess program. These programs are more complex than thermostats but they lack the generality of scope of humans or even dogs; they are concerned with solving very particular problems, such as playing a particular game or recognizing human faces or diagnosing diseases based on symptoms or recognizing messages that include plans for terrorist acts. These narrow-AI programs may be very good at what they do, but one thing the history of AI has taught us is that such very-narrow-scope tasks can often be effectively achieved via software systems that lack any of the properties we normally associate with minds: free will, self-awareness, emotion, autonomous decision-making, and so forth. On the other hand, some tasks that seem very simple to us humans have proved very difficult for narrowly-constructed AI systems – things such as holding a conversation in a "natural language" like English or walking down a crowded street. AI researchers have by and large shied away from these more contextual, embodied tasks and focused on the ones that can be solved by tricky algorithms acting in relative isolation. As I write these words I am earning my living in a similar way – I'm leading a team of software engineers and scientists involved with constructing narrow-AI programs that do things like recognize patterns in genetics or financial data and extract semantic relationships from newspaper articles, e-mail messages and biological research abstracts. But while I find this work reasonably interesting, I consider it quite distinct from my work on "artificial general intelligence" – the Novamente software system that several colleagues and I are building, which is oriented toward eventually yielding intelligence at the human level and then beyond.

The presupposition of much of the "narrow AI" work being done is that solving specialized subproblems, in isolation, contributes significantly toward solving the overall problem of creating real general AI.   While this is of course true to a certain extent, both my patternist mind-theory and my practical experience with AI development suggests that it is not nearly so true as is commonly believed. In many cases, the best approach to implementing an aspect of mind in isolation, is very different from the best way to implement this same aspect of mind in the framework of an integrated, self-organizing AI system. This phenomenon is rooted in the notion of autopoiesis, according to which the different aspects of an intelligent system all tend to adapt to each other, to the extent that each aspect is nearly implicit in the nature and interactions of the other aspects. There are a lot of possible ways that one can conceive to make computers carry out logical reasoning, for example – but only a small percentage of these ways will fit in naturally with other aspects of an integrated intelligent system forming an autopoietic whole. Now, one might argue that even though autopoiesis characterizes human intelligence, artificial intelligence could be different, and could perhaps involve a loosely-coupled collection of specialized subcomponents as envisioned by the more forward-minded of traditional AI theorists (such as Marvin Minsky, whose "Society of Mind" (1998) concept basically envisions intelligence as consisting of a population of loosely-coupled, highly specialized actors, a vision similar to Danny Hillis's "a lot of little things" understanding). But I don't buy it. My suggestion is that, just as once one has a really large amount of information to organize a hierarchical structure becomes inevitable, similarly, once one has a really diverse population of specialized algorithms to integrate, an autopoietic dynamic becomes inevitable. And I suggest also that the "Society of Mind" label is highly inaccurate as applied to Minsky's vision of a mind as a collection of loosely-coupled agents, because human society doesn't really consist of a collection of loosely-coupled agents – human society, as discussed earlier, is in a strong sense a mind all its own. Human intelligence has its individually embodied aspect but also its distributed, cross-body aspect, according to which a large percentage of human mental patterns actually span numerous individuals rather than residing in any one mind/brain.   The real society of mind, like the real society of humans, is a holistically evolving autopoietic system, not a loose assemblage of individual specialized agents.

The Surprising Unpopularity of AGI

So, why does AGI get so little attention nowadays?   By and large, I suggest, it's *not* because AI researchers believe AGI is impossible. The philosophy literature contains a variety of arguments against the possibility of generally intelligent software, but none are very convincing. Perhaps the strongest counterargument is the Penrose/Hameroff speculation that human intelligence is based on human consciousness which in turn is based on unspecified quantum gravity based dynamics operating within brain dynamics (Hameroff, 1987; Penrose, 2002); but evidence in favor of this is nonexistent. My impression is that most contemporary scientists believe that AGI at the human level and beyond is possible in principle.

The most articulate argument so far created *in favor* of the in-principle possibility of AGI is Marcus Hutter's theoretical work on algorithmic information theory and decision theory, mentioned above several times, which involves positing a very general mathematical definition of intelligence and then proving rigorously that arbitrarily high degrees of intelligence are possible given arbitrarily large amounts of computational power.   In essence, this theoretical work rigorously shows what has been obvious to many researchers for a long time: that AGI is at bottom a problem of processing and memory efficiency. With enough computing power, making AGI is trivial and can be done in a few dozen lines of easily-formulated LISP code. But this insight doesn't help very much in creating practical AGI systems using tractable amounts of computational power. In effect, the human brain consists of a collection of more or less clever tricks for achieving various sorts of more or less general intelligence within rather strict computation-power constraints.

Many AI researchers seem to take the position that, while AGI is in principle possible, it lies far beyond our current technological capability. This is a reasonable enough contention, since according to the best available estimates (which are admittedly very speculative), current computing hardware falls significantly short of the computing power of a single human brain (Merkle, 1998). Furthermore, existing operating systems and programming languages are arguably ill-suited to the task of creating general intelligence; and current compilers for languages that appear to be best-suited (LISP, Haskell and the like) are not pragmatically capable of robustly supporting applications that make the most of multiprocessor architectures and multi-gigabyte RAM. But still, these arguments are not terribly convincing. Even if it's true that current computers are much less powerful than the human brain, this isn't necessarily an obstacle to creating powerful AGI on current computers using fundamentally nonbrainlike architectures. And the shortcomings of contemporary software frameworks are surely just an inconvenience rather than a fundamental obstacle.

In essence, I believe, the reason there has been so little detailed research work on AGI is that there have been so few even moderately convincing general ideas in the area of AGI design. The paucity of plausible AGI designs has been so severe that a number of highly knowledgeable researchers have effectively given up hope, opining that the only or most likely path to AGI is going to be the emulation of the human brain. Eric Baum (2004) has presented this perspective very articulately in terms of the concept of "inductive bias." Much of human intelligence, he argues, is based on tacit knowledge accumulated over generations of evolution, which cannot feasibly be explicitly encoded in software – for similar reasons to those underlying the inability of linguists to fully articulate the tacit rules we use in processing language, in spite of decades spent trying to spell out formal grammars for natural language. Ray Kurzweil (1999, 2005) has argued that brain imaging will yield a reasonably complete understanding of human brain structure and dynamics by the middle of this century, and that the achievement of AGI via human brain emulation will likely follow not long thereafter.

I find the Baum/Kurzweil perspective a plausible one, and I currently possess no extremely strong argument against it, and yet intuitively I don't believe it. I have spent a large percentage of my research career (in parallel with pursuing various narrow AI projects with greater academic or commercial appeal and carrying out other sorts of research) working toward refuting this perspective via providing a detailed, thorough, high-quality design for a non-human-like AGI system. This has been a very difficult task and has given me a much better appreciation for why there are so few reasonably well fleshed out AGI designs out there. Intelligence has many different aspects and creating a viable design that addresses them all in a concrete and plausibly computationally efficient way is a very large job, and furthermore a job that gives few interim rewards.

Other Approaches to AGI

As far as I know, my Novamente project is the most serious attempt to create an artificial general intelligence (AGI) underway at the moment. However, there are a few other projects around that also have real AGI ambitions and that seem to me to be proceeding in a reasonable manner. Most of this chapter is going to be fairly specific to the SMEPH approach to AI that underlies Novamente, but before going in that direction, I'll briefly mention some other extant AGI projects just for sake of completeness.

Probably the largest and best-funded vaguely-AGI-oriented effort in existence today is the Doug Lenat's CYC project [59](Lenat and Guha, 1990). This began in the mid-80s as an attempt to create true AI by encoding all common sense knowledge in first-order predicate logic. So far they have produced a somewhat useful knowledge database and an interesting, highly complex and specialized inference engine, but they do not have a systematic R&D program aimed at creating autonomous, creative interactive intelligence. Their belief is that the largest subtask required for creating AGI is the creation of a knowledge base containing all human common-sense knowledge, in explicit logical form (they use a variant of predicate logic called CycL). They have a large group of highly-trained knowledge encoders typing in knowledge, using CycL syntax.

The Cyc knowledge base may potentially be useful one day to a mature Novamente, or another AGI system founded on experiential learning rather than deductive logic and knowledge databases. But the kind of reasoning, and the kind of knowledge, embodied in Cyc, just scratches the surface of the dynamics knowledge required to form an intelligent mind.

On a smaller scale, one group that I was sorry to see dissolve was Artificial Intelligence Enterprises[60], a small Israeli company whose engineering group was run by Jason Hutchens, a former colleague of mine from the University of Western Australia in Perth. This firm, before its effective dissolution in mid-2001, was seeking to create a conversational AI system. Unlike the Novamente project, however, their work focused on statistical learning based language comprehension and generation rather than on deep cognition, semantics, and so forth. Hutchens' research work will presumably continue in a different context.

One deep and powerful thinker who seems to have abandoned the quest for true AI is Danny Hillis, founder of the company Thinking Machines, Inc. This firm focused on the creation of an adequate hardware platform for building real artificial intelligence – a massively parallel, quasi-brain-like machine called the Connection Machine (Hillis, 1987). However, their pioneering hardware work was not matched with a systematic effort to implement a truly intelligent program embodying all the aspects of the mind. The magnificent hardware design vision was not correlated with an equally grand and detailed mind design vision. And at this point, of course, the Connection Machine hardware has been rendered obsolete by developments in conventional computer hardware and network computing.

Another interesting project is Hugo de Garis's Artificial Brain project (deGaris and Korkin, 2002) initiated at ATR in Japan, continued at Starlab in Brussels and Genotype Inc. in Boulder, Colorado, and now pursued by Hugo in the context of his professorship at the University of Utah. This is an attempt to create a hardware platform (the CBM, or CAM-Brain Machine) for real AI using Field-Programmable Gate Arrays to implement genetic programming evolution of neural networks. I view this fascinating work as somewhat similar to the work on the Connection Machine undertaken at Danny Hillis's Thinking Machines Corp. – the focus is on the hardware

---

[59] www.cyc.com
[60] www.a-i.com

platform, and there is not a well-articulated understanding of how to use this hardware platform to give rise to real intelligence. It is highly possible that the CBM could be used inside Novamente, as a special-purpose learning server; but CBM and the conceptual framework underlying it appear to me not to be adequate to support the full diversity of processing needed to create an artificial mind.

In the vein of "traditional AI", the well-known SOAR project (Laird, J.E., A. Newell, and P. S. Rosenbloom, 1987) is another project that once appeared to be grasping at the goal of real AI, but seems to have retreated into a role of an interesting system for experimenting with limited-domain cognitive science theories. Newell tried to build "Unified Theories of Cognition", based on ideas that have now become fairly standard: logic-style knowledge representation, mental activity as problem-solving carried out by an assemblage of heuristics, etc. The system was by no means a total failure, but it was not constructed to have a real autonomy or self-understanding. Rather, it's a disembodied problem-solving tool. But it's a fascinating software system and there's a small but still-growing community of SOAR enthusiasts in various American universities (see Nason and Laird, 2005 for some interesting recent work integrating SOAR with reinforcement learning).

The ACT-R framework (Anderson et al, 1997; Anderson, 2000), though different from SOAR in nearly all details, is similar in that it's an ambitious attempt to model human psychology in it various aspects, focused largely on cognition. ACT-R uses probabilistic ideas and is generally close in spirit to Novamente than SOAR is. But like SOAR, we feel that it does not contain adequate mechanisms for large-scale creative cognition, focusing instead on the modeling of human performance on relatively narrow and simple tasks.

In a completely different vein, the well-known Cog project at MIT is aiming toward building real AI in the long run, but their path to real AI involves gradually building up to cognition after first getting animal-like perception and action to work via "subsumption architecture robotics." This approach might eventually yield success, but only after decades.

California futurist and businessman Peter Voss, in his recent startup A2I2, is seeking to create an AGI based on a neural-net-like framework, aiming initially at a system with a qualitatively "dog-level" intelligence, and building from there. Stephen Omohundro, an early contributor to Mathematica and a veteran of numerous AI and complex systems research projects, has launched his own small venture called "Self-Aware Systems" aimed at creating program code that analyzes and modifies itself using Bayesian reasoning.

Pei Wang, a former research collaborator of mine, has his own AGI framework, NARS (Non-Axiomatic Reasoning System; Wang, 1995, 2005, 2006), which bears some resemblance to Novamente's inference component, but uses very different approaches for other things such as attention allocation, association-finding, schema learning, concept formation, and so forth. Also, NARS is not based on probability theory in any way, whereas Novamente's inference rules are probabilistically based.

Other projects worthy of mention include John Weng's (2004) SAIL architecture, Nick Cassimatis's PolyScheme (2004), Stuart Shapiro's SnEPs (Santore and Shapiro, 2003), and Robert Hecht-Nielsen's confabulation approach (2005). All these are currently active projects which actively address AGI.

To fully explore the relation between these other projects and Novamente would take us too far afield, but suffice it to say that the relationships exist and are interesting. For instance: NARS is based on an uncertain logic closely related to Novamente's PLN inference system. SnEPs is based on paraconsistent logic, whereas Novamente's PLN logic is also paraconsistent; furthermore, both SnEPs and Novamente have been used to control an agent in a simulation world based on the CrystalSpace game engine, based on somewhat similar approaches to embodied perception and action. Hecht-Nielsen's "confabulation" operation occurs naturally within Novamente as a consequence of PLN inference.

In spite of the various similarities, however, there are also significant differences between these other recent approaches and Novamente; and some of these differences are foundational and conceptual rather than technical. Novamente embodies a particular conceptual understanding of mind and intelligence; in this brief overview my goal is to get across a few important aspects of this conceptual understanding and explain how they manifest themselves in the AI architecture and in our plan for teaching Novamente.

Psynet Principles

How to make a software program that achieves a high degree of intelligence relative to its limited computational resources is not an easy problem. I feel certain there are many different solutions. However, patternist philosophy does provide some guidance as to what sort of solutions may be most likely. Most of all, if one adopts some other ideas about AI as a platform, patternist philosophy is extremely useful for helping one to sculpt these other ideas into a plausible and comprehensive approach to AI. In my own work, I have used patternist philosophy to help sculpt and refine a "complex dynamical systems" approach to AI; and then, on a more specific level, I have used it to guide the definition and development of a particular mathematical/conceptual approach to AI called SMEPH (Self-Modifying, Evolving Probabilistic Hypergraphs); and finally it has also been used to guide the development of the Novamente AI system, a very particular AI approach.

I have chosen to think about mind/brains as complex dynamical systems – systems that consist of many small parts, each acting with their own autonomy, and then cooperating explicitly and implicitly to give rise to holistic patterns of emergent activity. The question I have asked is: According to patternist philosophy, what kind of complex dynamical system may be likely to give rise to intelligent minds? The first answers one gets to this question are fairly broad, and then by introducing additional assumptions one can narrow things down, identifying particular classes of complex dynamical systems that are reasonably likely to give rise to minds.

In Chapter 7, I have discussed complex systems ideas from a more general perspective, looking at several potential "general laws of complex systems behavior." The identification of such general laws is one of the stated goals of the interdisciplinary complex systems research community, and up till now the quest for these laws has not been all that successful. It may be that my ideas in this direction will be of some use. However, from an AI perspective, I have found that the complex dynamical systems approach is valuable even in the absence of a set of unifying laws, simply via providing a deeply insightful way of thinking, and organizing one's thoughts, about artificial intelligence.

It may seem almost trite to say "AI systems may be thought of as complex dynamical systems" – and it *is* a bit obvious, in the sense that any AI system will necessarily be dynamic (in the sense of changing over time) and complex (in the sense of having a lot of different, interacting things going on within it). But even though it's obvious that any highly intelligent system will have these properties, it's not so obvious that these properties should be taken as foundational and that other aspects of intelligence should be framed in terms of them. A whole community of deep-thinking AI theorists has proposed that "mind as deductive symbol-manipulator" is a more important perspective, and that other aspects of intelligence should be framed in such terms. The SMEPH and Novamente frameworks do make use of deductive symbol manipulation, but the latter is interpreted in the context of a complex-dynamical-systems framework; which is different than embedding complexity and change over time within a logical-theorem-proving framework. So there is real meaning in which of the necessary aspects of intelligence one takes as more foundational.

Evolutionarily, it's clear that the complex dynamical systems aspect of intelligent systems is foundational, in the sense that intelligent biological systems evolved out of much less intelligent complex dynamical systems. However, this evolutionary heritage doesn't imply that the complex dynamical systems approach is a good one for thinking about AI. A number of contemporary AI theorists (e.g. Yudkowsky, 2005) argue that logical inference rather than complex self-organization should be taken as the foundational aspect of AI, because complex self-organization is inevitably tied to uncontrollability and unpredictability, and it's important that the superhuman AI's we'll eventually create should be able to rationally and predictably chart their own growth and evolution.

I agree that it's important that powerful AI's be more rational than humans, with a greater level of self-understanding than we humans display. But, I don't think the way to achieve this is to consider logical deduction as the foundational aspect of intelligence. Rather, I think one needs to consider intelligent systems as complex, self-organizing dynamical systems giving rise to complex, self-organizing systems of patterns on the emergent level – and then solve the problem of how a complex, self-organizing pattern system may learn to rationally control itself. I think this is a hard problem but almost surely a solvable one. The logical theorem-proving, decision-theoretic framework is, I contend, simply not rich enough to support powerful intelligence given severely limited computational power. The kind of logic that's most critical to AI is uncertain logic with probability and related quantifications of uncertainty at its center, and this kind of logic is sufficiently complex that its effective context-appropriate control requires the integration of logical reasoning into an overarching framework involving complex, self-organizing pattern-recognition and formation dynamics.

Putting together patternist philosophy with the general idea of minds as emerging from complex dynamical systems, in my prior writings I have created a way of thinking about minds and their physical embodiments that I call the "psynet model." The psynet model is a fluid entity, which I have presented in a variety of different forms in different publications – the different versions don't contradict each other, but they have differing levels of detail pertinent to different contexts. The core of the psynet model has always been the same, however. While the model was created largely with AI in mind, it was also intended to be applicable to human intelligence, and the applicability to the human mind/brain was fleshed out somewhat in *From Complexity to Creativity*.

A minimal list of "psynet principles" may be roughly stated as follows:

- An intelligent system must be a dynamical system, consisting of entities (processes) which are able to act on each other (transform each other) in a variety of ways, and most of whose role is to recognize or create patterns.
- This dynamical system must be sufficiently flexible to allow the emergence of system dynamics embodying the phenomena of autopoiesis and evolution; and to enable the crystallization of a dual network structure, with emergent, synergetic hierarchical and heterarchical subnets.
- This dynamical system must contain a mechanism for the spreading of attention from one part of the system to another based on shared relationship.

- This dynamical system must have access to a rich stream of perceptual data, so as to be able to build up a decent-sized pool of grounded patterns, leading ultimately to the recognition of the self.
- This dynamical system must contain entities that can reason based on uncertain information (so that, among other things, it can transfer information from grounded to ungrounded patterns).
- This dynamical system must be contain entities that can manipulate categories (hierarchical subnets) and transformations involving categories in a sophisticated way, so as to enable (among other phenomena) syntax and semantics.
- This dynamical system must contain processes able to recognize patterns involving symmetric, asymmetric and emergent meaning sharing, and temporal and spatial relatedness.
- This dynamical system must have a specific mechanism for paying extra attention to recently perceived data ("short-term memory").
- This dynamical system must be embedded in a community of similar dynamical systems, so as to be able to properly understand itself.

I stress that these loosely stated "psynet principles" don't exactly follow deductively from the patternist philosophy of mind. However, they are extremely conceptually harmonious with the patternist philosophy of mind, a statement that is certainly not true of every possible foundation for AI. Conceptually, they follow naturally from putting patternist philosophy together with the view of minds as complex dynamical systems. I have gone over the psynet model fairly lightly here because I've treated these ideas so extensively in prior publications, but also because I believe the best way to expound them in the present context will be to elucidate their implications in the context of the Novamente AI system in particular.

Knowledge Representation, Learning and Memory in Novamente

The Novamente design begins with a specific decision about knowledge representation. Information is represented inside Novamente using a network of sorts, but not a "neural network" as in contemporary AI approaches that seek to loosely model the brain — rather, it uses a special kind of mathematical network called a "weighted, labeled hypergraph," in which pieces of information are represented as nodes and links and patterns of activity of nodes and links. Whereas a link in a formal neural network has a numerical weight indicating its "synaptic conductance," and a formal neural network node has a weight indicating its "activation," Novamente nodes and links carry weights with different semantics. Each node or link is associated with a *truth value*, indicating, roughly, the degree to which it correctly describes the world. The term "Atom" is used to describe bothe nodes and links inclusively.

Novamente has been designed with several different types of truth values in mind; the simplest of these consists of a pair of values denoting a probability and the amount of evidence used to arrive at the probability. All nodes and links also have an associated *attention value*, indicating how much computational effort should be expended on them. These consist of two values, specifying short and long term importance levels. Truth and attention values are updated continuously by cognitive processes and maintenance algorithms.

Nodes and links in Novamente have a variety of different types, each of which comes with its own semantics. Novamente node types include tokens which derive their meaning via interrelationships with other nodes; nodes representing perceptual inputs into the system (e.g., pixels, points in time, etc.); nodes representing moments and intervals of time; and procedures. Links represent relationships between atoms (nodes *or* links), such as fuzzy set membership, probabilistic logical relationships, implication, hypotheticality and context. Executable programs carrying out actions are represented as special procedure objects that wrap up small networks containing special kinds of nodes and links. For the interested reader, Appendix 4 contains some tables giving a fairly complete enumeration of the various node and link types used in the current version of the Novamente system.

Next, the dynamics of the psynet model are enacted in Novamente via two primary learning algorithms acting on the node/link level: Probabilistic Logic Networks (PLN: Goertzel, Iklé and Goertzel, forthcoming), and a novel evolutionary learning approach that is a descendant of the Bayesian Optimization Algorithm (BOA: Looks et al, 2005; Pelikan, 2002).

PLN is a flexible "logical inference" framework, applicable to many different situations, including inference involving uncertain, dynamic data and/or data of mixed type, and inference involving autonomous agents in complex environments. It was designed specifically for use in Novamente, yet also has applicability beyond the Novamente framework. It acts on Novamente links representing declarative knowledge (e.g. inheritance links representing probabilistic inheritance relationships), building new links from old using rules derived from probability theory and related heuristics. PTL is context-aware, able to reason across different domains and to deal with multivariate truth values. It is capable of toggling between more rigorous and more speculative inference, and also of making inference consistent within a given context even when a system's overall knowledge base is not entirely consistent.

On the other hand, BOA was developed by computer scientist Martin Pelikan as an improvement over ordinary genetic algorithms (GA: an AI technique we mentioned briefly above, which solves problems via simulating evolution by natural selection). BOA significantly outperforms the traditional GA by using probability theory to model the population of candidate solutions. That is, instead of just taking the "fittest" candidate solutions to a problem and letting them reproduce to form new candidate solutions, it does a probabilistic study of which good candidate solutions' features make them good, and then tries to create new candidate solutions embodying these features. This, we feel, combines the evolutionary power of the GA with the analytical precision of probability theory—and provides a nice bridge between evolutionary procedure learning and probabilistic inference (PLN, the other main AI algorithm within Novamente). We have extended Pelikan's original BOA idea into a powerful procedure learning algorithm (currently named MOSES), which adaptively learns complex procedures satisfying specified goals. It can be used to recognize patterns (a pattern being formally representable as a procedure for calculating or producing something in a simple way, or else a procedure for controlling actuators, or a procedure for controlling patterns of reasoning or perceiving, etc.).

Cognitive processes such as large-scale inference, perception, action, goal-directed behavior, attention allocation, pattern and concept discovery, and even some aspects of system maintenance are implemented in Novamente as specific combinations of these two key algorithms, which are highly flexible and generic in their applicability. The idea is that, via these probabilistic processes, the patterns embodied in the nodes and links in the system's knowledge store will implicitly enact the more abstract dynamics described in the psynet model.

Nodes and links are important on their own, but also important as components of self-organized sets of nodes and links called "maps." A map is a collection of nodes and links that tend to get utilized together within cognitive processing – for instance, as well as a "cat" node, there may be a "cat map" consisting of all the nodes and links in a Novamente system's memory that are closely related to "cat" and tend to come up virtually every time cats are thought about.

In practice, a Novamente system consists of a collection of functionally specialized units called "Units," each one of which deals with a particular domain or type of cognition (procedure learning, language learning, focused attention on important items, etc.). Within each lobe there is a table of nodes and links, and also a collection of software objects called MindAgents, which carry out particular AI processes (visual perception, language parsing, abstract reasoning, etc.) using specialized combinations of PTL and BOA and in some cases other simpler heuristic AI algorithms. The particular assemblage of node and link types, MindAgents and lobes has been painstakingly created with a view toward properly giving rise to psynet-ish dynamics in the whole system.

Node Variety	Description
Perceptual Nodes	These correspond to perceived items, like WordInstanceNode, CharacterInstanceNode, NumberInstanceNode, PixelInstanceNode, PolygonInstanceNode
Procedure Nodes	These contain small programs called "schema," and are called SchemaNodes. Action Nodes that carry out logical evaluations are called PredicateNodes.
ConceptNodes	This is a "generic Node" used for two purposes. An individual ConceptNode may represent a category of Nodes. Or, a Map of ConceptNodes may represent a concept.
Psyche Nodes	These are GoalNodes and FeelingNodes, which are special PredicateNodes that play a special role in overall system control, in terms of monitoring system health, and orienting overall system behavior.

Table 1. Novamente Node Varieties

Link Variety	Description
Logical links	These represent symmetric or asymmetric logical relationships , either among Nodes (InheritanceLink, SimilarityLink), or among links and PredicateNodes (e.g. ImplicationLink, EquivalenceLink)
MemberLink	These denote fuzzy set membership
Associative links	These denote generic relatedness, including HebbianLink learned via Hebbian learning, and a simple AssociativeLink representing relationships derived from natural language or from databases.
ExecutionOutputLink	These indicate input-output relationships among SchemaNodes and PredicateNodes and their arguments
Action-Concept links	Called ExecutionLinks and EvaluationLinks, these form a conceptual record of the actions taken by SchemaNodes or PredicateNodes
ListLink and concatListLink	These represent internally-created or externally-observed lists, respectively

Table 2. Novamente Link Varieties

Map Type	Description
Concept map	a map consisting primarily of conceptual Nodes
Percept map	a map consisting primarily of perceptual Nodes, which arises habitually when the system is presented with environmental stimuli of a certain sort
Schema map	a distributed schema
Predicate map	a distributed predicate
Memory map	a map consisting largely of Nodes denoting specific entities (hence related via MemberLinks and their kin to more abstract Nodes) and their relationships
Concept-percept map	a map consisting primarily of perceptual and conceptual Nodes
Concept-schema map	a map consisting primarily of conceptual Nodes and SchemaNodes
Percept-concept-schema map	a map consisting substantially of perceptual, conceptual and SchemaNodes
Event map	a map containing many links denoting temporal relationships
Feeling map	a map containing FeelingNodes as a significant component
Goal map	a map containing PredicateNodes marked as current goals as a significant component

Table 3. Example Novamente Map Types

Novamente Design Aspect	Primary Functions
Nodes	Nodes may symbolize entities in the external world, simple executable processes abstract concepts, or components in relationship-webs signifying complex concepts or procedures
Links	Links may be n-ary, and may link Nodes or other Links; they embody various types of relationships between concepts, percepts or actions. The network of Links is a web of relationships.
MindAgents	A MindAgent is a software object embodying a dynamic process such as activation spreading or first-order logical inference. It acts directly on individual Atoms, but is intended to induce and guide dynamic system-wide patterns.
Mind OS	The Mind OS builds on a distributed processing framework to enable distributed MindAgents to act efficiently on large populations of Nodes and Links
Maps	A Map represents declarative or procedural knowledge as a pattern of many Nodes and Links
Units	A Unit is a collection of Nodes, Links and MindAgents devoted to carrying out a particular function such as vision processing, language generation, or a specific information processing style such as highly-focused concentration

Table 4. Major Aspects of the Novamente AGI Design

MindAgent	Function
Spontaneous Inference	Uses PLN inference to infer new links from existing ones, driven by a general "fitness function" that aims to create surprising or useful information
Goal-Directed Inference	Uses PLN inference to figure out how to achieve current goals
Goal Refinement	Uses PLN inference and heuristics to create new goals refining existing ones
Predicate Schematization	Transforms logical knowledge regarding goal achievement into schemata that can be executed to achieve goals
Schema Predicatization	Transforms schemata into declarative predicate representation for reasoning purposes
LogicalLinkMining	Creates logical links out of nonlogical links (a form of pattern recognition)
Evolutionary Predicate Learning	Creates PredicateNodes containing predicates that predict membership in ConceptNodes
Clustering	Creates ConceptNodes representing clusters of existing ConceptNodes
Importance Updating	Updates Atom "importance" variables and other related quantities
Hebbian Association Formation	Builds and modifies HebbianLinks between Atoms, based on a PLN-derived Hebbian reinforcement learning rule
Evolutionary Schema Learning	Creates SchemaNodes that fulfill criteria, e.g. that are expected to satisfy given GoalNodes
Concept Formation	Creates speculative, potentially interesting new ConceptNodes via blending existing ones
Predicate/Schema Formation	Creates speculative, potentially interesting new SchemaNodes and PredicateNodes by blending existing ones
Schema Execution	Enacts active SchemaNodes, allowing the system to carry out coordinated trains of action
Map Encapsulation	Scans the AtomTable for patterns and creates new Atoms embodying these patterns
Map Expansion	Takes schemata and predicates embodied in nodes, and expands them into multiple Nodes and links in the AtomTable (thus transforming complex Atoms into Maps of simple Atoms)
Homeostatic Parameter Adaptation	Applies evolutionary programming to adaptively tune the parameters of the system

Table 5. Example Novamente MindAgents

Maps and Emergent Knowledge Representation

Much of the meaning of Novamente's cognitive algorithms lies in the implications they have for dynamics on the map level. Here the relation between Novamente Maps and the concepts of mathematical dynamical systems theory (Devaney, 1988) is highly pertinent. Generally speaking there are two kinds of maps: map attractors, and map transients. Schema and predicate maps generally give rise to map transients, whereas concepts and percepts generally give rise to map attractors; but this is not a hard and fast rule. Other kinds of maps have more intrinsic dynamic variety, for instance there will be some feeling maps associated with transient dynamics, and others associated with attractor dynamics.

Many concept maps will correspond to fixed point map attractors – meaning that they are sets of Atoms which, once they become important, will tend to stay important for a while due to mutual reinforcement. However, some concept maps may correspond to more complex map dynamic patterns. And event maps may sometimes manifest a dynamical pattern imitating the event they represent. This kind of knowledge representation is well known in the attractor neural networks literature.

Schemata, on the other hand, generally correspond to transient maps. An individual SchemaNode does not necessarily represent an entire cognitive procedure of any significance – it may do so, especially in the case of a large encapsulated schema; but more often it will be part of a distributed schema. A distributed schema is a kind of mind map, and its map dynamic pattern is simply the system behavior that ensues when it is executes – behavior that may go beyond the actions explicitly embodied in the SchemaNodes contained in the distributed schema.

The maps in the system build up to form larger and more complex maps, ultimately yielding very large-scale emergent patterns, including patterns like the "dual network" (a combined hierarchical/heterarchical control structure) and the "self" (a fractal pattern in which a subnetwork of the hypergraph comes to resemble the hypergraph itself), which are posited in the psynet model of mind.

The concept of "map encapsulation" is also key to Novamente. Mind, in the patternist perspective, is viewed as a system for recognizing and creating patterns in the external world and in itself. In the Novamente approach, one aspect of this is "encapsulation," the creation of new *individual processes* that embody *networks of processes*. Along with spontaneous map formation, this is one of the most important ways that mind carries out "creation of coherent wholes." In Novamente, encapsulation takes a very explicit form: a pattern mining algorithm is used to form PredicateNodes containing within them small networks of Nodes and Links, which have been found to constitute recurrent patterns or "maps" in system dynamics.

Finally, there is an interesting potential relationship between Novamente maps and the theory of emotion outlined in Chapter 10 above. A specific prediction emerging from this combination is that complex map dynamics will be more associated with emotions than other aspects of Novamente cognition. Complex map dynamics involve temporal patterns that are hard to control, and that present sufficiently subtle patterns that the present is much better understood once one knows the immediate future. One may infer from this a possible major feature of the difference between Novamente psychology and human psychology: the strongest emotions of a Novamente system may be associated with the most complexly unpredictable cognitions it has — rather than, in humans, with phenomena that evoke the activities of powerful, primordial, opaque-to-cognition subsystems.

Novamente's Cognitive Architecture

The knowledge representations and learning mechanisms discussed above can be used in a variety of different ways. They can be used within narrow AI systems for purposes such as data mining or partial natural language understanding or automated theorem-proving – or they can be used within an overall architecture aimed at Artificial General Intelligence. The utilization of Novamente's representations and mechanisms for particular purposes is carried out via the creation of functionally specialized "Units", each one of which deals with information of a particular type and processes this information with a particular combination of cognition processes. The choice of Units and their arrangement is what we I refer to as "cognitive architecture" or "cognitive configuration."

Figure 1 depicts a specific Novamente configuration, intended for "experiential learning" based AGI – more specifically, for a Novamente system that controls a real or simulated body that is perceiving and acting in some world. Currently, in the Novamente project, we are not working with physical robotics but are rather using Novamente to control a simple simulated body in a 3D simulation world called AGI-SIM[61]. As I will argue extensively in the following chapter, I believe that pursuing some form of embodiment is likely the best way to approach AGI in practice. This is not because intelligence intrinsically requires embodiment, but rather because physical environments present a host of useful cognitive problems at various levels of complexity, and also because understanding of human beings and human language will probably be much easier for AI's that share humans' grounding in physical environments.

---

[61] see **sourceforge**.net/projects/**agisim**/

Figure 1: An Experiential Learning Oriented Cognitive Architecture for Novamente

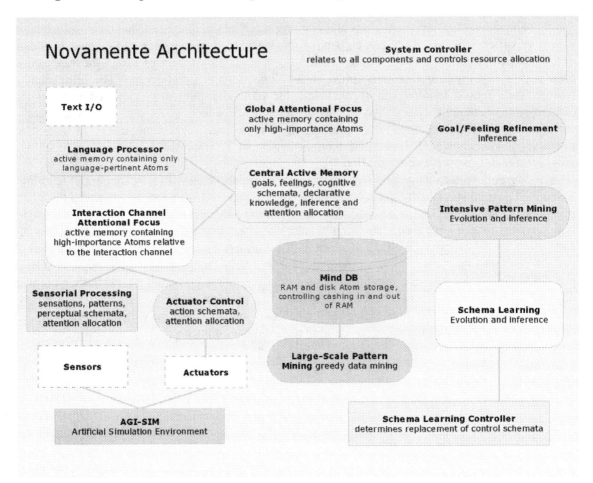

It must be emphasized that a diagram like Figure 1, on its own, doesn't tell you very much. It's very easy to make a fancy-looking block diagram describing an AI system, without any real content whatsoever. What gives Figure 1 its meaning is the specific knowledge representations and (above all) learning mechanisms that fit inside the boxes – boxes and lines like these are useful only as a way of structuring the intelligent dynamics of an appropriately designed underlayer.

The experiential learning configuration shown in Figure 1 centers around a Unit called the Central Active Memory, which is the primary cognitive engine of the system. There is also a Unit called the Global Attentional Focus, which deals with Atoms that have been judged particularly important and subjects them to intensive cognitive processing. There are Units dealing with sensory processing and motor control; and then Units dealing with highly intensive PLN or PEL based pattern recognition, using control mechanisms that are not friendly about ceding processor time to other cognitive processes. Each Unit may potentially span multiple machines; the idea is that communication within a Unit must be very rapid, whereas communication among Units may be slower.

Goal-Driven and Ambient Learning

Psychologically, at a big-picture level, one may think of the Novamente system's activities as falling into two categories: goal-driven and ambient. Ambient cognitive activity includes, for instance:

- MindAgents that carry out basic PLN operations on the AtomTable, deriving obvious conclusions from existing knowledge;
- MindAgents that carry out basic perceptual activity, e.g. recognizing coherent objects in the perceptual stimuli coming into the system;
- MindAgents related to attention allocation and assignment of credit;
- MindAgents involved in moving Atoms between disk and RAM.

All this happens in the Central Active Memory, and is supported by processes occurring in other Units.

Goal-driven activity, on the other hand, involves an explicitly maintained list of goals that is stored in the Global Attentional Focus and Central Active Memory, in the form of special PredicateNodes marked as current goals. Two key processes are involved:

- Learning SchemaNodes that, if activated, are expected to lead to goal achievement;[62]
- Activating SchemaNodes that, if activated, are expected to lead to goal achievement.

Explicit and Implicit Goals

The role of explicitly formulated goals in Novamente's cognitive architecture merits some discussion. My working definition of intelligence, as described at length in Chapter 4, is "achieving complex goals in complex environments." It does not follow immediately from this definition that intelligent systems need to explicitly represent goals. In general, goals can be either implicit or explicit. The Novamente design, however, does in fact express goals internally in a somewhat explicit manner, via at each point in time nominating specific PredicateNodes as current goals.

The notion of an "implicit goal" is fairly difficult to pin down. Almost any system's dynamics can be mathematically formulated to look like the pursuit of some goal, even if the system contains no knowledge of the goal in any concrete sense. If any quantity increases during the course of a system's dynamics, one can view the system as if it had the goal of increasing that quantity. The goal of life on Earth was to create humans, the goal of childhood is adulthood, the goal of adulthood is old age, etc.

---

[62] Novamente's goal-driven learning process is ultimately a form of "backward-chaining learning" (Russell and Norvig, 1995), but subtler than usual instances of backward chaining due to its interweaving of PLN and PEL and its reliance on multiple cognitive Units.

There is a difference, however, between implicit goal-seeking behavior as can be read into almost any complex system, and explicit goal-seeking behavior as is observed in intelligent systems like humans. An intelligent system generally contains some kind of representation of desired states of the world – "goal states" – and a way of comparing observed or experienced or hypothesized states of the world to these goal states. There are many forms that such a representation can take; the important thing, however, is that it's flexible enough to be acted upon by logical inference and other cognitive processes.

In Novamente this role is filled by PredicateNodes nominated to serve as current goals, which explicitly represent goal states, via processes that specifically reward schemata (embodying processes) that have been observed to lead to goal fulfillment. There are also implicit goals – states that the system implicitly tends to favor, although it lacks any explicit representation of them. In many cases, however, Novamente's self-study processes will gradually cause these implicit goals to become represented as explicit goals.

Why do I believe that an intelligent system needs explicit goals (along with its implicit ones)? For the same reason that it needs explicit representation of other kinds of knowledge. Remember, what we mean by "explicit" is simply "represented in a sufficiently flexible way as to be amenable to logical inference, association-finding, concept formation and other cognitive processes" – or, yet more simply, "represented as a mind-process that can be fairly freely acted upon by other mind-processes."

The Perception-Cognition-Action Loop

Implicit in the Novamente cognitive architecture is a close relationship between perception, cognition and action. It is key to recall that these processes are not as distinct as has sometimes been conceived in cognitive psychology. Each of these activities requires the other ones, not only in general but often in its particular mechanics.

Perception involves cognition for the perception of complex gestalts, which guides the perception of lower-level features; and it requires action, among other reasons, to guide the perceptual organs (e.g. the movements of the human eye must be tightly coupled with the brain processes receiving data from the eye, in order to correctly perceive a complex situation).

Action involves perception in order to monitor the intermediate phases of action execution – for instance, moving one's arm with precision is easier if one can feel and watch what one's arm is doing and make dynamic adjustments. Action also requires cognition, except in the most routinized cases. In playing tennis, one does not need to cognize to swing one's racket at the ball and hit it. But if one is returning a shot with an odd kind of spin one hasn't seen before, some quick cognition regarding the best strategy for returning it may be in order.

Finally, cognition sometimes is divorced from perception and action in the external world, but even in these cases it may make use of perception and action apparatus. Abstract mental imagery is a strong example of this. We "perceive" abstract ideas and "manipulate" these abstract perceptions, even in domains where perceptual metaphors are really not directly applicable. For instance, we may visualize relationships between people, or between entities in some abstract mathematical space. We do this because we have a highly effective mechanism for cognizing perceptual entities, and metaphorically extending this mechanism to other cognitive domains often yields good results.

Short-Term Memory

One important component of the above-described cognitive architecture is the "Global Attentional Focus" – which is, basically, the *explicit* manifestation within Novamente of the general concept of "short-term memory" or STM. Generally speaking, the "attentional focus" of a mind is defined as the set of knowledge and processes within the mind that is getting the most attention at a given point in time. It need not be located in one physical place; it may be distributed. In the human brain, attentional focus seems to be associated with some particular brain structures, and also with dynamical feedback loops spanning large regions of the brain (Edelman, 1990).

In Novamente, the attentional focus is generically associated with the set of maps that are highly active at any given time. Because of the nature of maps, it is not possible for too many maps to be active all at once: maps interfere with each other, so simultaneous activation of too many maps just leads to a jumble. Control of how many maps are simultaneously active is done implicitly via the parameters of the importance updating process.

Novamente dynamics naturally leads to an implicit attentional focus, a "moving bubble of attention." However, architecturally, we have also chosen to incorporate an explicit STM, in the form of a Unit that contains the Atoms found most important recently in the Global Active Memory. This allows intensive cognitive processing to focus on these Atoms alone, without processor competition.

However, there is one subtlety here. So far I have been speaking mainly in terms of importance as a single number associated with Atoms, but when dealing with perceptual interactions, it is useful to introduce instead a pair of importances:

- General importance
- Interaction-channel-specific importance

There are also two kinds of long-term importance in the system – general and interaction-channel-specific. This distinction arises from the fact that it's possible for a Novamente system to be hooked up to more than one interaction channel at a time. For instance, a complex Novamente might carry out financial trading activity at the same time as holding a conversation, but without confusing the two interactions, keeping the two totally separate. In that case, one considers multiple interaction-channel-specific activations and importances, one for each channel.

The generic STM of Novamente, then, consists of those Atoms that are in the top N, ranked in terms of either ordinary importance or channel-specific importance. What is the value N? This doesn't really matter, because in Novamente the distinction between "attentional focus" and "the unconscious" is fuzzy rather than crisp. There is not a separate data structure for attentional focus. One may conceive it as a fuzzy set, to which an Atom belongs at a given time with a degree equal to the max of its importance and channel-specific importance.

To oversimplify a bit, we may think in terms of three hierarchical levels:

- **Level 1**: the perceptual nodes which recognize patterns in texts and data sets – CharacterNodes, NumberNodes, and so forth; and SchemaNodes that project data into the outside world, or gather data from particular locations in the outside world.

- **Level 2**: a panoply of learning processes that create conceptual Atoms of various sorts, grouping the perceptual atoms. These act through the mind as a whole, but they act particularly avidly among the most important Atoms in the network.
- **Level 3**: the sprawling network of the Atomspace, with its humongous network of similarity and associative links.

The distinction between these three levels is not explicitly there in the Novamente codebase, but this tripartite division is a convenient way to think about attention in Novamente.

Level 2 is the essence here. In this analysis, focused attention within Novamente is revealed as a process that takes highly important Atomsets and makes them more coherent, more insular, more bounded and solid. Furthermore, it uses schema, as appropriate, to actively gather new data needed to do this whole-building. In other words, to support this process, Novamente dynamics must have:

1. Perceptual schemata that recognize features in the environment, and make them highly important;
2. A dynamic that encourages the formation of links within the Atomset that's the target of attentional focus;
3. A dynamic that discourages the formation of links crossing the boundary of the Atomset that's the target of attentional focus;
4. Action schemata that carry out actions estimated likely to increase the "wholeness" of the highly important Atoms.

Item 1 is basically a given. Items 2 and 3, on the other hand, are direct consequences of the built-in dynamics of the Novamente system. The most important Atoms are most likely to be chosen as inputs to various cognitive processes; thus links among these Atoms are very likely to form. On the other hand, *low-strength* AssociativeLinks are likely to form between Atoms in the focus and Atoms outside of it. These form inhibitory connections, which mean that any process that tends to act on sets of associated Atoms, will tend not to act on sets of Atoms bridging the focused-on Atom set and the rest of the Atomspace.

Item 4, on the other hand, occurs as a result of importance-driven schema execution. In other words, action schema that are associated to important things are more likely to execute. So if a whole bunch of related things are important, then schemata that are judged to be overall related to this set are very likely to get executed.

The Role of Logic in Novamente

Next, the role of logical reasoning in Novamente is philosophically somewhat subtle, and relies critically on the way that *representation* is handled. Logical inference acts on various types of links between ConceptNodes and PredicateNodes, including ExtensionalInheritanceLinks and IntensionalInheritanceLinks and other related link types. Its pertinence is reliant on the fact that many Novamente Nodes represent meaningful concepts and patterns.

Now, you may recall that not all concepts, in a Novamente system's mind, are going to be directly represented by Nodes. Some will be represented more complexly, e.g. by "maps" (fuzzy sets of processes that are often simultaneously active, or are commonly enacted according to a certain emergent temporal pattern). But the psynet model proposes that concept-representing maps arise largely from concept-representing processes: an hypothesis that distinguishes the psynet approach from, e.g., neural network models in which all concepts are represented as "distributed maps" arising from semantically empty lower-level entities such as neurons.

The crux of a concept-representing Node lies in its relationships to other Nodes – other concept-representing Nodes, or Nodes representing percepts or actions, goals, feelings, etc. And key among these relationships are *logical* relationships.

The term "logic" must be used with care; we use it in Novamente in a particular way, broader than some usages and narrower than others. In modern mathematics, "logic" is often used to refer to a highly certain, precise type of abstract thought. However, there is also a different sense of the term, which has a long history, dating back at least to Charles S. Peirce. As Peirce used it, "logic" encompassed various forms of speculative and uncertain reasoning, along with the absolutely certain deductive inference that most of the "logic" field deals with today.

In Novamente, "logic" and "reasoning" are used to refer to a set of Atom types whose TruthValues are defined using conditional probabilities, and some simple processes ("reasoning rules") for creating Atoms of this type. Of all the Novamente Atom types, these are the ones with the subtlest semantics. ConceptNodes refer to sets, and there are links such as ExtensionalInheritanceLink, InheritanceLink and so forth whose truth values represent conditional probabilities between sets. There are also higher-order logical links like ImplicationLink and EquivalenceLink, whose truth values represent conditional probabilities between links and/or PredicateNodes. This probabilistic reasoning framework yields conventional crisp logic as a special case.

This is a subtle aspect of Novamente – in fact, in parallel with this book, I (together with three colleagues) have written a whole book specifically on the Probabilistic Logic Networks approach to logical reasoning used inside Novamente. The mathematics of PLN contains many subtleties, and there are relations to prior approaches to uncertain inference including NARS (1995) and Walley's theory of imprecise probabilities (1991).

From a general AI/cognitive-science perspective, this whole logic framework should be viewed as a set of mechanisms oriented toward representing, recognizing and manipulating patterns of a certain specialized type – patterns that can be simply embodied as conditional probabilities. The reason for the inclusion of an explicit reasoning module in the Novamente system is that this sort of pattern seems to pack a particular wallop. The world and the mind itself seem to include many significant patterns that can be compactly represented in this way. And it happens that these patterns are susceptible to particularly simple manipulations. The Novamente schema module gives a means for constructing arbitrary complex actions and perceptual and cognitive patterns out of simple components; the logical reasoning rules give the system ways to speculatively understand something about some of these complex constructs in advance, by extrapolating knowledge obtained by experimentation with other similar constructs in the past.

Logical relationships are not the only kind of relationships between concept-denoting processes. Another important type is the associative relationship, which simply denotes a generic association between two processes – its strength records the extent to which, when one of the processes is active, the other is also active. For instance, "cat" and "sock" may be associated with each other; this fact however does not tell you the manner in which the two are associated (which might be, for instance, the fact that cats like to play with socks). However, logical relationships seem uniquely powerful in the sense that they come along with a very useful algebra for reasoning – for estimating new relationships from existing ones, with generally reasonable accuracy even in the face of uncertainty.

The Failures of Logic-Based AI

Logical reasoning is often taken as the most uniquely human cognitive characteristic. Language is the most unique human faculty in everyday terms, but logic is in a sense the apex of human achievement. It's logical reasoning that's led us to modern civilization – to such things as mathematics, science, and the institutions of democratic governance. Furthermore, inference is closely related to the issue of consciousness and awareness, since in humans conscious supervision and control of thought seems to be necessary for confronting a new problem with logical inference tools.

Also, logical reasoning is also something that is fairly close in some ways to the internal operations of computers. So, given all this, it's not surprising that a lot of AI research has focused on the area of reasoning. However, automated reasoning systems have performed very poorly to date, unable to carry out either:

- Everyday commonsense reasoning in the manner of a small child (though this was the goal of Cyc from the start, after a couple decades Cyc is still nowhere near achieving this goal), or
- Mathematical theorem-proving without detailed human guidance, beyond the level of simple theorems in set theory (Robinson and Voronkov, 2001).

The reason for this poor performance is moderately subtle. It's not that the AI programs are using bad reasoning rules. Their reasoning rules are correct, in fact more so than the reasoning rules implicitly used by humans in many cases. Humans are prone to stupid reasoning errors (Pietelli-Palmarini, 1996), and this often harms us in practical situations. Rather, the problem is that AI systems are applying the reasoning rules to the wrong sorts of entities, and they don't understand in what order to apply their reasoning rules – how to design a contextually-appropriate inference trajectory.

Regarding the "wrong sorts of entities" problem, there is a large literature on the nature of human concepts. This pertains to the intension/extension distinction discussed above in a PLN context. The classical school of thought held that concepts were defined by necessary and sufficient condition, but this has largely given way to a theory holding that concepts are defined mainly by prototypes and exemplars (Hunt and Ellis, 1999). Novamente and SMEPH suggest that both of these theories have some truth to them – that both necessary/sufficient conditions and prototypes/exemplars may be considered as probabilistic relationships in a concept hypergraph. Novamente theory also suggests a third aspect to the definition of concepts, which may be important for human as well as AI cognition: pattern-based intensional definition, wherein a concept is defined partly by the set of patterns associated with it.

Humans seem to have good heuristics for figuring out which concepts to use to describe a given situation. For instance, when perceiving a watch, a human must decide whether to think about it as a "watch", an "object", a "self-winding ladies' analog wristwatch", etc. The choice depends on context – i.e., it depends on which classification is going to be most useful for the inferences one wants to draw. One heuristic humans often use is that objects that are atypical of basic level objects tend to be named and identified at a subordinate level (Jolicoeur, Gluck and Kosslyn, 1984). In general, this "most useful level of categorization" problem is a subcase of the overall inference control problem – the problem of knowing which inference steps to carry out in which order.

The Novamente approach to this issue is conceptually simple: inference control strategies are represented as schema, and may be learned just like any other kind of procedural knowledge. The trick is that this is a very difficult learning problem. And this is where experiential learning comes in: humans learn to reason by starting out in very simple situations, and once their inference control strategies are thus honed, they are ready to deal with slightly more complex situations, etc. It may well be that in order to learn to reason effectively, an AI must go through the same sort of series of steps.

Reasoning and Forgetting

One of the many reasons that reasoning is important to intelligence is the necessity of *forgetting*. Memory is about remembering but equally importantly, it is about forgetting. Forgetting has profound consequences for mind. It means that, for example, a mind can retain the datum that birds fly, without retaining much of the specific evidence that led it to this conclusion. The generalization "birds fly" is a pattern A in a large collection of observations B is retained, but the observations B are not.

One hears far more in the AI and cognitive science literature about learning than about forgetting, but in fact, the latter is just as important. Without an intelligent forgetting mechanism, an intelligent learning mechanism in a memory-limited AI system will effectively be useless.

And the importance of intelligent forgetting implies that mind is intrinsically evolutionary: a system which is creating new processes, and then forgetting processes based on relative uselessness, is evolving by natural selection. This evolution is the creative force opposing the conservative force of self-production.

Forgetting is closely related to the notion of "grounding": A pattern A is "grounded" to the extent that the mind contains entities in which A is in fact a pattern. "Logical reasoning" is a system of process transformations largely specialized for producing incompletely grounded patterns from other incompletely grounded patterns.

For instance, the pattern "birds fly" is grounded to the extent that the mind contains specific memories of birds flying. Few concepts are completely grounded in the mind, because of the need for drastic forgetting of particular experiences. But in order to derive conclusions about incompletely grounded patterns, some kind of "abstract calculus of patterns" is required; one cannot proceed solely by evaluating patterns directly based on evidence. This is one of the main roles of "logical reasoning," as we use this term in Novamente.

Consider, for example, the reasoning "Birds fly, flying objects can fall, so birds can fall." Given extremely complete groundings for the observations "birds fly" and "flying objects can fall", the reasoning would be unnecessary – because the mind would contain specific instances of birds falling, and could therefore get to the conclusion "birds can fall" directly without going through two ancillary observations. But, if specific memories of birds falling do not exist in the mind, because they have been forgotten or because they have never been observed in the mind's incomplete experience, then reasoning must be relied upon to yield the conclusion.

In this example, it is more general knowledge that has been remembered, and more specific knowledge that has been forgotten. This is not universally the case, but it is a common overall pattern. The generality of a process may be defined, roughly, as the variety of situations in which it tends to become active. The processes in a mind will have a spectrum of degrees of generality/specialization, frequently with more specialized processes associated with maps residing lower in the map hierarchy.

One aspect of the hierarchical nature of the mind is the passage from specificity to generality. Lower levels of the hierarchy tend to refer to more specific percepts, concepts or actions; higher levels tend to refer to more abstract entities. And the necessity for forgetting is particularly intense at the lower levels of the system. In particular, most of the patterns picked up by the perceptual-cognitive-active loop are of ephemeral interest only and are not worthy of long-term retention in a resource-bounded system. The fact that most of the information coming into the system is going to be quickly discarded, however, means that the emergent information contained in perceptual input should be mined as rapidly as possible, which gives rise to the phenomenon of "short-term memory."

Intension, Extension and Pattern

One place where patternist thinking enters deep into the technical details of Novamente regards the distinction between intensional and extensional logical relationships. Inheritance links in Novamente represent relationships of specialization – e.g. cat inherits from animal – but the distinction between different kinds of inheritance is important. The two main kinds are called extensional and intensional: the former is more standard, and the latter is defined in terms of pattern theory.

The ExtensionalInheritanceLink in Novamente relates two sets according to their members. American inherits extensionally from Terran because things that are members of the set of American-things are also members of the set of Terran-things. The strength of the ExtensionalInheritanceLink between A and B denotes the percentage of A's that are also B's. PLN handles ExtensionalInheritance relationships but also deals with intensional relationships – relationships that relate sets according to their properties, or in other words, according to the patterns that are associated with them.

Conceptually, the intension/extension distinction is very similar to that between a word's *denotation* and *connotation*. For instance, consider the concept "bachelor." The extension of "bachelor" is typically taken to be *all and only the bachelors in the world* (a very large set). In practical terms, it means all bachelors that are known to a given reasoning system, or specifically hypothesized by that system. On the other hand, the *intension* of "bachelor" is the set of properties of "bachelor," including principally the property of being a *man*, and the property of being *unmarried*.

Some theorists would have it that the intension of "bachelor" consists solely of these two properties, which are "necessary and sufficient conditions" for bachelorhood; PLN's notion of intension is more flexible, it may include necessary and sufficient conditions but also other properties, such as the fact that most bachelors have legs, that they frequently eat in restaurants, etc. These other properties allow us to understand how the concept of "bachelor" might be stretched in some contexts – for instance, if one read the sentence "Jane Smith was a more of a bachelor than any of the men in her apartment building," one could make a lot more sense of it using the concept "bachelor"'s full PLN intension, than one could make using only the necessary-and-sufficient-condition intension.

To understand the relation between intensional and extensional inheritance in practice, consider the example of fish and whales. Extensionally whales are not fish, i.e:

```
ExtensionalInheritanceLink whale fish <.0001>
```

But intensionally, the two share a lot of properties, so we may say perhaps:

```
IntensionalInheritanceLink whale fish <.7>
```

The essential idea underlying PLN's treatment of intension is to associate both fish and whale with sets of patterns – $fish_{PAT}$ and $whale_{PAT}$, the sets of patterns associated with fish and whales. We then interpret:

```
IntensionalInheritanceLink whale fish <.7>
```

as

```
ExtensionalInheritanceLink whalePAT fishPAT
```

And we then define Inheritance proper as the disjunction of intensional and extensional inheritance, i.e:

```
InheritanceLink A B
```

is defined as

```
        ExtensionalInheritanceLink A B
OR
        IntensionalInheritanceLink A B
```

I hypothesize that most human inference is done not using ExtensionalInheritance relationships, but rather using composite Inheritance relationships. And, consistent with this claim, we suggest that, in most cases, the natural language relation "is a" should be interpreted as an Inheritance relation between individuals and sets of individuals, or between sets of individuals – not as a ExtensionalInheritance relationship. For instance:

```
"Fluffy is a cat"
```

as conventionally interpreted is a combination extensional/intensional statement, as is:

```
"Cats are animals."
```

This statement means not only that examples of cats are examples of animals, but also that patterns in cats tend to be patterns in animals.

Philosophically, one may ask why a pattern-based approach to intensional inference makes sense. Why isn't straightforward probability theory enough? The problem is – to wax poetic for a moment — that the world we live in is a special place, and accurately reasoning about it requires making special assumptions that are very difficult and computationally expensive to explicitly encode into probability theory. One special aspect of our world is what Charles Peirce referred to as "the tendency to take habits": the fact that "patterns tend to spread," i.e. if two things are somehow related to each other, the odds are that there are a bunch of other patterns relating the two things. To encode this tendency observed by Peirce in probabilistic reasoning one must calculate $P(A|B)$ in each case based on looking at the number of other conditional probabilities that are related to it via various patterns. But this is exactly what intensional inference, as defined in PLN, does. This philosophical explanation may seem somewhat abstruse – until one realizes how closely it ties in with human commonsense inference, and with the notion of inheritance as utilized in natural language.

To see the psychological naturalness of fusing extensional and intensional inheritance, let's consider a very simple example of probabilistic inference in Novamente. PLN is divided into two portions: first-order and higher-order. First-order PLN deals with probabilistic inference on (asymmetric) inheritance and (symmetric) similarity relationships, where different Novamente link types are used to represent intensional versus extensional relationships (Wang, 1995).   Example inference rules are deduction (A→B, B→C |- A→ C), induction and abduction (shown in Figure 2), inversion (Bayes rule), similarity-to-inheritance-conversion, and revision (which merges different estimates of the truth value of the same atom).

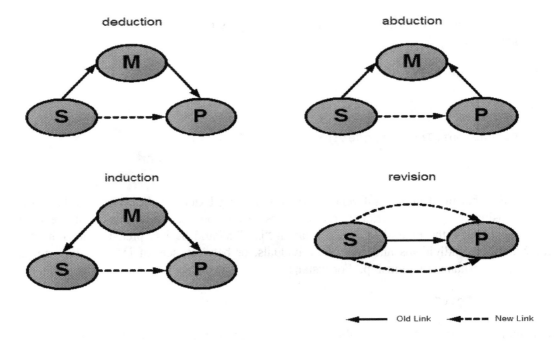

Figure 2. PLN First-Order Inference Rules

In the notation we typically use for describing Novamente Atoms, the PLN deduction rule looks like:[63]

---

[63] Here and there in this chapter I will lapse into Novamente technical notation, in which

X
Y
|–
Z

denotes that the premises X and Y lead to the conclusion Z. and

R
    A
    B

denotes that the relationship R holds between the argument A and B.

```
InheritanceLink A B
InheritanceLink B C
|-
InheritanceLink A C
```

where the deduction truth value formula (and there are a couple variants) tells you how to compute the truth value of the conclusion from those of the premises. Generally, each inference rule comes with its own quantitative truth value formula, derived using probability theory and related considerations.

A simple quantitative example of first-order PLN inference is:

```
InheritanceLink mud dangerous (.8, .7)
SimilarityLink sand mud (.6, .99)
|-
InheritanceLink sand dangerous (.31, .98)
```

The number-pairs such as (.8, .7) refer to the two components of a PLN truth value – the probability is .8, and the .7 represents the amount of evidence on which this probability estimate was based. This seemingly simple inference is actually carried out within PLN as a combination of two inferences. First, the similarity relation is converted to inheritance, yielding:

```
InheritanceLink sand mud (0.38, 0.98)
Then the deduction

InheritanceLink sand mud (0.38, 0.98)
InheritanceLink mud dangerous (.8, .7)
|-
InheritanceLink sand dangerous (0.31, 0.98)
```

is performed.[64] It's worth noting that the truth values in PLN combine both intensional and extensional information: the inheritance between mud and dangerous may be extensional, in the sense of deriving from actual observed instances of mud being dangerous; whereas the similarity between sand and mud is intensional, because it doesn't derive from there being a lot of instances that are both sand and mud, but rather from there being a lot of properties shared by sand and mud.

Attention Allocation in Novamente

Another critical aspect of Novamente that pertains particularly closely to pattern theory is what we call "attention allocation" or "assignment of credit." This has to do with regulating the system's own cognitive activities, an issue that has many different aspects. Firstly, as noted above, in practice a Novamente instance can't maintain an arbitrarily large collection of nodes and links in memory, so prioritization decisions must be made regarding which Nodes and Links to remove from RAM and save to disk. Next, among those Atoms remaining in RAM, decisions

---

[64] PLN requires "node probabilities" for this inference, which are defined relative to a relevant context. The example uses the values mud: 0.001, sand: 0.05, dangerous: 0.015.

must be made regarding which ones to think about: which ones to feed to PLN reasoning, and which ones to consider as goals for the guidance of evolutionary learning. When a goal is successfully achieved, credit must be assigned to the mind-components that helped achieve it (even indirectly) so that proper attention may be allocated to them in the future when a similar goal needs to be achieved.

It's worth taking a step back and interpreting these concrete issues in the general context of patternist philosophy. On a very abstract level, I have proposed, the dynamics of mind may be understood as "mind patterns extending themselves over each other and thus creating new synergetic patterns." This is Peirce's "One Law of Mind." Another way to view this is to state that the dynamics of mind can be largely understood in terms of the passage of "activation" among mind-patterns. Each mind-pattern may be viewed as having a certain amount of activation at each point in time, the activation value reflecting the total amount of system resources the intelligent system will spend dealing with that mind-pattern. The dynamics of this kind of activation is what we refer to as "attention allocation" and "credit assignment."

In principle, attention allocation could be treated as a kind of logical reasoning problem. What to focus attention on at a given time, in accordance with one's given goals, is a logical problem that can be approximately solved by probabilistic inference. However, approaching attention allocation as a logical inference problem, except in special cases, is not a tractable plan, because of our old enemy computational inefficiency: This would require a mind to spend nearly all its time figuring out what to pay attention to. And the processes involved in doing this figuring-out would need to have attention differentially allocated as well, leading to an absurdly useless mind that does hardly anything but figure out how much attention to pay to various ways of figuring out how much attention to pay to….

The human brain allocates attention to various processes by a complex combination of chemical and electrical neurodynamics. At any given time, a certain subset of the brain is highly active, and PET and fMRI scans allow us to visualize which parts of the brain are active during various activities. The regions of the brain that become active depend upon what the brain is sensing, what it's doing, what it's thinking, and what its current goals are.

High-level control of attention allocation is also a key part of human thought – such as when we consciously "force" ourselves to think about useful things rather than just letting our minds wander – but this is done against a substrate of more basic attention allocation. The processes that allow us to have the feeling of consciously forcing ourselves in certain thought-directions, are themselves generally allocated attention by simpler mechanisms.

The Novamente approach to attention allocation centers on the "importance," a specific number that is a component of each Atom's AttentionValue. Maps also have importances, defined in terms of the importances of their component Atoms. Importance directs system activity in several overlapping ways. First, the execution of complex procedures is importance-driven, with each Atom forming a component of a procedure chosen for execution based on its importance. Secondly, most Atom activities are "importance-driven", in that cognitive processes (embodied as MindAgents) select Atoms to act upon with probability proportional to Atom importance, where an Atom's importance (a component of its AttentionValue) is a quantity determined partially as a weighted average of the Atom's activation over the recent past. The result of this is that, on the whole, Atoms that are ranked more important over a period of time, will get thought about more. Similarly, on the whole maps that are implicitly ranked more important over a period of time will get more attention.

There are also some more sophisticated aspects to Novamente attention allocation, aimed at more aggressively pushing the system to concentrate on useful things. For instance, Atoms that have received a lot of recent attention but have not been recently involved in much useful knowledge-creation are at risk of having their allocation of attention decreased. But Atoms that have at some point in the past been very useful won't have their attention decreased below a certain minimum level (which is associated with a quantity called "long-term importance," a number that along with ordinary importance is contained in an Atom's AttentionValue).

The feedback between attention allocation and cognition is manifested in Novamente via the phenomenon of maps. A map is an interconnected set of Atoms whose dynamic unity is enforced by the system's attention-allocation mechanisms – so that when part of the map is highly important, the rest is likely to become highly important as well.

But the key question is then: How are these importance and long-term importance values, associated with Atoms, updated? A variety of schemes for making these sorts of decisions exist in the AI literature, but Novamente takes a somewhat novel approach. Special Links called HebbianLinks are created, indicating the degree to which the utility of one Atom implies the utility of another. PLN and evolutionary learning are then used to infer new HebbianLinks and new PredicateNodes involving HebbianLinks, from the original HebbianLinks learned via direct experience. In short, these "meta-level" learning processes are handled via the same cognitive mechanisms used for ordinary learning. This may seem simplistic but I believe it is actually a very elegant solution.

In a sense attention allocation and assignment of credit are being posed as inference problems, but the trick is that by and large they can be approached as very simple first-order inference problems. For instance, if A and B have often been important together, and so have B and C, then we may conclude:

```
HebbianLink A B
HebbianLink B C
|-
HebbianLink A C
```

via a mathematically direct porting of first-order PLN deduction to the probabilistic truth value stored in HebbianLinks. This tells us that maybe A and C will be important together.

Concretely Implemented Mind versus Emergent Mind

With learning, reasoning and cognitive architecture under our belts, we now return to the topic of knowledge representation. There is a critical distinction to be drawn between *concretely-implemented mind (CIM)* and *emergent mind*. This distinction has to do with the relationship between mind-patterns and mind's physical or software substrate. Some of the patterns making up a mind will be directly reflected in the physical substrate or software program underlying the mind (these are the "concretely implemented" ones), others will be more abstract and won't be easily derivable from this underlying layer (these are the "emergent" ones). The mind of a Novamente system is the set of patterns emergent among the bits in the RAM and registers of the set of computers running Novamente, and emergent between these bits and Novamente's environment. The CIM level – the nodes and links and process explicitly implemented in the Novamente software — form only a part of Novamente mind.

The real subtlety of the relation between the CIM and emergent levels in Novamente lies in the close parallels that exist between these levels. These parallels form a major conceptual difference between Novamente and most other AI systems. Generally speaking, other AI systems fall into one of two categories:

1. Subsymbolic systems, consisting of basic units ("tokens") that, taken individually, have no direct relationship to mind-level semantics (but that create mind-level semantics collectively);
2. Symbolic systems, whose basic tokens have mind-level semantics – and whose overall semantics are largely directly comprehensible in terms of the semantics of the basic tokens.

The archetypal subsymbolic AI system is the formal neural network. Formal neurons aren't "mind stuff" in any direct way – they don't represent concepts, memories, percepts or actions. They combine to form overall activation patterns that constitute mind-stuff. The dynamic processes of the neural network aren't directly thought processes, but rather processes for updating neural activations and synaptic weights. These are intended serve as the ultimate root of thought processes, but they act on a lower level than mental dynamics.

On the other hand, the archetypal symbolic AI system is the logic-based system; for instance a semantic network enhanced with formal logic axioms implemented as graph rewriting rules. In this case, each node of the network has a meaning such as "fork" or "chair" or "say X to user" or "edge of length roughly 1 inch at an angle of 45 degrees in the visual field." The thought processes of the system are conceived as basically identical with the graph rewriting rules, and the explicitly encoded control mechanisms that drive the application of these rules. The philosophy here is that the point of brain-level structures and dynamics is to give rise to mind-level structures and dynamics. In software, it is posited, we may create these mind-level structures and dynamics in a different way, by writing high-level code rather than by wiring neurons or simulated neurons together.

Of course, the line between these two approaches blurs somewhat in practice. For instance, a large multimodular neural network system may embody mind level structures on the architectural level, in its choice of modules and its sculpting of the interactions between modules. And a complex logic-based system like the production system ACT-R may have somewhat complex dynamics, giving it mind patterns beyond those explicitly coded into it.

However, there is no existing AI system that spans the two approaches as thoroughly as Novamente does. This is an aspect of the Novamente design that may be confusing on first encounter. Is Novamente a logic-based system with some subsymbolic-style representation and control? Or is it a complex, subsymbolically based dynamical system with some logic-based substructure guiding its dynamics? You decide! This ambiguity is essential to the system's power.

Our experimentation with Novamente to date has not been sufficiently sophisticated to fully test our intuitions about the potential complexity of its dynamics. Thus, the discussion of emergent mind in Novamente presented in this chapter is at present somewhat speculative. At time of writing, we have not yet seen most of the emergent phenomena that we predict will arise in the system — because until very recently, the Novamente implementation was too incomplete. If all goes well, then this paragraph will be completely obsolete by the time you read it!

Philosophically, the emergent level of representation in Novamente ties in with the discussion of autopoiesis above. It is critical that Novamente maps are not just any patterns of activity, but that many of them are rather "self-perpetuating" patterns of activity: patterns that, once they're invoked by some external or internal stimulus, have a tendency to maintain themselves for a while. What is special about the representation of thoughts and feelings as self-producing, self-reinforcing process systems? I have already enlarged on this in detail above. First is the extreme expressive power that this representation gives to a set of processes. Given N mind-processes, constructed in a reasonably flexible way, the number of possible self-reinforcing process system that can be formed by minor modifications of process parameters. Next there is the naturalness of this representation for the two primary "forces" of mental dynamics: autopoiesis and evolution. Maps display powerful autopoiesis, and they also evolve creatively and efficiently even under relatively weak evolutionary pressures. They are easy to combine with each other, easy to mutate, and relatively easy to reason about. They provide real-time responsiveness when needed, and also have the persistence required for long-term mental phenomena.

Novamente and the Symbolic/Subsymbolic Distinction

Daniel Dennett has framed the symbolic versus connectionist dichotomy in a clearer and more systematic way than any other author I've encountered. First consider Dennett's essay "Two Contrasts" (Dennett, 1998), which contains a very revealing discussion of Good Old-Fashioned AI (GOFAI), which he also calls "symbolic AI," and its weaknesses as compared to connectionist AI:

> *Suppose you have a GOFAI nonconnectionist AI theory: It postulates a certain level at which there are symbolic structures in something like a language of thought, and it has some mechanism for computing over these. Then, indeed, it makes little difference how you implement that. It makes no difference whether you use a VAX or a Cray, a compiled or interpreted language. It makes no difference how you determine the implementation, because all of the transitions are already explicitly stated at the higher level. That is to say, in technical terms, you have a flow graph and not merely a flow chart, which means that all the transition regularities are stipulated at that level, leaving nothing further to design, and it is simply a matter of engineering to make sure that the transition regularities are maintained. It makes no sense to look at different implementations, for the same reason it makes no sense to look at two different copies of the same newspaper. You might get some minor differences of implementation speed or something like that, but that is not apt to be interesting, whereas the relationship between the symbolic or cognitive level and the implementation level in connectionist networks is not that way. It really makes sense to look at different implementations of the cognitive-level sketch because you are counting on features of those implementations to fix details of the transitions that actually aren't fixed at the cognitive level. You haven't specified an algorithm or flow graph at that level. Another way of looking at this is that in contrast to a classical system, where the last thing you want is to have noise in your implementation (i.e., you want to protect the system from noise) in a connectionist implementation you plan on exploiting noise. You want the noise to be there because it is actually going to be magnified or amplified in ways that are going to effect the actual transitions described at the cognitive level.*
>
> *This becomes clear if you consider the hidden units in a connectionist network.... If you subject those hidden units to careful statistical analysis (it is made easier if you view... diagrams showing which nodes are active under which circumstances), you can discover that a certain node is always ON whenever the subject is (let us say) dogs, and never (or very weakly) ON when the subject is cats, whereas another node is ON for cats*

and not ON for dogs. Other nodes, however, seem to have no interpretation at all. They have no semantics; they're just there.   As far as semantics is concerned, they are just noise; sometimes they are strongly active and at other times weak, but these times don't seem to match up with any category of interest. As many skeptics about connectionism have urged, the former sorts of nodes are plausibly labeled the DOG node and the CAT node and so forth, and so it is tempting to say that we have symbols after all. Connectionism turns out to be just a disguised version of good old-fashioned, symbol-manipulating AI! Plausible as this is (and there must be some truth to the idea that certain nodes should be viewed as semantic specialists), there is another fact about such networks that undercuts the skeptics' claim in a most interesting way. The best reason for not calling the dog-active node the dog symbol is that you can "kill" or disable that node and the system will go right on discriminating dogs, remembering about dogs, and so forth, with at most a slight degradation in performance. It turns out, in other words, that all those other "noisy" nodes were carrying some of the load. What is more, if you keep the "symbol" nodes alive and kill the other, merely noisy nodes, the system doesn't work.

The point about this that seems to me most important is that at the computational level in a connectionist system, no distinction is made between symbols and nonsymbols. All are treated exactly alike at that level. The computational mechanism doesn't have to know which ones are the symbols. They are all the same. Some of them we (at a higher level) can see take on a role rather like symbols, but this is not a feature which makes a difference at the computational level. That is a very nice property. It's a property that is entirely contrary to the spirit of GOFAI, in which the difference between a symbol and a nonsymbol makes all the computational difference in the world.

Dennett, in the same essay, goes on to critique connectionist AI, with a focus on feedforward neural networks (rather than, say, attractor neural nets (Amit, 1999), or Grossberg-style (Grossberg, 1998; Grossberg et al, 2004) biologically more realistic models). He points out that such networks are "at best maybe an architecture for a little subcomponent of memory." They are subcomponents that can be trained to give certain outputs in response to certain inputs, but they can't answer queries like "Have you ever danced with a movie star?" because they don't have an adequately flexible representation of concepts. He likes the connectionist idea, but he considers the connectionist models available to be extremely oversimplistic and overspecialized. And of course, achieving simplified, specialized AI is nothing special: GOFAI has been doing that for decades.

According to Dennett's discussion in the above-quoted essay, it seems clear that Novamente is a connectionist AI system. Some ConceptNodes may be directly symbolic, others may be only indirectly symbolic, i.e. parts of concept maps that serve as symbols. If you remove the ConceptNode for "dog" (presuming there is one, and that "dog" isn't represented by a concept map), then it will be re-formable within a decent degree of approximation, using the system's node formation and inference mechanisms. Even without that particular node there, the system will probably be able to go on recognizing and reasoning about dogs, because nearly all the information in the "dog" node is also present in other nodes and relationships in the system, just less directly. Furthermore, in Novamente, it's definitely not true that the dynamics of the system is entirely specifiable at the symbol level. The nonsymbolic aspects of the system have a rich nonlinear coupling with the symbolic aspects, and if they're not engineered and tuned appropriately, the symbolic aspects of the system won't work.

So is Novamente connectionist? Wait a minute. In another interesting Dennett essay, "The Logical Geography of Computational Approaches," we find a description of "the defining dogmas of High Church Computationalism" (a relative of, though not quite a synonym for, GOFAI):

> **1) Thinking is information processing.** *That is, the terms of folk psychology are to be spruced up by the theorist and recast more rigorously; "thinking" will be analyzed into an amalgam of processes ("inference" and "problem solving" and "search" and so forth); "seeing" and "hearing" will be analyzed in terms of "perceptual analysis," which itself will involve inference, hypothesis-testing strategies, and the like.*
>
> **2) Information processing is computation** *(which is symbol manipulation). The information-processing systems and operations will themselves be analyzed in terms of processes of "computation," and since, as Fodor says, "no computation without representation," a medium of representation is posited, consisting of symbols belonging to a system which has a syntax (formation rules) and formal rules of symbol manipulation for deriving new symbolic complexes from old.*
>
> **3) The semantics of these symbols connects thinking to the external world.**

How does Novamente fare according to these criteria? It definitely adheres to the "Thinking is information processing" principle. In the Novamente design, thinking is analyzed into an amalgam of processes, as are perceiving and acting.

On the other hand, while Novamente does rely on the assumption that "information processing is computation," it doesn't do so in the narrow sense that Dennett means here. In Novamente, only some of the mind's information processing is symbol manipulation according to formal rules. Of course, on a different level, even Novamente's neural-nettish activation spreading consists of formal rules (after all it's based on probabilistic inference applied to HebbianLinks), but this is not what Dennett means; he's talking about formal rules that act on symbols, not formal rules that a computer program uses to implement the dynamics of subsymbolic entities.

And finally, in Novamente the semantics of symbols is one among several things that connects thinking to the external world. All in all, Novamente does not adhere to the dogma of the High Church of Computationalism as Dennett articulates it, but it adheres to a decent fragment thereof.

On the other hand, Dennett summarizes connectionism in terms of the following principles:

> 1. *"distributed" memory and processing, in which units play multiple, drastically equivocal roles, and in which disambiguation occurs only "globally."*
> 2. *no central control but rather a partially anarchic system of rather competitive elements.*
> 3. *no complex message-passing between modules or subsystems. "The fundamental premise of connectionism is that individual neurons do not transmit large amounts of symbolic information. Instead, they compute by being appropriately connected to large numbers of similar units."*
> 4. *A reliance on statistical properties of ensembles to achieve effects.*

5. *The relatively mindless and inefficient making and unmaking of many partial pathways or solutions, until the system settles down after a while.*

According to these principles, Novamente is largely connectionist. Its memory and processing are largely distributed in the sense Dennett articulates: units play multiple, very different roles, and most disambiguation occurs as a consequence of distributed activity. There is no central control; the Importance Updating Function which regulates attention allocation implements a competitive/cooperative nonlinear dynamic. There is no complex message-passing between modules or subsystems; there is a heavy but not universal reliance on statistical properties of ensembles, and there is a lot of trial-and-error exploration of partial pathways and partial solutions.

In all, our detailed consideration of Dennett's formulation of the symbolic/connectionist dichotomy reinforces our view that Novamente spans the two sides of the dichotomy. However, it does so not by simply piecing together aspects of the two paradigms, but by proposing a unified system within which connectionist and symbolic aspects are allowed to do what they do best. Procedure learning is not done using feedforward neural nets, but it's done using networks of small computational elements which have a similar distributed aspect, and are learned by a similarly noninferential mechanism. Memory is done using both maps and nodes, providing the crispness of symbolic memory and the flexibility of attractor-neural-net-style memory. Probabilistic logic rules are used to carry out precise inference steps, something that has not yet been successfully achieved in a realistic way within any purely connectionist AI system. But logic is carried out in a distributed network based way, so that logic is actually controlled using a "statistical ensemble based approach," and it is complemented by other less precise learning methods such as neural-net-like association formation and evolutionary concept formation.

Overall, the symbolic and connectionist approaches to AI each focuses on limited aspects of the mind. Each of them constitutes an interesting model of certain aspects of mental process, but in too many cases, very limited models have been interpreted more broadly, as models or metaphors for mental process overall. Novamente presents a model of mental process, in the form of a concrete AI system design, which incorporates all these aspects in a synthetic and intuitive way.

Complex Systems, AI and Novamente

In the next few sections I will explore the relationship of the Novamente design to some of the conceptual themes explored in the previous chapters. For starters: In Chapter 7 we discussed various possible principles of complex systems; it's now worth asking how these manifest themselves in the Novamente design.

From an AGI development perspective, the lack of rigorous mathematical/ scientific laws of complex system dynamics is majorly frustrating. If such laws did exist — i.e. if there were a systematic, rigorous science of complex systems, including rigorous versions of "principles" like the ones discussed in the previous section — then the ongoing development of the Novamente design would probably be a lot easier. However, the working out of specific complex systems like Novamente is a crucial part in developing complex systems science into a real science.

Conversely, complex systems science is useful to Novamente in two ways: as a general guiding philosophy, and to aid with particular decisions within particular Novamente components. On the general-guidance level, what complex systems science contributes are maxims such as the following:

- A system with fairly simple rules can give rise to extremely complex behaviors. Novamente's rules are not as simple as those of a neural net or cellular automaton, but nonetheless they are very simple compared to the behaviors that the system is supposed to display.
- It is important for system components to be interconnected, but in some cases, too much connectivity can be just as bad as too little.
- It's important to build systems that have large, not too wildly-shaped regions of parameter space corresponding to "complex" behavior, rather than periodic or chaotic behavior.
- It's important to build systems whose components cooperate with each other to yield intelligent emergent behavior.
- "In principle universal" computing power is easy to come by; the trick is to get universal computing power from relatively simple rules, which allow pragmatically useful behaviors to be displayed compactly and learnably.
- It may be useful to create systems in which emergent dynamics and component-level dynamics share similar patterns.

These are fairly simple lessons, but they combine well with related points that arise on a more detailed level. Although they may seem obvious, they nevertheless are not automatically followed by every plausible-looking AI mechanism that one might consider inserting in Novamente. In the course of Novamente design, careful attention has been required to be sure that they are followed, both on the high-level level and on the microlevel.

Now I will discuss a few of the specific complex systems principles mentioned above with an eye toward their Novamente implications.

My and my colleagues' experimentation with Novamente (and Webmind before it) has validated the edge-of-chaos principle in many particular ways. On the simplest level, the Webmind attention allocation component, which spreads "activation" values around between nodes and links as in a neural net, was seen to yield a variety of different attentional dynamics, including periodicity or convergence to a fixed point, and dynamics so complex as to be effectively quasi-random. Often setting parameters of this subsystem to intermediate-level values (e.g. setting the firing threshold for nodes to send activation neither too large nor too small) will yield acceptable behavior.

However, Novamente does not seem to depend in any critical way on the degree of validity of the Edge of Chaos Principle. What it depends on is that there exist subspaces of its parameter space in which the system will work reasonably well, even while it adds new nodes and links to its memory, deletes old ones, etc. The system adapts its own parameters over time in quest of optimum functionality, but if it had to adapt all its parameters constantly based on minor changes in the structure of its internal knowledge network, the design would be unworkable. Also, if slight changes in parameters generally led to huge changes in system behavior, the design would be unworkable. These problems seem not to occur in Novamente, at least based on our experimentation so far, and this is partly because minor adaptive tweaks to parameters seem to leave the system within the "complex behavior generating region," rather than sending it into the domains of chaos or unproductive, boring regularity.

The "principle of metasystem dynamics" – that successful complex systems are the ones in which the different parts work together to yield emergent patterns — is easy to see in

Novamente terms.   One wants the knowledge in different Atomspace subspaces to work together well, leading to useful emergent knowledge rather than contradictions.   And one wants the different AI processes used in the system to display cooperativity, in that their combined action results in patterns that they could not give rise to individually.

In building Novamente, one of our biggest challenges has been to create a set of specialized learning subsystems, that nonetheless cooperate with each other effectively to produce overall emergent intelligence.   Each subsystem is optimized to deal with the specific types of problems *that the other subsystems pose to it*. A system in which each component is optimized to deal with the inputs that the other components give it will naturally be more efficient than a system in which each component is unoptimized or is optimized in a more general sense.

Finally, Novamente has been built specifically to manifest the Structure-Dynamics Principle: the existence of a close (though not exact) mapping between structural and dynamical patterns.   What we call Novamente maps (sets of Atoms activated simultaneously or according to some other coordinated pattern) are basically sub-regions of system state space that are frequently visited.   Many Novamente maps are specifically triggered by one or a handful of particular Atoms, thus yielding a parallel between inter-map dynamic relationships and inter-Atom dynamic relationships.   I'll elaborate on this point in depth a little later.

A simple but pertinent example of the manifestation of the Structure-Dynamics Principle in the relation between Atoms and maps in Novamente is given by looking at the emergent and concretely-implemented aspects of *positive emotion*.

In Novamente we have FeelingNodes, which look like symbolic-AI-style representations of system feelings.   However, to pursue a human-mind analogy, these FeelingNodes are really more like basic limbic-system or otherwise chemically-induced brain stimuli than they are like richly texture high-level human feelings.   Novamente's Satisfaction FeelingNode is somewhat like the raw animal feeling of pleasure that we humans experience.

In the human mind, happiness is much more complex than pleasure.   It involves expectations of pleasure over various time scales, and it involves inferences about what may give pleasure, estimates of how happy others will be with a given course of action and thus how much pleasure one will derive from their happiness, etc.   Biological pleasure is in a sense the root of human happiness, but the relationship is not one of identity.   Changes in the biology of pleasure generally result in changes in the experience of happiness – witness the different texture of happiness in puberty as opposed to childhood, or maturity as opposed to early adulthood.   But the details of these changes are subtle and individually variant.

In Novamente, there will also be complex "feeling maps" centered around FeelingNodes but including many other Atoms as well.   Whether any of these will closely resemble the human feeling of "happiness" remains to be seen.   But for sake of discussion, let's assume for the next few paragraphs that it does, that there comes about some Novamente analogue of happiness defined as some complex map.

In this case, we have a parallel between:

- A concretely implemented mind structure, the Satisfaction FeelingNode;
- An emergent mind map, a metanode, the feeling of happiness (or its closest approximation thereof in Novamente psychology).

There is a substantial similarity between these two parallel entities existing on different levels, but not an identity.   Happiness, or its closest Novamente analogue, will be embodied in:

- A large, fuzzily defined collection of nodes and links (a "map")
- The dynamic patterns in the system that are induces when this collection becomes highly active (a "map dynamic pattern")

The Satisfaction FeelingNode is one element of the map associated with happiness. And it is a particularly critical element of this map, meaning that it has many high-weight connections to other elements of the map. This means that activation of pleasure is likely – but not guaranteed – to cause happiness.

This illustrates the Structure-Dynamics Principle, which is the heuristic rule stating that often, in complex systems, the set of *static patterns* overlaps greatly with the set of *dynamic patterns*. In other words, statics often encodes dynamics. In a Novamente context, maps encoding mental entities exist statically, but their purpose is to induce certain map dynamic patterns (often mind-wide in scope) when highly activated by the importance updating function. The concretely implemented mind exists to spawn the emergent mind.

Novamente as an Evolutionary System

What is the role of evolution in Novamente dynamics? There are really two separate aspects here:

- *Implicit evolution*, wherein the activity of the system as a whole is naturally interpretable as evolutionary.
- *Explicit evolution*, wherein evolutionary algorithms are used to create new Novamente Atoms.

One of the key goals of the Novamente design is to cause these two aspects of evolution to work harmoniously together.

Let us begin with implicit evolution. I will argue that the dynamics of long-term memory, as generally conceptually understood in the psynet model of mind and as specifically implemented in Novamente, fulfill the criteria for "natural selection" posited above. This observation gives a powerful motivation for using evolutionary computing in Novamente, because it means that when evolutionary computing works together with long-term memory in Novamente, what one has is *co-evolution*, a relatively well-understood and highly effective strategy for solving problems and creating structures.

First of all, one may argue on purely philosophical ground that memory must evolve by natural selection? Of course, this is not a rigorous proof, only an heuristic argument. But the argument is simple and, we believe, conceptually compelling. Suppose one assumes that the following three statements hold, regarding the nature of memory:

- Where X is an item in memory, the existence of relationships between X and other items in memory is a substantial determinant of the amount of attention that X gets.
- Many relationships involving X are emergent patterns between X and other things.

- Memory items X that don't get much attention may eventually be forgotten entirely, and will rarely be allowed to interact with other items and cooperatively create new items.

These three points, taken together, imply that the mental items X that get to survive in memory, and get to help generate new items, will tend to be the ones that have a lot of emergent pattern existing between themselves and other mind-actors. But this is precisely evolution by natural selection, in the "structural complexity" sense described above.

In the context of Novamente specifically, this argument manifests itself differently for different kinds of knowledge representation. For knowledge represented directly in Atoms, all one needs to show to demonstrate the existence of evolution is that the probability of a node surviving and prospering in the Atomspace is roughly correlated with the amount of emergent pattern generated between that node and other things in the Atomspace. But, this is almost obvious. Each link between one node and another embodies an emergent pattern between those two nodes. The probability of an Atom getting chosen to participate in some AI dynamic, that will create new Atoms or enhance the truth values of existing ones, is roughly given by its importance, which is directly tied to the weights of its links. Atom importance is not *exactly proportional* to the amount of emergent pattern embodied by the node's link set, but there is a strong relationship there. And the probability of a node or link surviving at all is proportional to its long-term importance, a quantity that is ultimately determined by its importance. Again, LTI is not directly proportional to the amount of emergent pattern involving an Atom, but there is a clear relationship of positive correlation.

We have not yet sought to measure the degree to which Novamente evolves by natural selection, according to the criteria given here, but it would certainly be interesting to do so. What needs to be done is to track, at each point in a Novamente's history:

- The amount of emergent pattern related to each Atom, using the patternist analyses of the various Novamente AI processes that are given throughout this book;
- The importance and LTI of each Atom over time.

Computing the correlation between these series, for each Atom, is a simple operation. Of course, the true amount of emergent pattern involving a Novamente Atom cannot be exactly estimated, due to fundamental and practical intractability issues. But reasonable estimates can be made; most simply, one can just look at emergent patterns that are explicitly embodied in links in the system.

Note that this interpretation of Novamente dynamics as evolutionary does not in any way require Atoms to reproduce via classical mutation and crossover operations. In fact some Atoms do reproduce this way, but such is not necessary in order for the system's overall dynamics to be evolutionary. What matters is mostly that more emergent-pattern-generating Atoms tend to survive longer, and secondarily that these emergent-pattern-generating Atoms tend to participate more in the creation of new Atoms, by one operation or another.

What about knowledge represented nonlocally, by maps in the Atom network? What determines the probability of survival of a map? In a continuous learning system like Novamente, this is determined by how often the map is accessed. Frequently accessed maps will be re-imprinted on the network, and will get to interact with other maps and with important Atoms, whereas infrequently accessed maps will be forgotten. A map is accessed if other processes habitually lead to patterns of activation that are in its basin, i.e. that are similar to it. Thus, *one* high-level system pattern is that maps survive if similarity relations (a simple kind of emergent pattern) obtain between them and other dynamic processes in the Atomspace. Here not between one data structure and another, but between one system state (the map) and other system states.

What do Edelman's neural maps correspond to in Novamente? Small neural maps intuitively correspond to individual Atoms, e.g. SchemaNodes or ConceptNodes or perceptual nodes. Larger neural maps intuitively correspond to Novamente maps — subgraphs of the Atom network that are typically activated as a whole, because of the particular pattern of link weights that they embody. According to Novamente dynamics, will these evolve by natural selection? The answer is yes, by exactly the same logic that Edelman has described in connection with the brain. Maps that are active when good results are obtained will have their importance reinforced by activation propagation from the GoalNodes embodying the good results. And pairs of active maps will sometimes be activated simultaneously, which (depending on the particular inter-Atom link strengths and activations involved) will sometimes lead to the creation of new maps fusing active maps or fusing parts of them.

So, all Atoms and maps are implicitly evolving, to some extent. But this is not strict neo-Darwinist evolution. There is self-organization both within and without the evolving entities.

The ecological aspect is the most obvious — survival in the dynamic Atomspace is not determined by a fixed fitness function; rather, the survival of a given Atom or map is based on its relationship with other Atoms and maps. Each one survives based on how well it "fits" with the others.

On the other hand, the epigenetic aspect is not there for simple Atoms, but it is there for maps and for Atoms containing compound schema or predicates. For a simple Atom, "genotype" and "phenotype" are the same: the mutative and combinatory operators that act on Atoms act on their links, which directly determine their fitness in the context of the system. But for a map, the genotype is the nodes & links contained in the map (and their links to other things), whereas the phenotype is the dynamic behavior that these Nodes and Links habitually follow together. Sometimes this dynamic behavior is very simple – just a simultaneous activation – and sometimes it can be quite complex, involving coordinated schema activity or complex temporal activation patterns. Similarly, for a compound schema or predicate node, the genotype consists of the node's external links, and the node's internal schema or predicate; and the phenotype has to do with the behavior of the internal schema or predicate, which may not depend in any simple, direct way on the elementary schema of which it is composed.

The explicit evolution that occurs in Novamente should be understood in the context of this implicit evolution. Explicit evolution is carried out by a process called the Evolution MindAgent, which is actually just a "utility" MindAgent that carries out tasks given to it by other MindAgents, via applying the MOSES algorithm. To give a task to the evolution MindAgent, one must specify a fitness function, and one may also specify some Atom type restrictions. The Evolution MindAgent then selects random Atoms from the Atomspace (consistent with the type restrictions), and causes them to mutate and to combine with each other (by whatever operations it knows that are relevant to Atoms of their type). It also constructs a probabilistic model of the evolving population, and selects or creates new Atoms based on this probabilistic model. In short, the Evolution MindAgent carries out *artificial selection* – it breeds Atoms fitting certain criteria.

The advantage of breeding is that it can relatively rapidly perfect specific traits of an organism. If one has a specific function that needs to be carried out, and a schema carrying out this function can be pieced together using a relatively small number of existing SchemaNodes, then breeding can do the trick. Probabilistic modeling techniques (e.g. Estimation of Distribution

Algorithms such as BOA and MOSES) appear to substantially accelerate the rate with which GA/GP-style breeding converges to reasonably good problem solutions.

The disadvantage of breeding is that, to be truly effective, it requires aa relatively narrow and rigid definition of what "the problem" is. Most of the problems that occur in an AGI system's experience can't easily be summed up in a compact fitness function. One often winds up with fitness functions like "the extent to which this schema makes the user happy." But assessing fitness according to such a criterion is very difficult, because it relies on extensive inference based on knowledge gained from experience. It is possible to use the Evolution MindAgent for this sort of problem – but one is incorporating so much ecology in the fitness function, that the simplicity and efficiency of the "breeding" paradigm is somewhat broken. In such cases may often preferable to simply let the implicit evolutionary process of Novamente dynamics do the trick. To take an animal breeding metaphor, one is no longer breeding chickens that are fat, one is breeding chickens that are good at surviving in a complex environment. And if one is going to do that, one may as well just let a bunch of chickens run wild in the environment and see which ones survive.

In conclusion, the complex-systems approach to evolutionary theory is part and parcel of the Novamente design. It is not merely a glib metaphor to say that ideas and patterns "evolve" within Novamente – it is a precise statement, to be considered in terms of the modern evolutionary synthesis that includes self-organizing epigenesis and ecology as key factors alongside selection. The degree to which Novamente, as whole and different parts of the system, is evolutionary can be quantitatively estimated most simply by estimating the degree to which survival is proportional to quantity of emergent pattern. Explicit evolutionary programming also plays a role, but it must be considered as an extreme case of the implicit evolution that is inextricable from Novamente dynamics. The continuity between evolutionary programming and intrinsic Novamente dynamics (which is implicitly evolutionary), is one reason why evolutionary programming is more natural than other optimization techniques for implementation in a Novamente context.

Novamente and the Cognitive Equation

Back in Chapter 12 on autopoiesis I proposed a general mathematical process and gave it the ambitious name "The Cognitive Equation." Novamente is supposed to be a cognitive system – so the question naturally arises: How specifically does the cognitive equation apply to Novamente? Given the content of the Cognitive Equation, what this question really boils down to is: To what extent is it the case that the components of Novamente, at a given point in time, represent patterns in Novamente's prior structure?

To answer this question in detail requires far more background on Novamente's data structures and dynamics than will be given in this book. The short answer, however, is that nearly every act of Atom creation in Novamente is in essence an act of pattern recognition. Relationships are created by MindAgents dedicated to inference, association formation, and so forth, precisely because these MindAgents have determined that these relationships are significant patterns in other Atoms in the system. New Atoms may be created purely speculatively, but if significant patterns are not recognized between them and other things in the system, they will perish rapidly; what keeps them around is if they, in conjunction with relationships that form involving them, prove to be patterns in the system's overall Atomspace. The same dynamic holds on the map level.

Novamente does not purely reflect the cognitive equation, because the Atomspace at any given time can contain some elements that do not embody patterns in the prior Atomspace. But the reason for this "cognitive imperfection" is a meaningful one. What Novamente does is to create, alongside with Atoms representing patterns in itself, Atoms representing possible patterns.

It then does some work to determine if they're actual patterns or not, and if not, it discards them (through a simple mechanism: their long-term-importance becomes so low that they are deleted from RAM). The imperfection is primarily due to the fact that the job of recognizing patterns in Novamente is being done by components of Novamente, and that the pattern recognition processes taking place often involve the creation of a large number of "conjectural" components that may or may not ultimately prove to constitute system patterns.

This kind of messiness is not present in the Old Cognitive Equation, which separated pattern recognition from self-generating dynamics. In a real mind, however, pattern recognition is part of self-generating dynamics, which is one reason I feel the New Improved Cognitive Equation is better. This fusing-together of self-generation and pattern-recognition reveals that the cognitive equation is less of a rigorous dynamical rule for cognitive systems, and more of a general dynamical pattern that will be obeyed by various intelligent systems to a greater or lesser extent. But nevertheless, the Cognitive Equation in its new form provides valuable conceptual guidance.

Developmental Stages

So far I have very loosely described a cognitive architecture, a knowledge representation and a set of learning mechanisms – and then related these entities to various philosophical concepts developed in prior chapters, and to known science regarding the human brain/mind. These entities are necessary ingredients for an artificial mind, but not sufficient. They merely set the stage for the self-organization and reflective learning processes that are what really make a mind. And this leads us to the fascinating and critical topic of AGI education. The basic principle underlying any reasonable AGI educational program must be the hierarchical composition of (conceptual, perceptual and behavioral) patterns. Advanced intelligence requires the recognition of complex patterns, but the search space of possible complex patterns is very large, and so a mind must work up to learning complex patterns via starting out with simple patterns and then incrementally building more and more complex patterns from the ones it already knows. PLN and MOSES are designed to be good at this kind of hierarchical building. The point of teaching an AGI is to present it with a series of learning problems that require it to learn to recognize more and more complex patterns, in an order that matches naturally with the logical buildup of more and more complex patterns from initially simple elements.

The teaching program we are planning to use for Novamente is based on an adaptation of Jean Piaget's classic development psychology ideas to the context of the AGISim simulation world (see Goertzel and Bugaj, 2005). It's understood that Piagetan theory is somewhat crude compared to more recent developmental psychology theories, and is also incomplete in many respects; but the truth of human development is highly complicated and there seems no other theory with the same mix of rough accuracy and extreme simplicity that Piaget's possesses. One of my own research foci in the near future will be the synthesis of modern developmental psychology insights with dynamical systems theory and pattern theory to form a more rigorous and general theory of developmental psychology, applicable to both humans and AI's.

Piaget conceived of child development as falling into four stages, each roughly identified with an age group: infantile, preoperational, concrete operational, and formal.

> **Infantile**: Here babies are exploring the world via observing what's around them and exercising their own actuators. They begin with some instinctive actions, and then start repeating their own actions when rewarded. Later, they begin repeating actions observed in others. Simple associations between words and object,

actions and images are made. One of the major learning achievements here is object permanence – infants learn that objects persist even when not being observed.

Preoperational: At this stage the child clearly has mental representations, but they're not well-organized using abstractions. Thinking tends to be intuitive not logical. This is where word-object and image-object associations become systematic rather than occasional. Simple syntax is mastered, including an understanding of subject-argument relationships (semantically: which types of events/actions tend to correspond with which arguments, lying in which categories) At this stage, the mind tends to classify objects by a single parameter (e.g. grouping all red objects together, or all square blocks together).

Concrete Operational: At this stage the child's mind progresses to the combination of symbolic and logical thought. Among the feats achieved here are: reversibility — the ability to undo steps already done; conservation — understanding that properties can persist in spite of appearances; theory of mind – an understanding of the distinction between what I know and what others know. (If I cover my eyes, can you still see me?) Concrete operations such as putting items in height order are easily achievable. Classification becomes more sophisticated, e.g. categorizing daisies versus roses. This kind of categorial distinction involves the combination of various types of properties.

Formal: At this stage we have abstract deductive reasoning, the process of forming then testing hypotheses, etc. This is full, adult human-level intelligence. Note that the capability for formal operations is intrinsic in the PLN component of Novamente, but in-principle capability is not the same as pragmatic, embodied, controllable capability.

Inspired by Piaget's general ideas I have created my own series of developmental stages, defined roughly as follows:

Infantile: Able to recognize patterns in and conduct inferences about the world, but only using simplistic hard-wired (not experientially adapted) inference control schemata.

Concrete Operational: Able to carry out more complex chains of reasoning regarding the world, via using inference control schemata that adapt their behavior based on experience (reasoning about a given case in a manner similar to what worked in prior similar cases).

Formal: Able to carry out arbitrarily complex inferences (constrained only by computational resources) via including inference control as an explicit subject of abstract learning.

Reflexive: Able to revise one's own cognitive processes rationally in accordance to one's own analyses of them. (Few humans achieve this stage at all, and the human brain architecture does not allow any humans to achieve it completely. But AI's will eventually get there!)

Careful study shows that the Piagetan learning tasks fit naturally into these categories, with Piaget's pre-operational phase appearing as transitional between the infantile and concrete operational phases. We suspect this approach to cognitive development may have general value beyond Novamente, though to argue this point would bring us too far afield here. We have designed specific Novamente /AGISim learning tasks based on all the key Piagetan themes. Currently our concrete work is near the beginning of this list, at Piaget's infantile stage.

We have designed specific Novamente /AGISim learning tasks based on a number of Piagetan themes, including: word-object associations, object permanence, subcategorization frames for common useful semantic constructs (this one is not a Piagetan theme exactly, but it's Piaget-related, and is critical in the context of language processing); classification – moving from single-parameter to multiple-parameter; reversibility of actions; conservation of properties; formation of abstract symbols; theory of mind (in several different senses); and, putting items in order according to height or other parameters. Currently our concrete work is near the beginning of this list, at Piaget's infantile stage.

Learning Object Permanence

Next I will discuss the specific task of learning object permanence, a topic which will require a brief digression into the simple visual system via which Novamente interfaces with the 3D simulation world, called AGISim, in which it is currently embodied. Rather than perceiving individual pixels or voxels within AGISim, Novamente perceives AGISim in terms of polygons. A Polygon Node represents a polygon observed at a point in time. A Persistent Polygon Node then represents a series of Polygon Nodes that are heuristically guessed to represent the same PolygonNode at different moments in time. Before object permanence is learned, the heuristics for recognizing Persistent Polygon Nodes will only work in the case of a persistent polygon that, over an interval of time, is experiencing relative motion within the visual field, but is never leaving the visual field. For example some useful heuristics are: If P1 occurs at time t, P2 occurs at time s where s is very close to t, and P1 are similar in shape, size and color and position, then P1 and P2 should be grouped together into the same Persistent Polygon Node.

Adjacency Links are created between Persistent Polygon Nodes, via a special formula that maps the relative positions of two polygons into a "strength" value in [0, 1]. Then a Clustering MindAgent looks for clusters in the graph of AdjacencyLinks between PersistentPolygonNodes: these clusters become AGISim Object Nodes. All this mechanism is relatively straightforward – but all it does is recognize an object as a set of persistent polygons that cohere together within the visual field during some continuous interval of time. If an observed object leaves the visual field and then re-enters, then these low-level in-built mechanisms don't tell Novamente anything about it. If a ball disappears behind a chair and then reappears, then upon reappearance it is classified as a new object! The Piagetan task of object permanence requires Novamente to learn that in fact it is still the same ball after it has reappeared.

This is not a very hard reasoning task. For instance, if the system is given multiple balls to play with, with different (unique) markings on them, then it can learn via experience that if a ball with marking X goes behind the chair, and it then goes behind the chair, it will find a ball with marking X rather than some other marking. Simple though it seems, this knowledge is represented in Novamente via a predicate involving a couple dozen different Nodes and Links, and learning it either requires a lengthy MOSES run or some fairly intensive backward-chaining inference. And a more interesting sort of inference occurs after this. Suppose the system has learned that balls retain markings: can it then extend this knowledge to infer the permanence of other sorts of objects? This requires what in PLN theory is called abductive inference.

This example very explicitly illustrates the difference between AGI research and narrow-AI research. In this case, we are making Novamente learn something that we could very easily tell it instead (information regarding what objects exist in AGISim is there explicitly in the AGISim server, and could merely be passed to Novamente). We are making it learn very simple and basic things because we believe that minds most naturally learn complex things via analogy to simple things and that analogies are most easily drawn to concepts and procedures about which a rich network of patterns has been formed. When Novamente reaches Piaget's concrete operational stage and needs to learn conservation laws, its job will be easier because it will be able to draw on its experience learning object permanence. Conservation of mass is basically "mass permanence," and the procedures it has developed for "learning about permanence" in the context of learning object permanence will be useful for it in learning about mass permanence. This simple example illustrates the general principle of composition of patterns, via which learning algorithms build complex patterns from simple ones.

Learning "Theory of Mind"

Next consider a more advanced example of experiential learning: recognizing that another agent has its own state of knowledge, with uncertain overlap with the Novamente agent's own state of knowledge. This occurs in the preoperational or concrete operational stages rather than the infantile stage of development; it has a fair number of preconditions. At time of writing, the Novamente system is not yet up to the task of learning this sort of thing: it is far too immature. But nonetheless, we can sketch out in detail how this sort of learning may occur.

One precondition to this kind of learning is the understanding that what the agent itself perceives may vary depending on its situation. A simple example of this is the recognition that "If I were in a different position or location, I would see something different." For instance, "If I rotated by pi radians, I would see the yellow block." It's not a big leap from this to the recognition that "You look like me, and you're rotated by pi radians relative to my orientation, therefore you probably see the yellow block." The only nontrivial aspect here is the "you look like me" premise.

Recognizing "robot" as a category, however, is a problem fairly similar to recognizing "block" or "insect" or "daisy" as a category. Assuming that:

- the agent can perceive most parts of its own "robot" body – its arm, its base, and so forth;
- the agent has figured out that physical objects like arms and bases look different depending upon how far away you are from them and what angle you're looking at them from;

the agent should be able to understand that it is naturally grouped together with other embodied agents (like its teacher) rather than with blocks or bugs. (Of course, a mirror would be a valuable though not necessary tool here. If the agent understood the principle of a mirror, which is not hard to learn using PLN, then it would be able to see that it has an eye similar to the eye of other robots.)

The only other major ingredient needed as a precursor to simple theory-of-mind is "reflection" – the ability of the system to explicitly recognize the existence of knowledge in its own mind. This may come for instance via the application of the atTime operator (an elementary

Schema provided to Novamente to use in its PredicateNode-building process) to the TruthValue operator. Observing that "at time T, the weight of evidence of link L increased from zero" is basically equivalent to observing that the link L was created at time T. It is quite simple to give Novamente the capability to form PredicateNodes like this, introspecting on its own state and structure.

Given these preliminary capabilities, the system may reason, for example, as follows:

```
Implication
        My eye is facing a block and it is not dark
        A relationship is created describing the color of the block

Similarity
        My body
        My teacher's body

|-

Implication
        My teacher's eye is facing a block and it is not dark
        A relationship is created describing the color of the
        block
```

This simple PLN inference is the essence of "theory of mind." (Note that in both of these implications the created relationship is represented as a variable rather than a specific relationship: thus this is definitely a "higher-order" inference, as would be very clear if it were written out in fully rigorous notation.)

The conceptual leap hidden here is that in the latter case the hypothesized relationship lives in someone else's mind, not the agent's own mind. This leap occurs via a number of inferential processes, including the recognition that some hypothetical relationships are associated with other bodies, whereas some are associated with the agent's own body. The hypothetical relationships associated with the teacher's body are the beginnings of Novababy's "theory of the teacher's mind."

Note that we don't need any "theory of the self" to proceed from observed relationships to elementary "theory of other agents' minds." Self is a complex topic and of course a mature mind needs to model its own self and those of others, but the point of view taken here is that a "self" is a complex system – a coherent self-organizing collection of relationships. The system's own self is a collection of interrelated relationships that are "reflective" in the above sense, and its image of another system's self is a collection of hypothetical reflective relationships, as in the above example. The prerequisite for the formation of self-systems, however, are basic theory-of-mind-oriented inferences as exemplified here.

Novamente and the Human Mind-Brain

In all details Novamente is very un-brainlike, yet much of the Novamente experiential learning architecture diagram (Figure 1 above) could be interpreted as a diagram of human brain processing, suggesting that at a high level some parallels may be drawn between Novamente and human brain structure. Table 5 draws several such parallels, in a speculative way. This table should be taken with several grains of salt – clearly, the brain is not sufficiently well understood for a table like this to be made with any reasonable degree of confidence. But it seems worthwhile to share the best guesses I have made in this regard based on the available evidence.

Human Brain Structure/Phenomenon	Primary Functions	Novamente Structure/Phenomena
Neurons	Impulse-conducting cells, whose electrical activity is a key part of brain activity	No direct correlate: Novamente's implementation level is different
Neuronal groups	Collections of tightly interconnected neurons, often numbering 10,000-50,000	Novamente nodes
Synapses	The junction across which a nerve impulse passes from one neuron to another; may be excitatory or inhibitory	Novamente links are like bundles of synapses joining neuronal groups
Synaptic Modification	Chemical dynamics that adapt the conductance of synapses based on experience; thought to be the basis of learning	The HebbianLearning MindAgent is a direct correlate. Other cognitive MindAgents (e.g. inference) may correspond to high-level patterns of synaptic modification
Dendritic Growth	Adaptive growth of new connections between neurons in a mature brain	Analogous to some heuristics in the ConceptFormation MindAgent
Neural attractors	Collections of neurons and/or neuronal groups that tend to be simultaneously active	Maps, e.g. concept and percept maps
"Neural Darwinist" map evolution	Creates new, context-appropriate maps	Schema learning via reinforcement learning, inference, evolution
Cerebrum	Perception, cognition, emotion	The majority of Units in a Novamente configuration
Specialized cerebral regions (Broca's area, temporal lobe, visual cortex, etc.)	Diverse functions such as language processing, visual processing, temporal information processing	Functionally-specialized Novamente Units
Cerebellum	Movement control, information integration	Action-oriented units, full of action schema-maps

Human Brain Structure/Phenomenon	Primary Functions	Novamente Structure/Phenomena
Midbrain	Relays and translates information from all of the senses, except smell, to higher levels in the brain	Schemata mapping perceptual Atoms into cognitive Atoms
Hypothalamus	(regulation of basic biological drives and controls autonormic functions such as hunger, thirst, and body temperature)	HomeostaticParameterAdaptation MindAgent, built-in GoalNodes
Limbic System	(control emotion, motivation, and memory)	FeelingNodes and GoalNodes, and associated maps

Table 5. Novamente vs. the Human Brain

One interesting thing to observe about the information in Table 5 is that the Novamente approach doesn't necessarily break the mind down into components in the same ways as the mainstream of modern cognitive science. For instance, memory and reasoning are typically considered as separate things, in the course of cognitive science research. Yet, in the Novamente approach, it is considered that most acts of "memory retrieval" are actually coordinated acts of reasoning, "constructing" memories from stored knowledge. Similarly, reasoning and perceptual pattern recognition are typically considered as different things, yet in the Novamenteapproach, perceptual pattern recognition is done via the same probabilistic equations used for abstract reasoning, deployed in simpler and more scalable ways.

These differences don't make it impossible to draw mappings between Novamente and the human mind/brain, but they do mean that these mappings must be drawn with care. In every case we've explored so far, when one probes deeply, one finds that the Novamente approach is harmonious with the ideas of some significant subset of cognitive science researchers. For instance (Riegler, 2005) advocates the constructive nature of memory, whereas (Goldstone et al, 2005) discusses parallels between perceptual learning and abstract cognition; etc. In many cases cognitive science divides mental process into categories based on convention and convenience; and when building an AGI one is confronted with a different notion of convenience – a division (like memory vs. reasoning) that's convenient for guiding the design of experiments on human subjects can be extremely inconvenient from the perspective of AGI design. In this regard Novamente is perhaps closer to neuroscience than cognitive science – neuroscientists are continually discovering feedback loops and dynamical and structural complexities that break through the simplistic divisions favored by many cognitive theorists. This is because neuroscientists, like AGI designers and engineers, are dealing with the necessary messiness of real complex systems, rather than with simplified theoretical abstractions.

Memory in Novamente and Humans

One area where Novamente clearly accords with cognitive neuroscience ideas is the division of memory into various subcomponents. The distinction between procedural, episodic and declarative memory is well-demonstrated both psychologically and neuroscientifically (Baddeley, 1999), and it is also quite natural in terms of Novamente's node and link knowledge representation.

Declarative knowledge is naturally represented in Novamente via probabilistic-logical link types, whereas procedural knowledge is naturally represented using links explicitly representing actions taken. Episodic memory, finally, is naturally represented via links joining probabilistic-logical relationships defining sequences of events with records stored in an "experience database." The distinction between the three memory types becomes, in Novamente terms, a matter of representational efficiency.

Experiential episodes can be stored in declarative logical terms but this is extremely inefficient; so for practical purposes it's better to store experiences in another, less flexible form, and map only their high-level structure in declarative form. As the mind and its goals change, the experience database will be repeatedly revisited and its contents re-represented declaratively in different ways based on different acts of pattern recognition.

On the other hand, procedural knowledge can also be stored in declarative form, and this is useful when one wants to reason about procedures. But it is generally the case that the most natural form of a procedure from the point of view of reasoning about the procedure is not the most efficient form from the point of view of actually executing the procedure. So for practical reasons one winds up with dual representations of procedures: an executable form that is compacted for execution efficiency, and an expanded form that's suitable for generalization and inference. A difference between AGI systems and the human brain comes up here: for humans it can be extremely difficult to create declarative forms for procedural knowledge. On the other hand, detail-level introspection is much easier for a software program than for a human brain, and procedural-to-declarative conversion doesn't need to be so problematic. This is one among many areas in which a slavish adherence to human neuropsychology is probably not clever AGI-design-wise.

Similarly, the distinction between short-term and long-term memory, fundamental to human psychology and well validated via neuroscience also turns out to be fundamental to AGI design, again for a combination of efficiency reasons. The "short-term memory" concept has undergone a number of name changes as cognitive theory has developed, but the basic idea is simple and solid. In a situation of limited processing power, not everything can be attended at once. The Novamente approach involves the assignment of "importance" parameters to nodes and links, and the use of inference and pattern recognition to update these importance values dynamically. Depending on the parameter values of the process, this dynamic updating will frequently lead to a situation in which a small percentage of nodes and links garner a vast majority of attention.

Furthermore, in the Novamente design this dynamic phenomenon (the emergence of STM) is structurally reified via the use of a separate lobe (the AttentionalFocus) for the most important nodes and links. This seems qualitatively similar to how, in the human brain, the focus of attention is structurally reified via separate brain structures devoted to STM in its various forms.

The refinements of the notion of STM that have occurred over the last few decades are also well reflected in the Novamente approach. One recent discovery is that, prior to the registration of sensations in the STM, there is a preliminary process in which fleeting connections are drawn between sensations and knowledge stored in LTM. This emerges naturally from the framework in which there is one overall knowledge-network and the "STM" is simply the "moving activation bubble" of most-important-entities.

Another discovery that has become gradually solidified since Baddeley (1989) first systematically presented it is the existence of multiple modality-specific STM's, such as a visual-perception STM, a linguistic STM, and a generic "mental workspace." This emerges naturally in the Novamente approach, because the specialized schemata that handle a process like visual perception or language processing will naturally gather separate bubbles of highly-active nodes and links around them. Again this dynamic process is reified in the Novamente architecture via the creation of specific lobes for modality-specific AttentionalFocus.

In cognitive science terms, Novamente embodies a flexible model of attention, as opposed to the more rigid filtering or late-selection based approaches that were popular among theorists in the past (Underwood, 1993). For instance, in the case where information comes in from two sensory channels and there are not enough resources to process both information-streams simultaneously, one achieves the result that: the earlier the stage at which selection between channels is possible (i.e. the closer the attentional focus is to the sensory level), the faster and more efficient the response to the attended channel, and the less is processed in the unattended channel. This accords with results from human psychology, for instance, research on the timing of accessing word meanings (Luck, Vogel and Shapiro, 1996).

The "binding problem" that is so critical in modern cognitive science (how do the physically disparate brain-activation patterns inspired by a percept get unified into a subjectively and pragmatically whole, unified perception?) is less critical in the AGI domain. Thinking about binding in the context of AGI impels one to decompose the problem into two parts: a conceptual part and a physiological part. Conceptually, there is a question of the logic and the cognitive dynamics by which disparate percepts are bound into a unified whole. Then, physiologically, there is the question of how the brain executes this logic and dynamics, which is a subtle issue because the parts of the brain representing different parts of a unified percept are often physically widely distributed. AGI systems give rise to the conceptual problem but not the physiological one. The solution to the conceptual problem seems straightforward in the Novamente context; it follows very closely Walter Freeman's (2001) ideas regarding the emergence of attractors in the brain. The linkages between sensation and LTM cause the *relationships between* the nodes and links involved in percepts corresponding to parts of a unified object to become important, which encourages the formation of predicates and concepts binding all of them together. From this perspective, the physiological "binding problem" then comes down to how the formation of dynamic associational and logical linkages occurs between percepts and concepts represented in distant brain regions – a very important question for neuroscience, but not directly relevant to AGI systems, except those that seek to emulate the way the brain uses three-dimensional geometry to help represent knowledge and guide cognitive and perceptual dynamics.

The nature of forgetting in human memory seems to an extent to represent general principles that are also applicable to Novamente and related AGI systems. It has recently become clear that a substantial amount of the forgetting that occurs in the human mind can be attributed to memory interference rather than simply "running out of space" (Wixted, 2004). This sort of phenomenon occurs naturally as a consequence of importance dynamics: if two pieces of knowledge contradict each other then when one gets attention the other will tend not to, and the loser in the rivalry will gradually get its importance downgraded until it's forgotten. In this view, running out of space is the ultimate reason for forgetting, but interference can explicitly cause memory items to be deprioritized. Also, the known fact that humans rarely truly forget anything that's been fully learned (Baddeley, 1989) is reflected in the notion of long-term importance. In Novamente, when an item is important on enough occasions, it achieves a high long-term importance, which basically guarantees that it will be preserved in the deep memory store (i.e. saved to disk) rather than being permanently forgotten.

Finally, regarding the particular structure of the contents of memory, cognitive science doesn't have anything definitive to say at the moment. The Novamente is reminiscent of semantic-network-based models of human memory, which originated with Quillian and have played a major role in many subsequent cognitive modeling approaches such as Anderson's work on ACT-R (1997). However, the nonlinear self-organizing dynamics in Novamente is also reminiscent of Walter Freeman style theories in which knowledge is represented in dynamical attractors. At the moment semantic-network-style models seem best able to deal with the human mind's treatment of abstract, linguistic or mathematical declarative knowledge, whereas attractor-style models seem better able to deal with knowledge directly related to perception and action, and also to issues such as binding and the creation of unified percepts and unified phenomenal selves from conceptually and physically disparate components. The Novamente approach unifies these two approaches by proposing a framework in which the importance levels of nodes and links in a semantic network display complex dynamics with attractors, and specific semantic network nodes in many cases "key" specific attractors.

Learning in Novamente and Humans

"Learning theory," in psychology, began in earnest with behaviorist studies of rote learning by pigeons, dogs and other animals. Commonalities between learning behavior in humans and these other animals were correctly observed, and some basic principles of learning were enounced: contiguity (spatiotemporally nearby things are associated), frequency (conditional probabilities are tabulated), contingency and blocking (e.g. after it's learned that a light predicts a shock, if a tone is introduced every time the light occurs, there is inhibition against learning that the tone predicts a shock). This kind of behaviorist learning has been shown to roughly obey probabilistic principles; e.g. the foraging behavior of wild birds automatically adapts itself to constitute a near-optimal solution to the relevant multi-armed bandit problem (Alexander, 1996).

Neurally, behaviorist-style learning ties in naturally with Hebbian learning – an old hypothesis which is increasingly substantiated by neurological research. We now know a fair bit about the chemical, genomic and proteomic dynamics underlying the processes of neuronal long-term potentiation that implement approximations of Hebb's basic learning rule (to increase the conductance of synapses that are repeatedly used).

However, the connection between neuron-level Hebbian learning and organism-level behavior learning is not quite so direct as many naively believed in the past. To achieve animal-level behaviorist learning via a Hebbian simulated neural network can require quite complex dynamics in a neural net of substantial size; see e.g. (Wilson, 2000) which uses a sophisticated Hebbian-style neural network model, implemented in terms of continuous-valued neurons and differential equations, to simulate behavior learning in Siamese fighting fish. This sort of work also indicates the amount of subtle tuning of Hebbian learning that is necessary to get it to give meaningful and useful results. It is interesting to contrast Wilson's work with the refinements of the Hebbian approach presented in Sutton and Barto's (1998) classic text on reinforcement learning. The former refines the basic Hebbian idea based on biological plausibility and quality of simulating biological learning behavior; the latter based on mathematical elegance and learning performance on computer science test problems. The clear message is that there are a lot of ways to tweak this basic learning mechanism, and the best way to tweak it for a given purpose is not at all obvious. In particular, we have little idea at present how Hebbian learning would be modified and adjusted to give rise to the type of intermediate-level learning we see in the human brain.

Clearly there is something powerful and valuable in the idea of Hebbian learning – something AGI designers should not ignore. However, from an AGI point of view, one is led to wonder whether it is sensible to implement Hebbian learning directly, or to try to figure out what higher-level learning dynamics neural Hebbian learning is giving rise to, and emulate *these* in software. In this regard the close connections between Hebbian learning and probability theory (Sutton and Barto, 1998) are highly interesting.

In (Goertzel, 2003), it is argued that Hebbian learning on the neuron level naturally gives rise to probabilistic reasoning on the level of neuronal clusters or sets of neuronal clusters. This suggests that if one hypothetically associates nodes in a semantic hypergraph with neuronal clusters or sets thereof, one can associate neural Hebbian learning with probabilistic inference on the probabilistic link weights in the hypergraph. I.e., it suggests that perhaps the reason Hebbian learning works so well in the brain is that it gives rise to approximate probabilistic inference on the level of the semantic hypergraph emergent from the brain. Of course, this is a speculative theory – cognitive neuroscience hasn't yet informed us how semantic-hypergraph-nodes are grounded in the brain, so we can't even ask detailed questions about how various sorts of inferences about their interrelationships are neurally grounded. But it is a speculation that accords with all available evidence, and is intuitively harmonious with existing knowledge in neuroscience, cognitive science and AI.

Hebbian learning is, on the face of it, a highly "local" learning method: it works by making incremental modifications to existing neurally based knowledge. In computer science local learning methods are valuable, but there's also a place for more global methods, which make big leaps to find answers far removed from existing knowledge. One of the most powerful global learning methods available is evolutionary learning, which takes many guises including genetic algorithms (Holland, 1992) and genetic programming (Koza, 1992), and roughly emulates the process of evolution by natural selection. It is known that the immune system adapts to new threats via a form of evolutionary learning, and Edelman (1987) has proposed that the brain does as well, evolving new "neuronal maps" – patterns of neural connection and activity spanning numerous neuronal clusters – that are highly "fit" in the sense of contributing usefully to system goals. He and his colleagues have run computer simulations showing that Hebbian-like neuronal dynamics, if properly tuned, can give rise to evolution-like dynamics on the neuronal map level ("neuronal group selection"). This is very interesting, and is something that could potentially be implemented in a Novamente context as well, where one could see emergent evolutionary learning arise from link-level probabilistic inference.

Novamente involves this sort of "emergent evolutionary map-level dynamics" phenomenon but, in a significant design decision, it also involves explicit evolutionary programming using MOSES, an algorithm related to but more efficient than genetic programming. Like the creation of specific Units for attentional focus and sensory modality focus, this is a case of explicitly choosing architectural features to match harmoniously with emergent phenomena.

So, Novamente's two key learning algorithms – probabilistic inference via PLN and evolutionary learning via PEL – may be seen to correspond to two key aspects of learning in the brain: Hebbian learning and neuronal group selection. And as with their neural correlates, these two learning algorithms fit naturally together – but for moderately different reasons. PEL and PLN fit together via their mutual reliance on probability theory, whereas Hebbian learning and neuronal group selection fit together because of their common reliance on the physiology and electrochemistry of neurons and other brain cells.

Conclusion

My main goal, in including this chapter in the book, has been to indicate the *sort of thing* that I think may be done to realize intelligence computationally. I think the Novamente AI design will work, in the sense of leading to advanced human-level-and-beyond AGI, but I think a lot of other things would probably work as well. To create an AGI a lot of hard technical problems must be solved right, but equally importantly, one must begin with a reasonably correct conceptual perspective. Via running through some aspects in which patternist philosophy can be used to interpret the Novamente design, I hope I have given some useful indications of what I think this correct perspective is.

Patternist philosophy teaches that mind is, in a strong sense, nothing special – all systems have patterns associated with them, since in the end everything (systems included) is made of patterns, and minds are "merely" the (fuzzy) sets of patterns associated with intelligent systems. But intelligence is just the ability to achieve complex (complexly patterned) goals in complex (highly patterned) environments. Experience is a different perspective than objective science, but, it's a perspective that may be attached to any system or any pattern regardless of intelligence. Of course, though, intelligent systems – and particular types of intelligent systems – will tend to have particular flavors of experience.

General intelligence requires a robust mechanism for the representation of general patterns, which gives compact representations to particular patterns of use to a particular system adapted to in a particular environment. It then requires learning algorithms for extrapolating new patterns from existing ones. General learning algorithms are needed, both incremental ones (like PLN) and global, speculative ones (like MOSES). Specialized learning algorithms are needed, in order to address frequently encountered resource-intensive learning problems in an efficient way (the specific heuristics for dealing with polygons mentioned above are an example of this). A flexible cognitive architecture is needed, able to incorporate ambient and goal-directed learning and to integrate various general and specialized learning mechanisms. Attention allocation and assignment of credit must be carried out effectively, which can be done if they are taken seriously and treated as difficult pattern recognition problems on par with other difficult pattern recognition problems. Finally, recognizing complex patterns right from the start is too hard — a mind must receive a sensible education that encourages the build-up of more and more complex patterns in a meaningful order; and one natural way to structure this educational process is to embed the mind in a body perceiving and acting in a world.

But all this is not to say that creating powerful AGI is straightforward in a pragmatic sense. The achievement of high levels of efficient intelligence is based on the use of algorithms that are particularly good at recognizing complex and appropriate patterns without using too many computational resources – and the discovery/creation of such algorithms is a difficult aspect of science. One point I have tried to make is that this discovery/creation process is best conceived *not* as a pure computer science or neuroscience problem, but rather at a problem living on the borderline between these technical disciplines and philosophy.

Chapter 16
Post-Embodied Mind

Human intelligence is hard to separate from human embodiment. We learn to think, as infants, largely in the course of learning to use our bodies. Even our most abstract goals can largely be viewed as sublimated versions of bodily goals. So many of our cognitive patterns and linguistic forms can be traced back to perception and action metaphors – from the inner visualizations with which many of us represent abstract knowledge, to the spatial relations implicit in pronouns like "above", "by" and "through."

A natural question, therefore, is: What about embodiment and *artificial* intelligence? More specifically, what about embodiment and artificial *general intelligence* (AGI)? It's clear that some cognitive capabilities currently considered under the loose heading of AI – for instance, chess-playing – can be carried out just fine by unembodied software programs. But these are "narrow AI" programs. What about software programs like Novamente is intended to ultimately be — programs that can autonomously learn about new problems and domains themselves, create new inventions and problem-solving strategies, reflect and spontaneously communicate. Suppose one accepts the "strong AI" claim that software programs, in principle, can achieve these things. Then the question still remains: Can these dreams be fulfilled by unembodied software programs, or do they require software programs embedded in sophisticated robot bodies, with sensors and actuators approaching, rivaling or exceeding the human body in diversity and precision?

Expert opinions on the importance of embodiment for AI have ranged all over the map. Some distinguished AI theorists and practitioners believe that embodiment is thoroughly unnecessary for AI; others argue that it's absolutely critical, and any AI system that's not embodied doesn't have a prayer of approaching true AGI.

In the Novamente project, I have opted for embodiment in a 3D simulation world (AGISim) and am considering later moving to embodiment in a physical robot body. However, I have made this choice not because I feel it is necessary for AGI's to be embodied — rather, it was largely a choice of convenience. Furthermore, there is no commitment to stick strictly to simulated embodiment as a way of getting knowledge into the system. If appropriate we are quite willing to feed various databases of information directly into Novamente's AtomTable. This is an approach I call *semi-embodiment*. This chapter is basically an extended description and defense of the semi-embodiment strategy, with some reference to Novamente but a focus on the more general issue.

I'll give a detailed analysis of exactly why embodiment is incredibly useful for AGI. But then, I'll argue that an exclusive and obsessive focus on embodiment is actually counterproductive for AGI. In fact, a lot of the technologies developed by the anti-embodiment crowd can be extremely useful for AGI – if they're integrated into an embodiment-encompassing framework. This leads me to the notion of *post-embodied mind* – intelligence that possesses a body (or more than one), but also possesses knowledge not derived from its body's sensorimotor capabilities in any way.

Post-embodied mind is not humanly natural, but I will argue that it's actually a more effective approach to intelligence than the one embodied in our brains. Furthermore, it may actually be the condition that we're evolving towards. A human with a special chip in their brains connecting them to the Internet and a massive network of interconnected databases – that would be a post-embodied human. It seems obvious that a post-embodied human would have a lot easier time learning and thinking than a plain old embodied human. Similarly, a post-embodied proto-AGI system is going to have a lot easier time getting trained to be really smart than a plain old embodied proto-AGI.

Embodiment: Necessary or Irrelevant for AGI?

Before launching into my own views, I'll give a quick overview of what others believe about embodiment and AI. I make no pretense to completeness – I'm just surveying some of the major perspectives, and giving a couple examples of well-known adherents to each.

First of all, among the "embodiment is unnecessary" crowd, there are several subspecies. The "knowledge encoding" crew believe that it's possible to write a long list of all the "common sense" facts about the world that every human learns through their embodiment, and simply supply this list to an AI, to fill up its brain with the basic knowledge it *would* get if it had a humanlike body. There are various opinions about how to write the list. Some, most notably Doug Lenat and the Cyc team (see www.cyc.com or Lenat, 1995), believe that it should be done in a formal mathematical language like predicate logic. Others, such as Chris McKinstry, founder of mindpixel.com, have posited that it should be done in a natural language like English.

On the other hand, some believe that explicit encoding of commonsense knowledge is unnecessary, and an AI system can pick up all it needs to know about the world through linguistic means – through conversations with humans. Jason Hutchens' HAL project, described at a-i.com (Graham-Rowe, 2001), was an example of this.

The most impassioned and articulate defense of the opposite position – that embodiment is critical for AI – was made by Hubert Dreyfus in his famous book "What Computers Can't Do" (1979). Dreyfus drew on Continental philosophy to argue that human intelligence is fundamentally situated in the body, and that considering intelligence as separated from embodied-ness is about as sensible as considering cognition as separate from memory. He considered the separation between intelligence and embodiment to be an artificial distinction made by a flawed research program, without basis in reality. In "The Embodied Mind" (1992), Francisco Varela and his collaborators deepened Dreyfus's critique by connecting it with Eastern philosophies of mind and being.

In terms of practical modern research, the embodiment-focused approach to AI is, not surprisingly, closely associated with robotics. Rodney Brooks (1999) and Hugo de Garis (De Garis and Korkin, 2002) are examples of well-known robotics researchers whose goal is to begin with simple robots carrying out simple embodied cognition, and then gradually make the robots and the corresponding cognitions more complex.

An alternative strategy, however, is to work with AI systems that control *simulated* bodies. At the present time, simulation worlds don't provide the kind of richness of sensation and action that the physical world does. On the other hand, they're a lot easier to play with: work on robot cognition has a dangerous habit of getting bogged down in the engineering and tuning of robotic systems, and never exploring the "cognitive" aspect all that fully. John Santore's (2003)

PhD work used Stuart Shapiro's SnEPs AI system to control an agent called "Crystal Cassie" in the "Crystal Space" simulated environment (an open-source environment created for 3D gaming). Similarly, John Laird (2002) has initiated a project using the SOAR AI system to control agents in 3D gaming environments.

As hinted above, my view is an intermediate one. As is usually the case in AI, the truth lies somewhere between the two extremes. Clearly, embodiment is not really *necessary* for AGI, in a theoretical sense. However, equally clearly, embodiment makes the task of teaching a proto-AGI system a heck of a lot easier – to such a great extent that trying to create a totally unembodied AGI would be a foolish thing. Hence the course I have taken with Novamente, which is somewhat similar to the approach prototyped by Santore with the SnEPs system.

I've just used the phrase "teaching a proto-AGI system." What I mean by this is as follows (and should be clear from the previous chapter). I divide the task of creating an AGI into two parts: first building the initial software system, and then teaching this "baby mind" how to think, reason, feel, etc. and filling its mind with knowledge. The proto-AGI is this initial software system.

This bipartite division of the AGI-creation problem is valid for almost all, but not all, approaches to AGI. In some approaches to AGI, the first part is almost trivial, because it's assumed that a very simple architecture can give rise to intelligence through repeated self-modification and self-analysis (Juergen Schmidhuber's (2004) OOPS system is an example of this). In most approaches to AGI, however, both of the stages are substantial and important, and the initial software system prior to learning has much of same the structure and dynamics that the system will have after a substantial amount of learning has taken place. Learning, at first, just provides content, refinement to the initially given components, and additional components existing within the initially given components. Eventually of course an AGI system may learn enough to rewrite all its source code and become something totally different than its creators intended. However, in most approaches to AGI, it is assumed that this kind of total self-modification will occur only after the system has become a highly sophisticated general intelligence using its original architecture. In this context, what I mean by a proto-AGI is a software system that has the overall software structures and dynamics needed to support general intelligence – but lacks the specific knowledge needed to operate in the world, and lacks the specific control structures needed to operate in various (practical and cognitive) situations.

How then can one turn one's proto-AGI into a genuine AGI? How does one turn an AI baby into an AI child? How critical is embodiment for this process? At the end of the previous chapter I discussed a quasi-Piagetan approach for training Novamente in the AGI-SIM world, and I think this sort of approach has a lot to recommend it. I will now give a few of the reasons why — and also some reasons why I believe this approach may viably be augmented by non-embodied information import into AGI systems' minds.

Knowledge Encoding and Natural Language

Firstly, I think it's clear that the knowledge encoding approach – taken on its own, without integration with other fundamentally different approaches — is intrinsically problematic. The Cyc knowledge base (the biggest one out there) has well over a million predicate logic relationships in it, each representing a piece of commonsense knowledge, but it's nowhere near complete enough to represent the knowledge in the mind of a small child (though it far exceeds the knowledge of most children or adults in various specialized areas such as geography, weights and measures, etc.). According to theoretical computer science, predicate logic has universal expressive power – so if the scope of human knowledge is finite (which is guaranteed, for instance, if the universe is computable or quantum computable), then it's possible in principle to encode all human knowledge in predicate logic. However, this "in principle" observation doesn't address the question of how to get all the knowledge out of human minds and into predicate logic. It may be that most of our commonsense knowledge is implicit, so that "we" (in the sense of our conscious linguistic minds) don't even know what we know. In that case, producing a commonsense knowledge base may not be possible until neuroscience can scan the human brain with tremendous accuracy, and use a highly refined theory of brain function to read the commonsense knowledge out of the scanned images. Most likely the scanning technology will be there in a few decades (see Broderick, 2002 and Kurzweil, 2000 for reasoned estimates), but when such a refined theory of brain function will emerge is anyone's guess.

On the other hand, the idea of teaching an AI system commonsense knowledge and cognition through the process of conversation is not fundamentally flawed, at least not in the obvious way that formal knowledge encoding is. Conversation may get across both explicit and implicit knowledge – often we say far more than we know we're saying. My suspicion, however, is that teaching a proto-AGI system via conversation alone will be *very, very painfully slow*.

One factor that could speed up the teaching-by-talking process is the judicious use of formally encoded knowledge. Databases built by formal knowledge encoding may be very helpful to AI's as they engage in conversations – providing explicit knowledge to help them anchor the implicit and explicit knowledge they obtain from their dialogues. For this purpose, it seems, commonsense knowledge bases built up using natural language will probably be more useful than knowledge bases like Cyc, created using formal logic. The reason is that a conversational AI system, to apply database knowledge in the context of conversation, must match conversationally-derived information with database information. If the database information is similar in form to the conversationally-derived information, then this matching problem will be relatively easy, and won't require complex, computationally expensive and hard-to-tune inferencing. On the other hand, if the database information is in a form very different from the conversationally-derived information, then the matching problem can be quite subtle.

This is an issue my colleagues and I have already encountered in practice, in our work with the Novamente AI system. We have a system component that maps natural language sentences into "Novamente nodes and links" – Novamente's internal knowledge representation. We can also load in knowledge from Cyc or other structured databases, translating their formal languages into Novamente nodes and links. However, the problem of matching natural language derived knowledge with database-derived knowledge turns out to be more irritating than one would initially think.

For instance, suppose one tells Novamente a simple sentence such as "Ben just gave Izabela the red ball." One would like Novamente to be able to infer that, shortly after that, Izabela had the red ball. In order to make that inference, Novamente requires the commonsense knowledge that "After X gives Y to Z, then shortly after that, Z has Y." This commonsense knowledge is implicit in Cyc, SUMO and some other available knowledge bases. However, extracting this knowledge from any of these databases in such a way as to provide value for language understanding requires a substantial amount of logical inference – nothing Novamente can't do, but more work than one would like one's AI system to need to do to figure out something so very simple. On the other hand, if one simply supplies Novamente with the relevant commonsense knowledge by telling it the sentence "After X gives Y to Z, then shortly after that, Z has Y", then Novamente will automatically map this sentence into nodes and links in a way that matches the way it maps the sentence "Be just gave Izabela the ball" into nodes and links. The commonsense background knowledge will match the conversational linguistic knowledge in form, making the matching between the two of them almost immediate, as it should be. For this reason, in the Novamente project, we would have much more use for *a Cyc-like compendium in simple English* than for Cyc itself.

Now, you might argue that a Cyc-like compendium in English is unnecessary, because all that information is implicitly there in the vast amount of text already present on the Internet. But, while this is true, it's also true that extracting commonsense knowledge from general texts requires sophisticated linguistic understanding – and sophisticated linguistic understanding, as much recent work in computational linguistics suggests, requires commonsense knowledge. So there is a chicken-and-egg problem here, which may potentially be resolved by formally encoding a small percentage of commonsense knowledge in simple English. The Novamente system already "understands" simple English (in the sense of successfully mapping it into internal nodes and links, using an interactive user interface that allows a human helper to correct its mistakes), so it can understand a simple English database. This commonsense knowledge can then help it to expand its knowledge of English, which enables it to better understand free text, which builds up its knowledge base, etc.

Symbol Grounding

One of the reasons often given for the necessity of embodiment in AI is the need for "symbol grounding" to aid with language understanding and cognition. The idea here is that, for a system without any sensors or actuators, the word "apple" is defined solely by its relations to other words, and to abstract entities such as database records. On the other hand, to a system with a body that can see, smell, hear, taste, grab and throw apples – or carry out some subset of these interactions with applies – the word "apple" will be associated with a host of nonlinguistically based patterns. The word "apple" may in other words be "grounded" (Harnad, 1990). This grounding, in essence, consists of a lot of commonsense facts about apples – not just abstract facts like "apples are usually red or green or yellow and "apples are edible fruits", but also a lot of more specific facts, such as information about the distribution of lengths and curvatures of apple stems, information about how bitter various apples have tasted at various stages of ripeness (and how similar their particular variety of bitterness has been to the bitterness of various other fruits), etc. In principle, all these facts could be encoded in a knowledge base using predicate logic or natural language, but it would be a heck of a pain – because most of these facts are implicit in the human mind, only articulated via great and unnatural effort. Most of these facts are things that emerge *naturally and pre-verbally* in any reasonably powerful pattern-recognition system supplied with a large number of examples of applies, and are cast into a verbal or otherwise easily-communicable form only awkwardly and difficultly.

Creating an AI system that can ground concrete terms like "apple" isn't such an amazingly difficult thing. At this stage, given the advanced state of various relevant sorts of narrow-AI research, it's basically an exercise in integrated systems design and the tuning of statistical learning: one has to take a perception component and a linguistics component and hook them up together, via some kind of cognition component that's able to recognize correlations between words and patterns among perceptions. There is nothing here that strains current technology. Research projects such as the *Robot Brain Project* (MacDorman et al, 2001) have sought to push in this direction.

A question here is: How can you tell if your AI system has correctly grounded "apple"? Being able to correctly distinguish apples from non-apples isn't really good enough. You need to test whether the system can draw useful conclusions about apples – and about other things, using apples as a metaphor – with roughly the same kind of fluidity as a human.

What's a bit subtler is the grounding of linguistic "glue words" like prepositions. Words like "through," "past", "over" and "near" embody subtle patterns of spatial relationship, which are hard for humans to articulate yet easy for us to manipulate implicitly. The semantics of these words is a classic case of implicit knowledge. Cyc is fairly weak in this area in spite of giving numerous different senses for each preposition and attaching predicate logic expressions to each. Cyc gives 14 different kinds of "in", each defined by different logic expressions, but the "essential meaning" of "in" seems not to be contained in any of these definitions exactly – it's a fuzzier kind of implicit knowledge. It's not that the "real definition" of "in" can't, in principle, be expressed in logical form – of course, it can be. But the real definition is a mixture of abstract formal concepts like those given in Cyc, with a load of specific examples of in-ness, familiar to us all from everyday life, and a load of perception-and-action-heavy patterns abstracted from these specific examples. Enumerating all these cases, for a system without a library of embodied experiences in its mind, would be extremely laborious. On the other hand, for a system with a library of embodied experiences, it's easier: there are thousands of examples of in-ness in storage, and plenty of patterns on different levels of abstraction emergent from these. A new in-ness situation may be understood by reference to these various examples and the low and mid-level patterns abstracted from them, not only by the very abstract patterns among these examples that are captured in formal, Cyc-style definitions of the senses of "in."

Note, I'm not arguing that Cyc-style abstract definitions don't exist in the human mind, nor that they shouldn't exist in an AI mind. The point is that these definitions, in an embodied mind, are only the top of a pyramid of more and more abstract patterns emergent from sensorimotor data. Sometimes questions are best resolved by reference to the top of the pyramid, and sometimes by some of the patterns lower down, even sometimes very-low-level patterns close to the level of the original sensorimotor data.

Reality as a Training Ground for Learning Cognitive Heuristics

And now we have reached another significant problem with unembodied AI. The network of patterns on various levels of abstraction, which emerges in an embodied system from sensorimotor data, is not only directly useful to a mind, it's also useful as a metaphor for thinking about non-directly-sensorimotor-related things – and as a *training ground* for teaching the system to manipulate hierarchical/heterarchical networks of patterns, aka dual networks.

Regarding the "metaphor" issue: It's no coincidence that so many of our prepositions are spatial metaphors – to a large extent, we've learned to think by learning to think about the spatiotemporal world all around us. But it's not only that we humans learn the meaning of "through" by reference to a database of experienced examples of through-ness, but also that we learn how to manipulate and interrelate abstract relationships like "through", "in" and "because" by learning to manipulate and interrelate their concrete instantiations in the perceived and enacted world. Without this "playground" for learning to interrelate such relationships, it will be hard for an AI system to get the knack. Not impossible, just *substantially more difficult*.

Cognition requires a host of heuristics, many which are too subtle and implicit for us to explicitly program into our AI systems – our AI systems have got to learn them. Reasoning about time and space, about other minds, about our own goals and actions, about plans over a short term versus a long time, and dozens of other examples – all these sorts of reasoning involve generic inferential methods, but they also involve case-specific inference control heuristics, which must be either learned by experience, or supplied by some very subtle sort of cognitive/computing science that is nowhere near existence yet. There is plenty of "AI" work in these areas at present, but it's all extremely crude compared to what human children do. Contemporary AI planning systems, for example (see e,g, Russell and Norvig, 2002), exceed most adults in planning the operations of a factory when all the factors and conditions involved are well-defined – but perform worse than the average young child in planning problems where the situations and entities involved are nebulously defined and/or poorly understood.

The example of planning hints at some other important points. Embodiment isn't just about getting a rich field of sensorimotor data in which to ground one's symbols, it's also about having a body to move around and generally control. The process of controlling one's body with simple multi-time-scale goals in mind, is an excellent training ground for learning how to achieve goals by controlling systems in the context of diverse data rich in patterns on multiple scales. Furthermore, the process of modeling one's own physical self is excellent practice for modeling one's own mental self – and modeling one's own mental self is critical, if one wishes to modify and improve oneself.

In none of these cases is embodiment *strictly necessary* for learning how to think – there are other approaches to learning planning, other approaches to learning introspection and self-control, other approaches to learning to balance patterns on multiple space and time scales, etc. The point is that embodiment provides a way of learning all these things, and more, in a richly interconnected way.

Human-like versus Non-human-like Embodiment

Another important point is that, since we are embodied, we have reasonable intuitions for the way various sorts of learning go in embodied systems. If a purely non-embodied learning mechanism is plausible – which I doubt, but I'm not sure about this – it would be more alien to us and hence will likely be harder for us to monitor and tune, and fill with our various human insights into how the world operates and how thinking may usefully be done. In fact, even an AGI embodied in a humanlike body would be fairly alien to us due to its different cognitive architecture. Given the current state of robotic technology, the most likely case to come about is an AGI embodied in a not-very-humanlike body – which will obviously have a psychology highly different from ours, but hopefully similar enough that a real psychological connection can be built between it and its human teachers.

One way to realize that it's a mistake to equate "embodiment" with "precisely human-style embodiment" is to think about cases like Helen Keller – a woman who was blind, deaf and had a very limited sense of smell. Mostly she experienced the world through touch, and her teacher "spoke" to her in language by tracing out the shapes of letters on her hand. Helen Keller was intelligent and articulate and, it would seem, basically developed a full understanding of the world. Of course, touch – together with kinesthesia and inner feelings like hunger, thirst, pain, sexuality and so forth – provides a lot of bits of data at every point in time. But there's nowhere near the complexity of the data coming in through vision. A key point to notice is that touch is the sense that truly provides the sense of embodiment. The skin divides the self from the other, and allows one to tell when one is touching some other object, to feel the nature of the environment in which one is immersed, etc. To embody an AI, I'd rather have a digital Helen Keller – a robot with a body with skin and sensations all around it – than a typical modern mobile robot with a camera eye and some bump sensors and sonar.

There are two aspects to "learning how to think" that are hard to fully distinguish from each other: learning how to think at all, versus learning how to think like a human. All humans share, to a certain extent, a common world-model. Parts of this world-model may be explicitly stored in the fetal brain, at an abstract level, but I doubt this is a very significant factor. More important, I suspect a lot of the world-model comes out of the relationships between our in-built human drives and feelings: hunger, thirst, pain, sexuality, movement, heat, cold, etc. Some of the rest comes out of the relations between our particular sensors and actuators – for instance, a mind with sonar would tend to build different world-models, as would a mind that controlled wheels or wings instead of legs. And some of the rest comes from the social relationships that ensue from our biology, e.g. parent-child relationships, sexual relationships, issues of trust and deceit and the signals that go along with these, etc. In order to share the human world-model in the sense that a human would, an AI really would need to have a human-like body. Helen Keller shared most of the basic human world-model, because she had a human body (and also note that her brain was wired to think in terms of the common human senses even though some of her sensory input devices didn't work – for instance, she still had her visual cortex to think with, though it couldn't be applied directly to visual sensations, it could still be applied to think about other tasks in a vision-like way).

Human patterns of cognition are closely tied to the human world-model, not just to the fact of embodiment. For this reason, an embodied AGI, unless it has a body modeled very closely after the human body, is not going to think like us – not even if its head is filled with Cyc-ishly or conversationally imparted "human common sense." It will take this human commonsense and integrate it with its own nonhumanly-embodied experience and come up with something different. The closer its embodiment is to ours, the more easily we'll be able to guide it as it learns how to think.

This notion of the "human world-model" is described by Eric Baum (2004), from a computational-learning-theory perspective, as *inductive bias*: a predisposition to recognize certain types of patterns in the world. For instance, it's fairly well demonstrated empirically that, although the human brain doesn't come with any specific linguistic *knowledge* at birth, humans do have some strong inbuilt biases to recognize certain kinds of linguistic patterns (Pinker 2000; Calvin and Bickerton, 2001). If we create an AGI system and give it neither linguistic knowledge nor linguistic-pattern-oriented biases, and then try to teach it human language, then we are placing it at a severe disadvantage relative to humans – even if we've given it an accurately human-like body to move around in! For this sort of reason, Baum argues that real AGI is a long way off: he believes that we need to wait for neuroscience to fully interpret the brain, so we can read the inductive biases out of it and program them into an AI system. On the other hand, I believe that we can work around the "inductive bias" problem via a messy, creative, integrative, post-embodied approach.

Post-Embodied AI

I believe there is a strong argument for creating embodied AI, as opposed to taking a wholly unembodied approach. However, I'm not a member of the "AI should be driven by robotics" camp. In fact I'm currently involved in a lot of work – on things like unembodied natural language processing – that most hard-core embodied-mind enthusiasts believe is worthless so far as AGI is concerned.

The practical fact is that, right now, given the technologies at our disposal, embodying AI systems is a pain. Furthermore, modern robot technology focuses on things like vision processing and arm movement, which are "far from the essence" where embodied intelligence is concerned – there's been disappointingly little work on robot skin, for example. For these reasons, there's a strong real-world motivation to go as far as one can possibly go with unembodied methods – and integrate embodiment only where avoiding it would lead to truly absurd inefficiencies. With this in mind, I like to distinguish two approaches to embodied AI: *pure embodiment*, in which the AI knows only what it learns via its body (though it may of course have *predispositions* toward certain types of knowledge programd into it from creation), and *impure embodiment*, in which the AI has one or more bodies but may also have other important sources of knowledge (generally, some sources present in it from creation, some interactively presenting it with new information as it learns). What I advocate is impure embodiment, or, to give it a more positively-spun name, *post-embodiment*.

Post-embodiment is a path of pragmatism – but interestingly (and this is one reason for the name), it's also perhaps a more forward-looking perspective than the classical embodied-mind view. We humans are embodied, and our minds exemplify the usefulness of embodiment for cognition – but, we humans are obviously far from optimal intelligent systems. More and more, as culture and technology advance, we voyage far beyond our bodies, and gain knowledge from remote regions of the outside world. It's not too long until we'll be able to graft computer chips into our brains and tap directly into quantitative, relational, textual and multimedia databases of various kinds – thus gaining vast amounts of information that have no direct connection to our physical bodies … information coming directly into our brains, not through our senses. Furthermore, virtual reality technology will in time mature and allow us to have the sense of occupying multiple "physical bodies" in multiple "places." In short, there is an argument to be made that human intelligence will become less embodied during the next decades as technology advances. And clearly this will be an enhancement rather than a regression of our intelligence.

Post-embodiment for humans is a futurist hypothesis, but for AGI's, it can be the initial condition. AGI's should have bodies, because embodiment provides a highly efficient medium for learning a lot of valuable things in an interconnected way. However, AGI's should also use every other means of knowledge-gathering and learning-stimulation at their disposal.

What does this mean in practice? The most sensible course for AGI seems to be threefold.

- *First*, one should embody one's proto-AGI system in one or more worlds so it can have rich sensorimotor experience. Simulated worlds may be useful for grounding of various relational concepts, and learning self-control and self-modeling. However, given the simplicity of current simulated worlds, real-world perceptual data seems to be very valuable as a complement, due to its greater richness, leading to a richer hierarchical/heterarchical network of emergent patterns.
- *Second*, one should converse with one's proto-AGI system regarding all sorts of things, including what it sees and does in the worlds it interacts with.
- *Third*, one should feed one's proto-AGI system as much prepared knowledge as possible — preferably in simple natural language, but also in the form of quantitative, relational and logical data, since this is more readily available than "simple natural language" based databases.

So for instance, if a post-embodied AI system is told "Ben gave Izabela the ball," then it may know that shortly afterwards, Izabela has the ball, for one of two reasons. It may know this because someone explicitly told it that "If X gives Y to Z, then shortly afterwards, Y has Z." Or it may know this because it observed a number of instances of real-world interaction associated with the word "give," and noticed this as a pattern in several cases with actual X's, Y's and Z's. The presence of formally-given commonsense knowledge may be useful, but this knowledge can only go so far and eventually experientially-based learning has got to take over. For instance, how does the system know that, after the giving is done, Ben no longer has the ball? This is not part of the intrinsic nature of giving, because if Ben gives Izabela a cold, he may still have the cold; and if a wife gives her husband a child, she still has the child. Again, it may be that the AI system has explicitly been given knowledge of the form "Balls are solid physical objects" and "If X gives Y to Z, and Y is a solid physical object, then shortly afterwards, Y has Z and X does not have Z." On the other hand, it may be that the system has learned this pattern from observing a number of situations with a number of real X's, Y's and Z's. In any particular example, one can retroactively say: "If the system had this and that pieces of commonsense knowledge fed explicitly into it, it could figure that out." But the problem is that there are far too many examples, and extracting all this implicit knowledge from the human mind is quite slow and difficult. The above formula regarding solid physical objects, though it may look fine-grained to the untutored, is in fact an overgeneralization, and needs to be replaced by more refined formulas – et cetera, ad nauseum.

The middle ground proposed here is well-illustrated by this simple example. A purely embodied approach doesn't give the system any explicit background knowledge, it lets the AI system learn things like the relation between giving and having on its own, at the same time as it learns linguistic syntax, learns to move its body, and so forth. A purely unembodied approach tries to encode every last rule in some formal way (using e.g. predicate logic or natural language). A post-embodied approach tries to have its cake and eat it too – for instance, in this example, this might involve explicitly encoding "If X gives Y to Z, then shortly afterwards, Y has Z" and letting the system figure out "If X gives Y to Z, and Y is a solid physical object, then shortly afterwards, Y has Z and X does not have Z" from experience. Note that the experiential learning problem here is significantly easier than in the purely-embodied case; because all the system needs to learn is a modification of a pattern it's already been told. Or, in the post-embodied approach, one might tell the system the latter, more complicated formula about embodiment and physical objects, and let it figure out the refinements on this based on experience. The important point isn't exactly where the fed-in knowledge leaves off and the experientially-learned knowledge begins, the point is that both kinds of knowledge exist and are allowed to synergetically feed off each other.

For another example, consider Cyc's 14 senses of the word "in" – which a careful analysis shows not to be sufficiently fine-grained to provide a thoroughly human-like understanding of the types of in-ness in the world. In a purely embodied approach, the AI would learn some rough equivalent of these senses – and also acquire a more refined understanding — through experience. In a purely unembodied approach, if the 14 senses are inadequate, they must be either improved by the creation of a larger database with 37 or 61 senses, or else improved via a series of conversations in which various examples of the many different shades of in-ness are discussed. In the post-embodied approach, we can begin with the notions of in-ness provided by resources like Cyc and online dictionaries, and allow them to be elaborated via insights into in-ness obtained by observing – and conversing with humans about — various examples of in-ness in the perceived, interacted-with world. It seems clear that, if it can be pulled off, the post-embodied approach provides a more efficient approach to learning – as it brings more information into the AGI's mind more quickly.

This approach is "impure" because it doesn't require that the proto-AGI learns everything from experience. Rather, it assumes that one's proto-AGI system has a flexible enough knowledge representation, and a powerful enough set of cognitive dynamics, that it is able to integrate diverse forms of knowledge (formal, experiential and linguistic) and use each form to assist its understanding of other forms. More specifically, for instance, it assumes that the perception module of one's proto-AGI system is set up so that the emergent patterns that form within it can be easily and naturally matched up (by the system's "inference" module, whatever form that may take) with the patterns that form in the system upon interpretation of linguistic utterances.

If there is a reasonable objection to this integrative approach, it's that it may be too hard to get the emergent perception-patterns to match up to the linguistic-understanding patterns. In a purely embodied mind, the linguistic-understanding patterns will have largely emerged from perception-patterns, therefore this matching is almost guaranteed. In a post-embodied mind, this matching is *not* guaranteed, and the system must be engineered carefully to ensure that it occurs. Otherwise the kind of linkage shown in the above example – where perception-based learning elaborates patterns explicitly told to the system – won't happen in practice, and the post-embodied approach doesn't work.

I believe the representations and dynamics in the Novamente system are adequate to this task – and, over the next few years, we'll find out if I'm right or not. At the moment we have a roughly half-complete proto-AGI Novamente, with the rest of the design worked out in moderate detail and awaiting implementation and testing. We're just beginning the process of feeding the system knowledge in natural language, and transforming knowledge from Cyc and other knowledge repositories into Novamente's knowledge representation. And a project to embody Novamente in a simulation world (based on CrystalSpace, the same technology used for Crystal Cassie) is well underway.

The perspective on embodiment I've outlined here is highly controversial when considered in the context of contemporary, conventional AI theory. Even so, though I consider it to be basically "just common sense." However, at least it's not sterile "common sense" – in the context of Novamente, it has led me to some fairly interesting insights regarding the useful fusion of linguistic, formal and perceptual knowledge – some of which I've hinted at above. So the ideas given here constitute a conceptual perspective that has proved useful for guiding practical work on at least one proto-AGI system.

But my main point right here isn't Novamente in particular – it's the general approach, the concept of *post-embodied AI*. Embodiment is important; it's incredibly useful as a learning mechanism for minds – but we shouldn't get carried away with it and assumes that all non-embodied mechanisms for getting information into AI's are Bad Things. Rather, in a sufficiently flexible AGI framework, it's possible to have embodiment and *also* utilize the approaches typically associated with anti-embodiment philosophies. This may have the effect of making both the pro- and anti-embodiment schools of thought unhappy with one's work. However, it may also provide the maximum rate of progress toward actually creating AGI.

Chapter 17
Causation[65]

with Jeff Pressing and Pei Wang

"Causation" is a concept created by human minds and cultures in order to provide pragmatic answers to various sorts of questions. The cognitive process called "causal inference" exists to address "why?" and "how?" questions, at both the conscious and unconscious levels. For instance, we use causal inference to address questions such as (to give a partial list):

1. Having observed A, what can we expect of B?
2. What is the most plausible explanation for the observation set C?
3. What will happen when we change our behavior from A to B?
4. How should we intervene to cause effect X?
5. Did/will Y cause Z?
6. What does our causal knowledge suggest about the effectiveness of a certain sequence of behaviors {A, B, C} in attaining goal G?

Humans have evolutionarily gained an intuitive sense of how to usefully answer these sorts of questions in various physical and social contexts, by combining various cognitive mechanisms to form a loose concept of "causality."

Though the causality concept is messy and heterogeneous, however, I do believe it has some simple concepts and processes at its core. One of my central contentions in this chapter is that the essence of human causal inference lies in two processes:

> *Predictive Implication* (e.g. is it the case that after A occurs, B usually occurs as well?),
> *Plausible Causal Mechanism* (is there a plausible mechanism via which occurrence of A might rationally be seen to lead to occurrence of B?).

---

[65] A substantial fraction of this chapter was originally written in 2000 or so by my collaborator-at-the-time Jeff Pressing, who tragically died in 2002 (in his early 50's). A few paragraphs were written by my colleague Pei Wang at around the same time. The text has been extensively edited since then and the current version has been shaped to represent my views and taste, though certain incorporating many of their ideas and turns of phrase. (In fact, I am pretty sure Jeff would agree with everything stated here, but there's no way to find out, at least not pre-Singularity. Pei agrees with the ideas presented modulo our ongoing disagreement about the role of probability theory in intelligence, as opposed to more heuristic forms of uncertain inference.)

Human causal inference, I propose, is a complex emergent phenomenon that is guided and triggered by these two core processes: finding predictive implications and finding plausible causal mechanisms.

The "plausible causal mechanism" aspect of causal inference is, I will argue, well-analyzed in terms of pattern theory. To oversimplify a little: a "plausible causal mechanism" underlying the hypothesis "A causes B" may be considered as a simple explanation of why A may cause B. In essence, such a simple explanation is a pattern in the set of pairs (instance of A, instance of B). Simplicity may be gauged here relative to a mind's body of background knowledge, which means that the plausibility of a posited mechanism is judged both by Occam's Razor and by inference based on the mind's knowledge about the world.

Considered generally, these ideas about causation are not entirely original; however, the specific formulations presented here are different from those given in the literature, and in my view are both more harmonious with pattern theory and more easily applicable in an AGI context. If you would like to explore the philosophical literature on causation, some reasonable reviews of 20'th century philosophical ideas on causation are given in (Bunge, 1959; French et al, 1984; Michotte, 1962) and more recently in (Woodward, 2003). For a summary of the modern probabilistic perspective on causation, see (Pearl, 2000) and (Dupre and Cartwright, 1988). Also, in an interesting psychological study, Leslie and Keeble (1987) showed that even human infants have a tendency toward causal attribution.

Next, another key contention of the chapter is that human causal inference is deeply wrapped up with free will. In essence, I suggest, when humans think "A causes B," part of what they are thinking is something like "If I were to spontaneously and freely choose to manipulate A, then this would incur changes in B." The "spontaneously and freely" here has to do with the difference between correlation and causation. It is often very difficult to tell whether a correlation between (for example) an earlier event-class A and a later event-class B is because "A causes B" or simply because there is a common cause of A and B. But if the human mind envisions itself freely enacting an event in class A, and then reasons that in this circumstance an event in class B would probably follow — then this is tantamount to reasoning that there is a causation rather than a correlation going on. To sum up, I suggest that in human psychology, "cause" exists largely as an unconscious metaphor for "I cause." The factors of predictive implication and plausible causal mechanism fit into this framework naturally.

Like free will, consciousness and many other aspects of human folk psychology, causality has both useful and misleading aspects. It is definitely useful for any mind to be able to learn implications of the form "If I do A, then B will happen" and "If I do A in context C, then B will happen." Direct mining of observed historical data is one mechanism by which minds can learn these predictive implications. And, establishing a plausible causal mechanism is another valuable means of increasing the strength of such implications – this is a case of abductive inference, which transfers truth-value to the proposed implication from other situations where the same or similar mechanisms have been observed. Human causal inference involves both of these factors, which are important ones.

However, human causal inference also involves systematic errors, some of which may be related to the illusory nature of the human concept of free will, and more generally to the rooting of "cause" in "I cause." We humans have a tendency to try to assign a single cause for an event when it would be more realistic to look for a complex combination of causes – this is probably because our causal inference procedures are tuned for choosing courses of action for ourselves, and we have more ability to enact single events than complex combinations of events. So we intrinsically look for causes of the type that we could enact ourselves, even in contexts that have nothing to do with our practical activity.

Finally, you may recall that causality played a key role in the above discussion of complex systems, in that nearly all of the "complex systems principles" proposed there relied essentially on the notion of causation. So, the discussion of complex systems given there is only complete after the ideas in this chapter are included within them. This is a somewhat subtle point, in that one of the main messages of this chapter is that causality is a messy and subjective concept; it follows from this and the above ideas that system complexity is also a messy and subjective concept. If ascription of causality depends on mindset and context, then ascription of system complexity and associated structures and dynamics does also. I realize this is not how people are accustomed to thinking about system complexity, but I do believe it is accurate.

Perspectives on Causation

There is a huge variance in the pragmatic concept of causality among various humans and human groups. On the group level, social psychologists have spoken of *causal palettes* that cultures or social groups use to provide ready candidates for causes used to explain events. In traditional societies, for instance gods, persons, animals, magical (e.g. healing) substances and laws and historical events are commonly invoked. On the other hand, Western societies have their own biases. For example, it has been well-established empirically that Western societies, being individual-centered rather than group-centered, are biased towards inferences of cause based on individual agency rather than situational forces (in social psychology, this is called the *fundamental attribution bias*).

The reason for this variation among human assessments of causality is the severe computational difficulty involved in assessing predictive implication and causal mechanism in real-world contexts. The human mind relies on an assemblage of dubiously-reliable heuristics, a fact which has caused the human intuitive concept of causality to become highly subtle and predictable. An AI like Novamente will use its own heuristics for the subprocesses of causal inference, different from the human heuristics, and will evolve its own style of causal ascription. AI's will doubtless have their own causal biases. In the long run, AI's are likely to exceed humans in their ability to modify their own causal and other biases; and we believe they will make significantly more accurate (though never perfect) causal inferences.

One way to think about causality is to introduce a distinction between *event, process* and *agent* oriented causation. This is not an entirely crisp categorization — these three types of causality overlap quite a lot, because processes can often be broken down into sequences of events, isolated events can be viewed as part of a process, most events and processes can be viewed as created by agents, and so forth. Nevertheless, in practical human psychology, these three aspects of causality are not neatly reducible to each other, and so it is worth considering them all as important perspectives.

A basic causal typology along these lines might be drawn as follows:

Cause type	Effect type	Example	Nature of Causal Explanation
Discrete event	Discrete event	Rock falls on a glass, breaking it	Physical laws & properties
Discrete event	Process	Signing of treaty sets up long-term occupation by military force	Mental models (in legal form) setting agreed-upon constraints on future behaviors of the parties
Process	Discrete	Accumulation of potential in nerve cell yields a spike output	Lawful axon properties, underpinned by physical law
Process	Process	Changes in interest rates affect equity markets	Quasi-lawful financial relationships emerging from the aggregate effects of the self-interested actions of many traders
Agent	Discrete Event	Person claps their hands, producing a sound	Physical properties of human hands, relevant motor programs, underpinned by agent intention
Agent	Process	Person balances a pole on its end	Agent's intention, mental models of the skill and physical powers to execute it

Next, in addition to these various types of causality, we must also consider the distinction between *local* and *distal* causes. For example, does the announcement by Greenspan cause the market to change, or is he just responding to changed economic conditions on interest rates, and they are the ultimate cause? Or, to take a further example, suppose a man named Bill drops a stone, breaking a car windshield. Do we want to blame (assign causal status to) Bill for dropping the stone that broke the car windshield, or his act of releasing the stone, or perhaps to the anger behind his action, or his childhood mistreatment by the owner of the car, or even the law of gravity pulling the rock down? . Most commonly we would cite Bill as the cause because he was a free agent. But different causal ascriptions will be optimal in different contexts: typically, childhood mistreatment would be a mitigating factor in legal proceedings in such a case.

Of course, in reality, the various types of causality are often mixed up, as are local and distal causes. Consider, for instance, a financial market such as the S&P 500. Here there is a variety of causes that can affect movements in this equity price index, including:

- changes in business conditions in significant component submarkets like technology (process),

- interest rate directions (process),
- government financial announcements (events),
- comments from significant individuals (events, agents),
- specific acts by operators with significant market share (events, agents),
- general attitudes of traders & investors (agents).

This complex process is evidently a mixture of event, process and agent-based causalities, at varying distances from the actual event in question. We need to identify all of these factors to get a complete "why" answer for the event. In general, there can be multiple causes between events; and we can sometimes even speak about circular or reciprocal causation.

Given all this particularity and complexity, it may seem remarkable that there are a few simple ideas underlying nearly all ascriptions of causality. But it would seem that there are; it would seem that the diversity of human psychological and cultural judgments of causality has some very simple processes at its core. Above I have described these simple processes in terms of "predictive implication" and "plausibility of causal mechanism. Another (slightly more precise) way to formulate the same observation is to observe that causal relationships usually presuppose the following conditions:

1. precedence (cause precedes effect) AND
2. A) regularity of relationship (between cause and effect) AND/OR
   B) plausibility of a causal mechanism between cause and effect.

The combination of points 1 and 2A leads to the pure "predictive implication" view of causality, which is basically the "classical logical" position on causality as enunciated by John Stuart Mill (in his classic work *A System of Logic*) and others. The addition of 2B incorporates more of the richness of actual human psychology as regards causal inference.

Mill's basic idea was to distinguish sufficient conditions as causally sufficient, and necessary conditions as causally necessary, when coupled with precedence or synchrony. This concept is most clearly defined in the case of discrete event causality. Suppose that we have two discrete events X and Y, and that X precedes Y. Then the classical logical position may be summarized as:

- If X is sufficient for Y then X may be a cause of Y, or it may be an effect of one or more common causes of effects X and Y.
- If X is necessary for Y then it is either a cause of Y or an enabling condition of Y or an effect of all existing causes of Y (discussion of the distinction between "causes" and "enabling conditions" will be given below).
- If X is necessary and sufficient for Y, X is the only cause of Y.

In Novamente this process is embodied by the formal and probabilistic process of creating software objects called PredictiveImplicationLinks. Although the discussion in this chapter is not really restricted to Novamente, I will use this lingo here for convenience's sake. A PredictiveImplicationLink pointing from X to Y is, basically, a symbol denoting that if X occurs, then it's likely that Y will occur afterwards. The "likely" here may be quantified by a probability value.

The probabilistic aspect of PredictiveImplicationLinks is important in many ways, for instance regarding the identification of temporal relationship. In many situations, if our knowledge of precise time relations is limited, we may be happy with having cause simply not clearly succeed effect. Hence, the required probabilistic degree of temporal precedence is weak (it can include simultaneity). And some senses of causality don't require strict causal precedence but merely simultaneity – for instance when we say that gravity causes objects to fall, no temporal precedence is implied.

Case 2A is most useful where we have a lot of data, as in examining process, event or agent relations that are instances of classes of many other similar cases. Here we can easily assess "regularity of relationship" by directly computing or simply inferring the strengths of the relevant PredictiveImplicationLink.

On the other hand, what if we want to assess causation between individual events or arising from unique agent actions? Then regularity cannot be used as a criterion. Causal inference is only possible then if we can identify a *credible mechanism of production* between cause and effect, Case 2B above. Such mechanisms typically follow physical or behavioral laws, and are based on abstractions from classes of prior experience. They often provide explanation on the basis of existing knowledge. This approach requires assessing the credibility of mechanisms, which is itself an inferential task, but a different one that the direct assessment of predictive implication. Among other subtleties involved, human minds have special heuristics for assessing the credibility of mechanisms involving conscious agents, as opposed to simple events.

Now, Case 2B does fit into the classical-logical approach to some extent. In cases one where has knowledge of the mechanism joining cause and effect, one can then use this knowledge of mechanism to infer a PredictiveImplicationLink joining cause and effect. However, the human concept of "causation" becomes subtle here, in that the strength of a human's belief in a given instance of causality is not necessarily identical to the strength of their belief in the associated PredictiveImplicationLink. Knowledge of mechanism is often valued highly in intuitive causality assessment, even more so than would be justified by the implications of knowledge of mechanism for "predictive implication between cause and effect."

In practice the human sense of causality involves contextually determined combinations of 2A and 2B – of assessments of predictive implication and plausibility of causal mechanism.

These contextually determined combinations are particularly subtle in the case of agent causality – the ascription of a conscious agent (human or AI) as the cause of an event? An agent is associated with a large set of events, and a large set of processes. Thus, in principle, agent causality reduces to event causality and process causality. But this is not in general a tractable approach to understanding agent causality. There are characteristic patterns to the sets of events and processes associated with agents, and these are obviously pertinent to causal inference – they help to direct the PredictiveImplicationLink search process, and even more strongly, the plausible causal mechanism search process. The identification of these patterns is a matter of experiential learning, by which a mind may learn cognitive schemata telling what kinds of implications and mechanisms to look for in what agent-causality-related contexts. In the case of human intelligence, this process seems to be largely a matter of implicit emergent intelligence, rather than any sort of low-level mechanism.

It's clear that human psychology contains many specialized heuristics regarding agent causality, and that these heuristics are to some extent culturally variable. Of all the types of causality, agent causality is the place where AI's are *least* likely to closely emulate human intuition. It seems more likely that AI's (such as Novamente) may emulate, in a general way, the manner in which humans dynamically create special heuristics for assessing agent causality — via the application of cognitive schemata learning to create special causal-inference schemata using basic event and process causal inference processes as ingredients.

Causes versus Enabling Conditions

A distinction is often made between causes and so-called *enabling conditions*. Enabling conditions satisfy condition 2A, and 1 above, so they are generally implicated in producing an effect; but they display no significant variation within the context considered pertinent. Example: oxygen is necessary to use a match to start a fire, but since it is normally always present, we usually ignore it as a cause, and it would be called an enabling condition. If it really is always present, we can ignore it in practice; the problem occurs when it is very often present but sometimes is not, as for example when new unforeseen conditions occur.

In Novamente, the difference between a cause and an enabling condition often has to do with the importance level of a PredictiveImplicationLink. The relationship:

```
PredictiveImplicationLink oxygen fire
```

may have a high strength, but it is going to have a very low importance unless one is dealing with some cases where there is insufficient oxygen available to light fires. This example indicates that the intuitive notion of causation has a somewhat subtle relationship with basic Novamente knowledge structures and learning dynamics.

From this perspective, enabling conditions are just an extreme case of distal causes. Whether local or distal causes are worth paying attention to in a given context, depends on all sort of particularities of the context, which will often well summed up by the importance of the given causal relations while the system is thinking about the context. These relationships may be captured by associative relationships between nodes representing contexts, and nodes representing causal relations, e.g:

```
HebbianLink
      space_travel
      PredictiveImplicationLink
            oxygen
            fire
```

Such a relationship means that when the space_travel context is important, the causal relation between oxygen and fire will likely also become important, reflecting the fact that in this context it is no longer a pragmatically irrelevant "enabling condition," but a very relevant cause. (When making a campfire on the moon, one may need to spray oxygen out of one's oxygen cannon to get the fire going.)

Distinguishing Correlation from Causation

Another well-known conceptual point regarding causality is that *correlation and causation are distinct*. To take the classic example, if a rooster regularly crows before dawn, we do not want to infer that he causes the sun to rise.

In general, if X appears to cause Y, it may be actually due to Z causing both X and Y, with Y appearing later than X. We can only be sure that this is not the case if we have a way to identify alternate causes and test them in comparison to the causes we think are real. Or, as in the rooster/dawn case, we may have background knowledge that makes the "X causes Y" scenario intrinsically implausible in terms of the existence of potential causal mechanisms.

In the case of roosters and dawn, clearly, we have both implication and temporal precedence. Hence there will be a PredictiveImplicationLink from "rooster crows" to "sun rises." But will the system conclude from this PredictiveImplicationLink that, if a rooster happens to crow at 1 AM, the sun is going to rise really early that morning – say, at 2 AM? How is this elementary error avoided (in the human mind or in an AI system)?

There are a couple answers here. One of them involves the contribution of the existence of potential mechanisms to the intuitive assessment of causality (in Novamente terms, the contribution of explicit assessment of the existence of causal mechanisms, to the strengths of CausalLinks). We will turn to that a little later. There is also a simpler, though perhaps less complete, answer, which involves nothing other than inference. This answer says: *The strength of this particular PredictiveImplicationLink may be set high by direct observation, but it will be drastically lowered by inference from more general background knowledge.*

The idea is as follows. If the system had never seen roosters crow except an hour before sunrise, and had never seen the sun rise except after rooster crowing, the posited causal relation might indeed be created. What would keep it from surviving for long would be some knowledge about the mechanisms underlying sunrise. If the system knows that the sun is very large and rooster crows are physically insignificant forces, then, this tells it that there are many possible contexts in which rooster crows would not precede the sun rising. Conjectural reasoning about these possible contexts leads to negative evidence in favor of the implication:

```
PredictiveImplicationLink rooster_crows sun_rises
```

which counterbalances – probably overwhelmingly – the positive evidence in favor of this link derived from empirical observation.

More concretely, one has the following pieces of evidence:

```
PredictiveImplicationLink <.00, .99>
     small_physical_force
     movement_of_large_object

PredictiveImplicationLink <.99,.99>
     rooster_crows
     small_physical_force

PredictiveImplicationLink <.99, .99>
     sun_rises
     movement_of_large_object

PredictiveImplicationLink <.00, .99>
     rooster_crows
     sun_rises
```

which must be merged with:

```
PredictiveImplicationLink rooster_crows sun_rises <1,c>
```

derived from direct observation. So it all comes down to: how much more confident is the system that a small force can't move a large object, than that rooster crows always precede the sunrise? How big is the parameter denoted c compared to the confidence we've arbitrarily set at .99?

Of course, for this illustrative example we've chosen only one of many general world-facts that contradicts the hypothesis that rooster crows cause the sunrise... in reality many, many such facts combine to effect this contradiction. This simple example just illustrates the general point that reasoning can invoke background knowledge to contradict the simplistic "correlation implies causation" conclusions that sometimes arise from direct empirical observation.

Contextual Variations in Causal Ascription

Next, let us return to the issue of contextual variations in habitual causal ascription. In the case of answering the question "what causes a stock market (S&P) movement?" as discussed above, there are several subtle things going on.

First of all, background knowledge is invoked to inferentially rule out correlations that are known not to be causal based on "common sense," knowledge of mechanism, etc. – as in the rooster/sunrise example.

Then, among the many different predictive implication links that point to the node indicating the stock market movement (and that have survived the previous phase), the implication that is most contextually relevant is selected. When a schoolteacher is speaking, the context is different than when Greenspan is speaking. This is partly a question of association spreading: some of the relevant temporal implication links are more associated with any given context than others. And it's partly a question of inferential user modeling: By ascribing a given cause, the speaker wants to affect the listener's mental model in a certain way, and can reason about how to do so, and choose a candidate cause to cite accordingly. All this is subtle in practice, but can be accomplished by the basic associational and inferential mechanisms provided within Novamente.

The full human notion of causality then emerges as a combination of contextual association, explicit inferential modeling of the context of discourse, and invocation of relevant background knowledge – but all within the basic framework of constructing predictive implication links. One may think of the initially, empirically identified predictive implications as "candidate causes", of the correlations that survive the invocation of background knowledge as "strong candidate causes," and of the correlations that are chosen by a system to explain an event in a social context as "culturally selected candidate causes." Note that some of the predictive implications ultimately selected may never have been observed in reality, but may be entirely inferred based on indirect means – although direct observation of predictive implication is of course a very valuable method for inferring causality.

In order for an AI system to adequately handle causality, it is not important that the system balance all the different factors involved in causal inference in exactly the same way that humans do. For the purpose of communicating with humans, it may be useful for an AI system to know how to ground the word "cause" in a purely human-like way in various contexts. But from the point of view of pure cognitive performance, it seems clear that there is nothing so special about the exact balance of factors that humans use for causal ascription – particularly since human individuals and cultures vary so much in their habitual ascriptions of causality. What is important is to one's AI easy ways of computing all the main factors underlying the human notion of causality. It will then combine these factors in its own way to form its own intuitive, context-variant measure of causality.

Plausible Causal Mechanisms

Now I'll dig a little deeper into the nature of causality – tying causality in with patternism, in the context of the logical of evaluating causality in an AI context.

I've discussed a variety of theoretical aspects of causal inference, but in terms of specific algorithms, so far the only concrete proposal I've discussed so far is the formation of PredictiveImplicationLinks via direct observation or inference. I've pointed out the crucial role played by the integration of diverse background knowledge into these links, and have also noted that the distinction between a cause and an enabling condition may be reflected in the importance of PredictiveImplicationLinks.

But, I've also noted that predictive implication is not enough to establish causality. The pragmatic, intuitive human notion of causality seems to involve an additional ingredient: establishment of a "plausible causal mechanism." This is a somewhat slippery idea but I will argue here that it can be meaningfully and generally formulated in patternist terms.

The simplest case is event causality. To find a causal mechanism between event-class A and event-class B, a mind may try to explicitly find hypothesized "mechanisms" S so that for each event Y in class B, there is some event X in class A so that X may plausibly be hypothesized to be produced by Y using mechanism S. For instance, suppose a stock trader hypothesizes that the event-class "decrease in the rate of increase of housing prices" may be causal for the event-class "decrease in the stock price of companies selling supplemental insurance policies." This predictive implication may be observed via analysis of historical financial data, but to establish a causal relation rather than just a predictive implication, a plausible causal mechanism must be established. This may be supplied, for instance, by hypothesizing a mechanism involving interest rates. When the housing market tightens, the Fed lowers interest rates to encourage home-buying, but lower interest rates make it harder for supplemental insurance companies to earn enough money from their savings to offset the cost of paying claims. This hypothetical mechanism makes the existence of a causal relationship more likely, according to the human intuitive notion of causation.

Similarly, for process causality, if we want to say that process A causes process B, then this is made much more plausible if we find a simple pattern via which "A at time t" can be used to approximately predict "B at time t+s". This is a dynamical pattern representing an ongoing causal mechanism determining B in terms of A. The simpler the pattern that we detect is, the more likely that it is real and important. For instance, the above example may be recast in terms of process causality. If one wants to claim that "insurance company stock prices" is causally affected by "housing prices", correlational analyses may be helpful, but it's also valuable to identify mechanisms by which month N's housing prices might affect month (N+1)'s insurance company stock prices. If these mechanisms are simple they may form patterns in these stock price histories according to the definition of pattern as "a representation as something simpler." The mechanism suggested above in the context of event causality may also be cited here.

In general, I suggest, a plausible causal mechanism underlying the hypothesis that A causes B is a *significant pattern* tying together specific instances in B with specific instances in A that precede them. When such a pattern is present, the likelihood that one is seeing a "real causal pattern" rather than just a correlation is more likely. Human judgments of causation involve a combination of assessments of predictive implication with assessments of plausible causal mechanism as judged by pattern recognition.

In Novamente formalism, we may say that if A causes B, there is likely to be a function ("schema") S, so that:

```
ImplicationLink
      AND
            InheritanceLink X A
            InheritanceLink Y B
      ExecutionLink S X Y
```

and so that S is *as simple as possible*. S is the "plausible causal mechanism" referred to in the above informal discussion. Of course, a schema S that is just a lookup table of ordered pairs (X,Y) satisfying *(Inheritance X A) AND (Inheritance Y B)*, isn't a meaningful "causal mechanism" mapping A into B. On the other hand, a small and compact schema that effectively summarizes a large set of ordered pairs (X,Y), gives a genuine reason to believe that *A* and *B* may be "really connected." In other words, S must be a pattern in the set of ordered pairs (X,Y). In this case, in Novamente, we may create a special link called a CausalLink pointing from A to B. Appendix A3 gives a more complete explanation of the formalism of CausalLinks, including an extension of the idea to deal with process causality explicitly. In the above financial example, the proposed mechanism concerning interest rates plays the role of the schema S.

In all, we see that causality is not a fundamental aspect of the universe, but is more of a pragmatic conceptual tool, which minds use to produce useful answers to certain questions that concern them. What is considered a cause of a given phenomenon at a given point in time depends on the context, and depends on what kinds of real or hypothesized actions are being considered as "causal effectors." Pattern-recognition is an important part of the causal inference process, and is in turn bolstered by causal inference which is in many cases a useful heuristic pattern recognition tool.

Chapter 18
Belief and Self Systems

Minds are collections of patterns, but they are not disorganized bags of patterns. They lack the perfect regular structure of crystals or quasicrystals – and they also lack the relatively simple emergent patterns of physical self-organizing systems like Benard cells or turbulent waterways — but they are structured nonetheless. They are structured in the manner of evolving, autopoietic systems. Among the most notable structures within minds are belief and self systems. A "belief system" is a network of patterns embodying propositions about some aspect of the world or the mind, which all interlock in the sense of autopoietic systems: the beliefs are about the world/mind, but they are also about each other, and if one removes a belief from the system it can easily be regenerated by the others in the system. A "self system" is a belief system whose subject is specifically the intelligent system that contains the belief system – these are among the most important, subtle, and complex of belief systems.

Belief systems are the source of a fairly large percentage of human irrationality. The self-creating aspect of belief systems often makes it difficult for humans to adapt their minds to new information, because the old information is part of an autopoietic network of ideas, and is continually regenerated by other patterns that had coadapted with it. Incorporating a single new fact may require a thoroughgoing alteration of the autopoietic attractor of a large part of the mind, and not all minds are structured so as to easily permit this.

On the other hand, in science-fictional treatments of AI, it is often implied that digital minds will be perfectly rational. But the patternist theory of mind suggests very strongly that this will not be even approximately the case, at least not until the very far future[66] when ultrapowerful AGI systems have essentially unlimited computing power at their disposal. No doubt, it will be possible for AI systems to be significantly more rational than humans are (not such a tall order!). But to achieve perfect logical consistency does not seem computationally tractable.[67] AI's whose intelligence is merely a few order of magnitude greater than human-level, will almost surely also be stuck with self-reinforcing belief systems that impede clear thought – though certainly not to the same level as us humans.

On the other hand, there are some important ways in which we believe nearly all AGI systems will be significantly more rational than humans. In particular, "belief systems" (tightly

---

[66] Potentially, this "very far future" could occur within the next century or even the next few decades, if a technological Singularity occurs; but I still count this as very far from the current state of things, conceptually if not temporally!

[67] To see this intractability, assume for the sake of argument that all beliefs are propositions in Boolean logic (the reality is more complex, not less so). Consider a significantly cross-referential belief system S — one in which most beliefs refer to a number of other beliefs. Then, the problem of determining whether a new belief is logically consistent with the belief system S is at least as hard as the well-known problem of "Boolean Satisfiability," or SAT (Garey and Johnson, 1979).

interconnected networks of beliefs), in the human mind, have a strong tendency to perpetuate themselves. This same tendency will exist in AI systems, but without as much power.

Self-Perpetuating Belief Systems and Nonlinear-Dynamical Attention Control

The dynamic by which autopoietic belief systems cause irrational thought and behavior is a subtle and important one, and it's closely related to the phenomenon of "rationalization" (I say this realizing, of course, that the natural language term "rationalization" does not have quite as crisp of a meaning as my interpretation of the term). In AI language, rationalization is a particular dynamic of inference control. It does not necessarily involve incorrect inferences (though it sometimes may); what it involves is the mind being very calculatedly careful about which inferences it chooses to make, because there are some conclusions that it *emotionally* wants to see. In other words: Suppose a certain conclusion C is strongly emotionally desired. There may be many logical inferences that lead to this conclusion, and many that weigh against this conclusion. Rationalization occurs when the mind focuses its attention almost exclusively on those inferences that lead to the desired conclusion C, ignoring the other ones.

There is a strong relationship between rationalization and circular inference. The effect of circular inference with large cycles is often to encourage the creation of "concept clusters" – sets of concepts, densely connected by relationships, each one of which is supported by a number of other relationships in the cluster. This kind of dynamical phenomenon can be favorable, in that it can encourage the recognition of actual clusters in the structure of the world. It can also help with the formation of the emergent dual network archetypal structure. However, such tightly-connected concept clusters can also be problematic, due to the phenomenon of inhibition. Often concepts in the (human or Novamente) mind display inhibitory relationships – two concepts or concept-sets representing mutually exclusive ways of looking at the world will inhibit each other. This means that a tightly connected concept cluster, once it begins to achieve prominence in the mind, can effectively shut out other sets of concepts from getting any attention at all. And this can result in rationalization, because inference operations, when trying to justify a concept, will tend to focus on inferences involving entities that are being significantly attended by the mind. Add to this the fact that tightly connected concept clusters *that provide emotional satisfaction* may be *extremely strongly attended* by the mind, and you have a recipe for irrational thinking:

1. Circular reasoning with large cycles causes coherent concept clusters to form
2. Some of these clusters involve strong emotions, which therefore become highly prominent over time
3. Reasoning tends to bolster relationships in these concept clusters because it tends to choose inference steps using premises that are prominent in the mind, and items contradictory to those in the prominent concept clusters are unlikely to be prominent, due to inhibition.

Rationalization is an example of what, in *Chaotic Logic*, I have called the "immunological behavior of belief systems." In effect, a belief system can sometimes treat contradictory ideas the same way a body treats foreign cells. In Novamente, an Atom-cluster representing a powerful belief system can build low-strength, high-importance AssociativeLinks to Atoms representing contradictory evidence, which has the result of inhibiting the contradictory evidence from getting any attention, so long as the belief-system cluster is highly active.

How is this kind of problem prevented? It can't be prevented entirely. But in an AI context, its dangerous effects can be palliated somewhat, by appropriate design of the "attention allocation" mechanism that determines which items in the mind get more attention at which times. In Novamente, the Long-Term Importance component of the AttentionValue associated with each Atom acts to mitigate against rationalization, and other pathologies associated with overly persistent belief systems. Suppose there is a tightly-connected cluster of Atoms, which has very high importance. Suppose that it's closely related to a powerful emotion, which constantly spreads it activation (factor 2 above). Even so, if this tightly-connected cluster is not generating new information, then it will gradually become less and less important, so that other Atoms can become more prominent, and be chosen more often by inference MindAgents and other cognitive processes.

Essentially, what the LTI quantity does is to wire in "learning and creating new knowledge" as a system-level goal of Novamente. There may well be explicit goals of this nature in a Novamente system, expressed in GoalNodes, which is used to help guide higher-level actions of the system. But the wiring-in of LTI at a low level ensures that these goals are adhered to throughout every single system process, in a much more efficient way than would be possible through explicit GoalNode activity. In a highly advanced, fully self-modifying Novamente, the GoalNode concerned with learning and knowledge creation might modify the attention allocation mechanism so as to make the LTI term or its future analogue operate yet more effectively.

Consider a simple example of emotionally-motivated circular inference. Joe has three beliefs regarding his girlfriend:

A: She is beautiful
B: I love her
C: She loves me

Each of these beliefs helps to *produce* the others. He loves her, in part, because she is beautiful. He believes in her love, in part, because *he* loves *her*. He finds her beautiful, in part, because of their mutual love relationship. All of these concepts are complex, multidimensional ones, represented by complex maps, not by single, crisp formulas.

Now, consider another example. When Joe looks at a chair obscured by shadows, he believes that the legs are there because he believes that the seat is there, and he believe that the seat is there because he believes that the legs are there. Thus sometimes he may perceive a chair where there is no chair. And thus, other times, he can perceive a chair that really *is* there far more effectively than a standard image-processing computer program equipped with a high-precision camera eye. The standard computer program understands a chair as a list of properties, related according to Boolean logic. But he understands a chair as a collection of *processes,* mutually activating one another.

The legs of a chair are defined partly by their *relation* with the seat of a chair. The seat of a chair is defined largely by its *relation* with the back and the legs. The back is defined partly by its relation to the legs and the seat. Each part of the chair is defined by a fuzzy set of patterns, some of which are patterns involving the *other parts* of the chair. The recognition of the chair involves the recognition of low-level patterns, then middle-level patterns among these low-level patterns, then higher-level patterns among these. And all these patterns are organized associatively, so that when one sees a certain pattern corresponding to a *folding* chair, other folding- chair-associated patterns become activated; or when one sees a certain pattern corresponding to an *armchair*, other armchair- associated patterns become activated. But, on top of these dual network dynamics, *some* patterns inspire one another, boosting one another beyond their "natural" state of activation. This circular action is the work of the self-generating nature of mind — and, we suggest, it is *necessary* for all aspects of intelligent perception, action and thought.

Each of these examples involves a fairly small set of beliefs, which reinforce each other in the mind – mutually and circularly, perhaps a little more than logic would dictate, based on the available evidence. In the first case, the circular reasoning is nudged along by strong emotions: the goal of achieving personal happiness is so important that inference is pushed along paths that lead to beliefs that increase this goal. In the second case, the circular reasoning is nudged along by the deep-seated desire to recognize *something* in the environment. In both cases, a lot of rationalization is occurring. The parameters of inference control are being adapted to enable inference to arrive at emotionally-satisfying, goal-satisfying conclusions. This is a dangerous process, which would be squelched by an effective attention-allocation process, but it can be a valuable one too. There can be some cases in which allowing more circular reasoning better helps the system to achieve its goals (though this can never be taken to an extreme; if it is, then the system will just fill up with meaningless "knowledge" derived through small inference cycles).

Good and Bad Self-Perpetuating Belief Systems

From an AI perspective, it is not necessarily desirable to design an attention allocation mechanism that will stomp out the self-reinforcing "coherence" of belief systems entirely. In the absence of near-infinite computational resources, the autopoietic aspect of belief systems serves a valuable role. What one wants an AI system's attention allocation mechanism to do is to stomp out belief systems that are no longer producing significantly useful ideas. In the next few pages I will run through some example belief systems, pointing out their utility or lack thereof and how this relates to the way they would be treated within Novamente.

An interesting feature of belief systems is that, very often, a belief that makes no sense on its own, is highly valuable as part of a system. This is tied in with the fact that, sometimes, a belief system that is held together partly by "irrational" means can extremely useful at producing pragmatically important logical conclusions.

To understand these phenomena, recall the old story about the farmer who hires an applied mathematician to help him optimize his productivity. The first part of the task involves figuring out how many cows can fit in certain parts of the barnyard in a given period of time. The mathematician thinks about the problem and begins "First, let us assume a spherical cow...," — and the farmer fires him.

The farmer thinks the mathematician is off his rocker, but all the mathematician is doing is applying an element of his belief system – a system which has a huge amount of overall coherence and productivity. This particular element, though absurd in the context of the farmer's belief system, is often quite effective when interpreted in terms of the belief system of modern science. The notion seems ridiculous "in itself", but it was invented for a certain purpose and it serves this purpose well. It is right that this belief (that cows may be for some purposes reasonably assumed spherical) is important, because it's part of a coherent system of beliefs that is useful. There may even be some circular reasoning in this system: the belief that objects may be assumed spherical may reinforce some abstract beliefs about the nature of objects, and vice versa. The point is that the overall system of beliefs is useful; it continues to generate valuable patterns, and valuable predictions (about things like how many cows will fit in a certain pasture!).

On the other hand, a concrete example of an unproductive, destructive belief system – the actual belief system of a specific paranoid schizophrenic woman, is given in *Chaotic Logic* and discussed there in some detail. The belief system is as follows:

*"Jane" almost never eats because she believes that "all her food" has been poisoned. She has a history of bulimia, and she has lost twenty-five pounds in the last month and a half; she is now 5'1" and eighty five pounds. She believes that any food she buys in a store or a restaurant, or receives at the home of a friend, has been poisoned; and when asked who is doing the poisoning, she generally either doesn't answer or says, accusingly, "**You** know!" She has recurrent leg pains, which she ascribes to food poisoning.*

Furthermore, she believes that the same people who are poisoning her food are following her everywhere she goes, even across distances of thousands of miles. When asked how she can tell that people are following her, she either says "I'm not stupid!" or explains that they give her subtle hints such as wearing the same color clothing as her. When she sees someone wearing the same color clothing as she is, she often assumes the person is a "follower," and sometimes confronts the person angrily. She has recently had a number of serious problems with the administration of the college that she attends, and she believes that this was due to the influence of the same people who are poisoning her food and following her.

To give a partial list, she believes that this conspiracy involves: 1) a self-help group that she joined several years ago, when attending a college in a different part of the country, for help with her eating problems; 2) professors at this school, from which she was suspended, and which she subsequently left; 3) one of her good friends from high school.

Her belief system is impressively resistant to test. If you suggest that perhaps food makes her feel ill because her long-term and short-term eating problems have altered her digestive system for the worse, she concludes that you must be either stupid or part of the conspiracy. If you remind her that five years ago doctors warned her that her leg problem would get worse unless she stopped running and otherwise putting extreme pressure on it, and suggest that perhaps her leg would be better if she stopped working as a dancer, she concludes that you must be either stupid or part of the conspiracy. If you suggest that her problems at school may have partly been due to the fact that she was convinced that people were conspiring against her, and consequently acted toward them in a hostile manner — she concludes that you must be either stupid or part of the conspiracy.

In *Chaotic Logic* I gave a formal model of Jane's belief system is given, involving the following set of beliefs:

```
C₀: There is a group conspiring against me
C₁: My food is poisoned by the conspiracy
C₂: My friends and co-workers are part of the conspiracy
C₃: My leg pain is caused by the conspiracy
C₄: My food tastes bad
C₅: My friends and co-workers are being unpleasant to me
C₆: My leg is in extreme pain
```

Of course, each of the brief statements listed above next to the labels C_i is only a shorthand way of referring to what is in reality a diverse collection of ideas and events (a "concept map" in Novamente lingo). For instance, the statement "my co-workers are being unpleasant to me" is shorthand for a conglomeration of memories of unpleasantness. Different processes related to C_5 may focus on different specific memories.

The interrelationships between these beliefs are not so difficult to unravel. Obviously, the belief C_0 is a pattern among the three beliefs which follow it. So, fairly simple inference processes may lead to the creation of C_0, as a response to the simultaneous activity of C_1, C_2 and C_3, or some binary subset thereof.

Next, what about C_1, C_2 and C_3? These three may be handled similarly; we will explicitly discuss only the first. Taking C_0, C_2, C_3 and C_4 as given, C_1 is a fairly natural inference. In an actively reasoning mind, C_1 may be produced by the cooperative action of these four beliefs. In fact, it is a *pattern* in a large set of individual events associated with these beliefs, because it *simplifies* the long list of events that are summarized in the simple statement "My food is being poisoned." This statement encapsulates a large number of different instances of apparent food poisoning, each with its own list of plausible explanations. Given that the concept of a conspiracy is already there, the attribution of the poisoning to the conspiracy provides a tremendous simplification; instead of a list of hypotheses regarding who did what, there is only the single explanation " *They* did it." Note that for someone without a bent toward conspiracy theories (without a strong C_0), the cost of supplying the concept "conspiracy" would sufficiently great that C_1 would *not* be a pattern in a handful of cases of apparent food poisoning. But for Jane, relative to the state of her mind (which includes C_1 and C_0), C_1 powerfully simplifies C_4.

Now let us turn to the last three belief-processes. What about C_5, the belief that her co-workers are acting unpleasantly toward her? First of all, it is plain that the belief C_2 works to produce the belief C_5. If one believes that one's co-workers are conspiring against one, one is far more likely to interpret their behavior as being unpleasant.

And furthermore, *given* C_2, the more unpleasant her co-workers are, the *simpler* the form C_2 can take. If the co-workers are acting pleasant, then C_2 has the task of explaining how this pleasantry is actually false, and is a form of conspiracy. But if the co-workers are acting unpleasant, then C_2 can be vastly simpler. So, in this sense, it may be said that C_5 is a pattern in C_2.

By similar reasoning, it may be seen that C_4 and C_6 are both *produced by* other beliefs in the list, and *patterns in or among* other beliefs in the list.

The arguments of the past few paragraphs (drawn from *Chaotic Logic*) are somewhat reminiscent of R.D. Laing's *Knots* (1972), which describes various self-perpetuating interpersonal and intrapersonal dynamics. Some of Laing's "knots" have been cast in mathematical form by Francisco Varela (1978). However, Laing's "knots" rather glibly treat self-referential dynamics in terms of prepositional logic, which as we have seen is of dubious psychological value. The present treatment draws on a far more carefully refined model of the mind.

It follows from the above arguments that Jane's conspiratorial belief system is in fact a "mind attractor." If Jane were a Novamente system, this belief system would be a "concept map" (however, there is reason to believe Novamentes will never be susceptible to this sort of delusion). It's not a fixed-point mind attractor, because the specific contents of the belief-processes C_i are constantly shifting. So the belief system is not exactly *fixed*: it is subject to change, but only within certain narrow bounds. It is a *strange attractor* for mind dynamics.

If Jane's brain allocated its attention according to Novamente's Importance Updating Function, how might this kind of problem be avoided? This is not a trivial question, because, in fact, Jane's belief system is far from sterile. It is constantly creating new ideas, new hypotheses. Each day it generates new hypotheses, explaining each thing that happens to her in terms of the conspiracy against her. However, there are some glaring problems with these explanations.

For one thing, her explanations display two disturbing properties:

1. **"Conservatism":** The explanations provided are incredibly stereotyped; no matter what the phenomenon, the explanation is always "the conspirators did it." New events never require new sorts of explanations.

2. **"Irrelevance":** The theories she creates to explain an event often don't have that much to do with the specific structure of the event. Formally, the collection of patterns that emerge between the event and the explanation is often surprisingly small.

What's wrong with this kind of conservatism and irrelevance? In some circumstances, it could be correct; there could be domains in which the same old explanation was actually correct for any phenomenon. But in the case of Jane's life, and in most real-world situations, this is not the case; the world is more dynamic than that.

The practical result of this conservatism and irrelevance – and the real problem with Jane's belief system – is that it has negligible predictive power. It can generate explanations for almost anything after the fact, but before the fact, it has almost no power to make correct predictions about what's going to happen. Similarly, if a friend tells her about some events, Jane can provide an explanation for these events within her belief system, quite readily. But she cannot predict, using her beliefs, what is going to happen to her friend – not even with a so-so degree of accuracy. The problem isn't just that her belief system doesn't predict "objective reality" – the problem is that it doesn't even predict the events in Jane's own *subjective reality*, it only provides them with *post hoc* explanations. Because explanations generated in such a conservative and irrelevant way cannot possibly predict a reality as complex as the one we live in, and even a highly delusional person cannot help but create a subjective reality mirroring much of the complexity of the "real world" they live in.

Next, for a rather different example of a self-reinforcing belief system, let us look back in history, and consider Galileo's belief that what one sees when one points a telescope out into space is actually there. This seems quite reasonable from today's perspective. After all, it is easy to check that when one points a telescope toward an earthbound object, what one sees is indeed there. But we are accustomed to the Newtonian insight that the same natural laws apply to the heavens and the earth; and the common intuition of Galileo's time was quite the opposite. Galileo was going against the commonsense logic of his time.

It was said at the time that he was making hypotheses that could not possibly be proven, merely dealing in speculation. Now we see that this objection is largely unfounded; we have measured the heavens with radio waves, we have sent men and robotic probes to nearby heavenly bodies, and the results agree with what our telescopes report. But to the common sense of Galileo's time, the idea of sending men into space was no less preposterous than the notion of building a time machine; no less ridiculous than the delusions of a paranoiac.

Furthermore, it is now known that Galileo's maps of the moon were drastically incorrect; so it is not exactly true that what he saw through his primitive telescopes was actually there!

Galileo argued that the telescope gave a correct view of space because it gave a correct view of earth; however, others argued that this analogy was incorrect, saying "when the telescope is pointed toward earth, everyone who looks through it saw the same thing; but when it's pointed toward space, we often see different things."

Now we know enough about lenses and the psychology of perception to make educated guesses as to the possible causes of this phenomenon, reported by many of those who looked through Galileo's telescopes. But at the time, the only arguments Galileo could offer were of the form "There must be something funny going on either in your eye or in this particular lens, because what is seen through the telescope in the absence of extraneous interference is indeed truly, objectively there." In a way, he reasoned dogmatically and ideologically rather than empirically.

How is Galileo's belief system intrinsically different from the paranoid belief system discussed above? Both ignore common sense and the results of tests, and both are founded on "wild" analogies. Was Galileo's train of thought just as crazy a speculation as Jane's, the only difference being that Galileo was lucky enough to be "right"? Clearly this is an overstatement – at very least, we many say that, whereas both of them blatantly ignored common logic in order to pursue their intuitions, Galileo's intuition was better than Jane's. But what made his intuition better?

Whereas Jane's belief system is conservative and irrelevant, Galileo's belief system was *productive* and *predictive*.

It was productive in that, once you assume that what you see through the telescope is really out there, you can look at all the different stars and planets and draw detailed maps; you can compare what you see through different telescopes; you can construct detailed theories as to why you see what you see. True, if it's *not* really out there then you're just constructing an elaborate network of theory and experiment about the workings of a particular gadget. But at least the assumption leads to a pursuit of some complexity: it produces *new pattern*.

It was predictive to the extent that, when Galileo looked at the moon and saw something there, someone else could look at it and see the same thing. Furthermore, telescopic observations about heavenly bodies allow one to make vastly more accurate predictions about their motions. The quality of the predictions Galileo obtained with his crude instruments was not so good, which is why he had trouble convincing his colleagues – but it was better than nothing.

Of course, the reason Galileo couldn't make great predictions based on his telescopic observations was that the technology was so young. This brings up a more general point: when a belief system is in the early stages of development, it should generally be given more slack, in terms of productivity and predictiveness. But once it's been around for a while, if it still isn't any good, it should be treated far more harshly. In human psychology, however, exactly the opposite dynamic often seems to happen: once a belief system has been around for a while, the mind tends to get used to it and has trouble eliminating it, whereas new belief systems have a hard time getting started. In Novamente, this kind of "slack for new ideas" can be introduced on two different levels:

- By giving some newly created concepts and relationships a high initial importance, so they get a chance to be thought about extensively, and prove themselves or not;
- By giving some newly created concepts and relationships a moderately high initial long-term importance, so they won't be deleted from memory before they've had a chance to integrate themselves into the mind and prove themselves worthy of retention.

An evocative term for a belief system that is both productive and predictive is: a "dialogical" system[68][69]. A dialogical system is one that engages in a dialogue with its context. The opposite of a dialogical system is a "monological" system, a belief system which speaks only to itself, ignoring its context in all but the shallowest respects. A system which is in an early stage of development, but will eventually be dialogical, may be called "predialogical." Of course, the subtlety here is that, in its early stage of development, a predialogical system may be indistinguishable from a monological one. Pre-dialogicality, almost by definition, can be established only in retrospect. Human minds and societies deal with the problem of distinguishing monologicality from predialogicality the same way they deal with everything else — by induction and analogy, by making educated guesses based on what they've seen in the past. And, of course, these analogies draw on certain belief systems, thus completing the circle and destroying any hope of gleaning a truly objective theory of "justification."

The Self-Creating Self

A particularly important example of a self-perpetuating belief system is — the self itself.

Psychology provides many different theories of the self. One of the clearest and simplest is the "synthetic personality theory" proposed by Seymour Epstein (1984). Epstein argues that the self is a *theory*. This is a particularly useful perspective for AI because theorization is something with which AI researchers have often been concerned.

Epstein's personality theory paints a refreshingly simple picture of the mind:

> *[T]he human mind is so constituted that it tends to organize experience into conceptual systems. Human brains make connections between events, and, having made connections, they connect the connections, and so on, until they have developed an organized system of higher- and lower-order constructs that is both differentiated and integrated.*

If we replace "connections" with "patterns" (remember the Metapattern, "it is pattern which connects") then we see that Epstein's theory of self is highly patternist indeed.

And, in addition to making connections between events, human brains have centers of pleasure and pain. The entire history of research on learning indicates that human and other higher-order animals are motivated to behave in a manner that brings pleasure and avoids pain. The human being thus has an interesting task cut out simply because of his or her biological structure: it is to construct a conceptual system in such a manner as to account for reality in a way that will produce the most favorable pleasure/pain ratio over the foreseeable future. This is obviously no simple matter, for the pursuit of pleasure *over many different time scales simultaneously* presents many contradictions – which means that the experience of pleasure and the accurate modeling of reality will often contradict each other as well, particularly when reality-modeling indicates that short-term pleasurable activities are not going to effectively build towards long-term pleasure.

---

[68] The terms "dialogical" and "monological" are defined here slightly differently than in Chaotic Logic, but the conceptual meaning is the same

[69] The terms "dialogical" and "monological" are not original; they were used by Mikhail Bakhtin in his analysis of Dostoevsky (1984). Bakhtin calls the reality of Dostoevsky's novels "dialogical," meaning that it is the result of significant interaction between different world-views.

And a hypothetical future AGI, in spite of its many differences from human beings, will be in a similar situation. It will lacks some of the problems that humans possess – for instance, we humans have basically evolved to live in hunter-gatherer societies, but we are now stuck living in civilization, which makes many of our evolved instinctive behaviors a lot less productive than they were 10,000 years ago. Violent urges were a lot more useful in the distant past than they are currently, yet we're still stuck with them, and for some people, the need to be violent sometimes is a serious component of happiness. But even without the burden of legacy goals and associated legacy behaviors, Novamente still has the problem of achieving its goals over many different time scales at once; and the problem that sometimes accurate modeling of reality may make it less happy in the short run.

Epstein divides the human conceptual system into three categories: a self-theory, reality-theory, and connections between self-theory and reality-theory. And he notes that these theories may be judged by the same standards as theories in any other domain:

> *[Since] all individuals require theories in order to structure their experiences and to direct their lives, it follows that the adequacy of their adjustment can be determined by the adequacy of their theories. Like a theory in science, a personal theory of reality can be evaluated by the following attributes: extensivity [breadth or range], parsimony, empirical validity, internal consistency, testability and usefulness.*

A person's self-theory consists of their best guesses about what kind of entity they are. In large part it consists of ideas about the relationship between oneself and other things, or oneself and other people. Some of these ideas may be wrong; but that is not the point. The point is that the theory as a whole must have the same qualities required of scientific theories. It must be able to explain familiar situations. It must be able to generate new explanations for unfamiliar situations. Its explanations must be detailed, sufficiently detailed to provide practical guidance for action. Insofar as possible, it should be concise and self-consistent, productive and predictive

The acquisition of a self-theory, in the development of the human mind, is intimately tied up with the body and the social network. The infant must learn to distinguish her body from the remainder of the world. By systematically using the sense of touch — a sense which has never been reliably simulated in an AI program — she grows to understand the relation between herself and other things. Next, by watching other people she learns about people; inferring that she herself is a person, she learns about herself. She learns to guess what others are thinking about her, and then incorporates these opinions into her self-theory.

Crucially, a large part of a person's self- theory is also a *meta-self-theory*: a theory about how to acquire information for one's self-theory. For instance, an insecure person learns to adjust her self-theory by incorporating only negative information. A person continually thrust into novel situations learns to revise her self-theory rapidly and extensively based on the changing opinions of others — or else, perhaps, learns not to revise her self-theory based on the fickle evaluations of society.

The interpenetration between self-theories and meta-self- theories is absolutely crucial. The fact that a self-theory contains heuristics for exploring the world, for learning and gathering information, suggests that a person's self- and reality-theories are directly related to their *cognitive style,* to their mode of thinking.

And indeed, we find evidence for this. For instance, Ernest Hartmann (1991) has studied the differences between "thick-boundaried" and "thin-boundaried" people. The prototypical thick-boundaried person is an engineer, an accountant, a businessperson, a strict and well-organized housewife. Perceiving a rigid separation between herself and the outside world, the thick-boundaried person is pragmatic and rational in her approach to life. On the other hand, the prototypical thin-boundaried person is an artist, a musician, a writer.... The thin-boundaried person is prone to spirituality and flights of fancy, and tends to be relatively sensitive, perceiving only a tenuous separation between her interior world and the world around her. The intriguing thing is that "thin- boundaried" and "thick-boundaried" are *self-theoretic* concepts; they have to do with the way a person conceives herself and the relation between herself and the world. But, according to Hartmann's studies, these concepts tie in with the way a person *thinks* about concrete problems. Thick-boundaried people are better at sustained and orderly logical thinking; thin-boundaried people are better at coming up with original, intuitive, "wild" ideas. This connection is evidence for a deep relation between self-theory and creative intelligence.

What Hartmann's results indicate is that the way we *think* cannot be separated from the way our *selves* operate. This is so for at least two reasons: one reason to do with the hierarchical network, another to do with the heterarchical network. First of all, every time we encapsulate a new bit of knowledge, we do so by analogy to other, related bits of knowledge. The self is a big structure, which relates to nearly everything in the mind; and for this reason alone, it has a broad and deep effect on our knowledge representation. This is the importance of the self in the heterarchical network.

But, because of the hierarchical nature of knowledge representation, the importance of self goes beyond mere analogy. Self does not have to do with arbitrary bits of information: it has to do, in large part, with the *simplest* bits of information, bits of information pertaining to the immediate perceptual and active world. The self sprawls out broadly at the lower levels of the dual network, and thus its influence propagates upward even more widely than it does.

This self/knowledge connection is important in our daily lives, and it is even more important developmentally. For, obviously, people do not learn to get oriented all at once. They start out, as small children, by learning to orient themselves in relatively simple situations. By the time they build up to complicated social situations and abstract intellectual problems they have a good amount of experience behind them. Coming into a new situation, they are able to reason associatively: "What similar situations have I seen before?" And they are able to reason hierarchically: "What simpler situations is this one built out of?" By thus using the information gained from orienting themselves to previous situations, they are able to make reasonable guesses regarding the appropriate conceptual representations for the new situation. In other words, they build up a dynamic data structure consisting of new situations and the appropriate conceptual representations. This data structure is continually revised as new information that comes in, and it is used as a basis for acquiring new information. This data structure contains information about specific situation and also, more abstractly, about how to get oriented to new situations.

We suspect that it is not possible to learn how to get oriented to complex situations, without first having learned how to get oriented to simpler situations. This regress only bottoms out with the very simplest situations, the ones confronted by every human being by virtue of having a body and interacting with other humans. And it is these very simple structures that are dealt with, most centrally, by the self-theory. There is a natural order of learning here, which is, due to various psychological and social factors, automatically followed by the normal human child. This natural order of learning is reflected, in the mind, by a hierarchical data structure in which more and more complex situations are comprehended in terms of simpler ones. But those who write traditional AI programs have made little or no attempt to respect this natural order.

Instead, traditional AI programs are provided with concepts that "make no sense" to them, which they are intended to consider as given, *a priori* entities. On the other hand, to a human being, there are no given, a priori entities; everything bottoms out with the phenomenological and perceptual, with those very factors that play a central role in the initial formation of self- and reality-theories. To us, complex concepts and situations are made of simpler, related concepts and situations to which we already know how to orient ourselves; and this reduction continues down to the lowest level of sensations and feelings. To traditional AI programs, the hierarchy bottoms out prematurely, and thus there can be no functioning dynamic data structure for getting oriented, no creative adaptability, no true intelligence.

Self- and reality- theories, in the psynet model, arise as mind attractors *within the context of the emergent dual network.* This means that they cannot become sophisticated until the dual network itself has self-organized to an acceptable degree. The dual network provides routines for building complex structures from simple structures, and for relating structures to similar structures. It provides a body of knowledge, stored in this way, for use in the understanding of practical situations that occur. Without these routines and this knowledge, complex self- and reality- theories cannot come to be. But on the other hand, the dual network itself cannot become fully fleshed out without the assistance of self- and reality-theories. Self- and reality- theories are necessary components of creative intelligence, and hence are indispensible in gaining information about the world. Thus one may envision the dual network and self- and reality- theories evolving together, symbiotically leading each other toward maturity.

Subselves

So far we have discussed self-systems only in a very abstract way. To some extent, this is inevitable, because selves are extremely complex things. A human or mature Novamente's self is not something that could be written down in a few simple formulas and analyzed! However, it is possible to proceed a little further in the detailed analysis of self. In this section and the next we will take some small steps in this direction; further steps still will be taken later in the book in a more technical Novamente context.

The main point to be made in this section is that:

- Human selves tend to be dissociated, divided into "subselves."
- There are quite likely solid information-processing-based reasons for this, so that we may tentatively expect selves within other (computational resource constrained) AGI systems to have a similar property.

The subself phenomenon makes the dynamics of the self interesting in quite particular ways. We will explore briefly here the notion of "subself dynamics," and describe a specific hypothesis as to what kinds of subself dynamics are associated with healthy personality and a generally well-functioning mind.

The concept of subselves has become popular in psychotherapy circles in recent years (Rowan, 1990). There are techniques for "letting one's subpersonalities speak," for coming into contact with one's lover's subpersonalities. A woman might have a "scared little girl" subpersonality, a man might have a "neighborhood bully" subpersonality. A straight-laced society woman might have a repressed "slut" subpersonality, denied expression since her teenage years. In this type of therapy one deals with subpersonalities on an entirely individual basis: each person must discover, with the therapist's help, what their subpersonalities actually are.

There should be nothing counterintuitive about this notion of subselves: A self is just a mind attractor, consisting of a network of concepts and procedures for mediating between mind and world. There is no reason whatsoever why a mind should not contain many such networks, overlapping and interpenetrating. Different situations require fundamentally different mental processes, perhaps even mutually contradictory mental processes — it is foolish to expect that this diversity of aims should be accomplished by a unified system.[70]

Of course, excessive separation of subselves can be a major psychological problem. One does not want subselves with different names, entirely different identities, mutually exclusive memories. But one does want subselves with slightly different identities, and slightly different memories. One's memory of a situation when at work may be significantly different from one's memory of that situation when at home — and for good reason, because memory is in large part a constructive process. One builds memories from the raw materials provided by the mind, in order to serve specific purposes. Different subselves will have different purposes and thus different memories.

In Epsteinian terms, it should be understood that a "subself" contains a self system, a world system, and a system of interrelations between them. Insofar as they bring to mind multiple personality disorder, the words "subself" and "subpersonality" may seem too strong. But if this association is put aside, one finds that, if they exaggerate the situation at all, it is only by a very small margin. For in truth, the word "personality" as it is generally understood would seem to be a perversion of the facts. "Personality" is too individualistic; it implies that the qualities of a person are fundamentally a property of that person alone, when in fact these qualities are in every respect *social*. They are formed through social interaction, and they are also *defined* by society, in the sense that, to a person from a substantially different society, they would be largely incomprehensible. So a given person's "personality" does not really exist except as part of a *network of personalities* — just as a self/reality subsystem does not really exist except as part of a network of self/reality subsystems. And in certain cases two subselves residing in different bodies may be more closely related than two subselves residing in the *same* body; for instance, the roles that two lovers assume when with one another often have little to do with their other subpersonalities. This fact is exploited frequently in fiction.

What can we say about the way in which subselves in a single mind should interact, in order that the self as a whole (and the mind as a whole) should function effectively? In From *Complexity to Creativity*, I put forth a hypothesis in this regard, and (somewhat immodestly) called it "The Fundamental Principle of Personality Dynamics."

This principle was inspired by the ideas of the philosopher Martin Buber, who in his famous book *Ich und Du* (Buber, 1971) distinguishes two fundamentally different ways of relating: the "I-It" and the "I-You" (sometimes called "I-Thou"). What the Fundamental Principle states is that a healthy mind, as a rule, consists of a population of subselves carrying out mutual I-You relationships with each other. This kind of subself community leads to robust, adaptive belief systems. On the other hand, a subself community containing a preponderance of I-It relationships will be characterized by self-sustaining belief systems of minimal adaptive value. This is purely a hypothesis, supported by intuitive observations of human personalities; but we

---

[70] A priori, it would not be unreasonable to extend the subself idea even further, to obtain subsubpersonalities, and so forth. I propose, however, that in the case of human beings this is not necessary. In other words, I suggest that, just as there is a magic number 7+/-2 for human short term memory capacity, and a magic number 3+/-1 for levels of learning (see Bateson, 1980, or *The Evolving Mind*), there is a *magic number 2+/-1* for personality system depth.

believe it has the potential to serve as a valuable guide for Novamente parameter-tuning and teaching.

What is the difference between I-It and I-You? Buber phrases it in quite mystical terms, basically saying that an I-It relationship involves treating the other as an object, whereas an I-You relationship involves non-judgmentally and non-analytically accepting the other as a whole. Our own take on these notions is a little more reductionist, but conceptually not too far off from the original.

In our view:

- The mechanism underlying an I-It relationship is a recognition of the particular patterns displayed by another individual, another self.
- The mechanism underlying an I- You relationship is an *implicit* recognition of the *overall emergent structure* of the other.

The overall emergent structure of another will, in general, be too complex and subtle to enter into consciousness as a whole. Thus it will be experienced as broad, featureless, abstract. But the mind can nevertheless experience it.

According to Buber, the experience of the You does not depend on the details of the individual being experienced. But that does not contradict the equivalence posed here, between the You and the overall emergent structure of another self. For it is the key insight of complexity science that the high-level emergent structures observed in a system *need not depend in detail on the lower-level details of the system.*

If one subself understands another subself in an I-You manner, then it is bound to be tolerant of the particular patterns making up this other subself, even if, taken individually, they might not seem agreeable. The humanistic subself will happily let the scientific subself have its spherical cows. On the other hand, in the I-It relationship, the holistic integrity of the other is not immediately perceived, and thus the intuitive reaction will often be to reject those aspects of the other's being that are not agreeable in themselves.

So, in a network of I-You relationships, there will be a tendency of selves not to interfere with each other's actions on a microscopic basis. There will tend to be a certain respect — based on a mutual understanding of the contextuality of actions, on a direct perception of the roles of individual patterns in autopoietic self-and reality-systems. What effect does this respect, this leeway, have on mental functioning?

Clearly, an atmosphere of tolerance and respect will lead to a decreased need for *defensiveness* on the part of individual thought-systems. If each individual thought-system proposed by a certain self is immediately going to be shot down by other selves, then the self in question will have to be very careful to protect its thought-systems, to make them self-sufficient and resistant to attack. On the other hand, if there is an atmosphere of relative leniency and tolerance, then resilience is no longer so important, and other aspects of thought-systems may be emphasized.

This line of reasoning leads up to the conclusion that a system of selves, characterized by I-It relationships, will tend to produce overly dogmatic, entrenched thought systems. A system of selves characterized by I-You relationships will, on the other hand, tend to produce more well-balanced thought systems, which are adaptively effective as well as adequately self-preserving. In scientific terms, a statistical correlation is posited: between the adaptivity and productivity of thought systems, and the I-You nature of inter-subself relationships. This correlation is what I call the "Fundamental Principle of Personality Dynamics"

Obviously, I-It interself dynamics is not the **only** explanation for dogmatic, inflexible the previous section we gave an example of an inflexible thought system, the conspiracy theory of a paranoid individual. Everything said to the paranoid person becomes "part of the conspiracy." The basic principles of the system never adapt, and the new explanations that these principles generate bear little relation to the situations they allegedly respond to. The system generates little emergent pattern in conjunction with its environment — it is "un-fit." However, this system did not arise primarily as a consequence of subself dynamics, but rather as a consequence of *unvaryingly I-It interactions with the outside world.* The paranoid individual, more than anyone else perhaps, does not accept anyone else as a You. Their suspicion holds them back from this intimacy. Their world is an object to be manipulated and analyzed.

It is interesting to explore these ideas via envisioning thought-experiments involving AGI systems involved in experiential interactive learning.

Consider, for instance, a baby AGI system that carries out three kinds of interactions:

1. Conversations with random individuals over the Web, who chat with the baby just for fun;
2. Conversations with AI programrs, who chat with Novamente because they are trying to teach it something;
3. Interactions with the financial markets, in the form of continual trading on a wide variety of financial instruments.

Each of these interactions is going to have fundamentally different characteristics.

In the first case, Novamente will probably be rewarded for saying silly and outrageous things, or for displaying knowledge of obscure facts, or formulating complex (even if unoriginal) sentences. Lengthy conversations will be unusual, and the user will probably not often be willing to put significant effort into the task of understanding what Novamente means, in cases where finding the right words proves difficult.

In the second case, Novamente will probably not be rewarded for continual humorousness and frivolousness, or for utterances that display the "surface signs" of intelligence (long sentences, obscure facts) but will rather be rewarded for having conversations that indicate it is fundamentally increasing its understanding of the world.

In the third case, verbal conversation is not the focus at all, although conversations reporting its trading activity or asking humans specific questions regarding its trading activity may occur. Rather the focus is on highly rapid, highly precise interactions with other software programs.

Each of these "interaction domains" has different properties, and the types of actions for which Novamente gets rewarded in the different worlds will be different. The consequence of this will likely be that Novamente develops separate subselves – separate complexes of concepts, actions and self-and-world-models – corresponding to each of these interaction domains.

Of course, there will be some spillover from one to the other. On a deep level, the system will be learning some general lessons from all three sorts of interaction. And on a shallow level, it is inevitable that, now and then, specific behaviors learned in one interaction domain will be used within one of the others. But this is exactly what one expects with subselves – not a complete dissociation of subselves associated with different domains (that's multiple personality disorder), but a probabilistic dissociation.

Artificial Intersubjectivity

These ideas have interesting consequences for AGI design, bringing us back to the concerns of Chapter 15. In *From Complexity to Creativity*, I introduced the concept of Artificial Intersubjectivity or A-IS: namely, that a collection of artificially intelligent agents, in order to achieve a high level of intelligence via interacting in a simulated world, must collude in the modification of that world, so as to produce a mutually more useful simulated reality. In this way they may evolve interrelated self- and reality-theories, and thus artificial intersubjectivity.

The key question, from an AGI design perspective, is whether this can be expected to happen spontaneously or not. This ties in with the human-psychology question of how much in-built mechanism we have for social modeling. While the jury is still out on the details, the correct answer seems to be "quite a lot" (Calvin and Bickerton, 2001).

So, it would seem that, speaking practically, spontaneously and automatic intersubjectivity cannot be counted on. Unless the different interacting AI agents are in some sense "wired for cooperativity," they may well never see the value of collaborative subjective-world-creation. We humans became intelligent in the context of collaborative world-creation, of intersubjectivity (even apes are intensely intersubjective). Unless one is dealing with AI agents that evolved their intelligence in a social context — a theoretically possible but pragmatically tricky solution — there is no reason to expect significant intersubjectivity to spontaneously emerge through interaction.

Fortunately, there is an alternative, which is the design strategy called "explicit socialization," which involves explicitly programming each AI agent in a community, from the start, with:

- An a priori knowledge of the existence and autonomy of the other programs in its environment, and
- An a priori inclination to model the behavior of these other programs.

In other words, in this strategy, one enforces A-IS from the outside, rather than, as in natural "implicit socialization," letting it evolve by itself. This approach is, to a certain extent, philosophically disappointing; but this may be the kind of sacrifice one must make in order to bridge the gap between theory and practice. In a Novamente context, what this boils down to is creating special Unitss that are explicitly purposed to serve as models for other minds: architecturally quite simple once one decides to do it. This aspect of Novamente is very well-suited for experimentation within the AGI-SIM simulation world.

Creative Subselves

One of the more interesting consequences of the theory of subself dynamics is the model of creativity to which it leads. Creativity, of course, is a topic highly relevant to Novamente. We wish our AGI system not only to have an acute logical problem-solving ability and to learn from experience, but to create original concepts and external structures pulled out of its own digital depths.

Creativity is a subtle and multifaceted phenomenon, involving the integration of essentially all of a mind's cognitive functions (Goertzel, 1997). On a very high level, however, I suggest that it is fostered by the existence, in highly creative individuals, of a "creative subself" mind attractor, a self-system whose sole reason for being is the creative construction of forms.

A creative subself, I suggest, has an unusual "shape" — it interfaces significantly with only a very limited range of lower-level perceptual/motor processes. For instance, a creative subself focused on musical composition and performance would have next to no ability to control processes concerned with walking, speaking, lovemaking, etc. On the other hand, in the domain in which it does act, a creative subself has an unusually high degree of autocratic control. For the creative act to proceed successfully, the creative subself must be allowed to act in a basically unmonitored way, i.e., without continual interference from other, more broadly reality-based subselves.

A creative subself makes speculative, wide-ranging and inventive use of the mind's associative memory network. It knows how to obey constraints, but it also knows how to let forms flow into one another freely. It does a minimum of "judgment." Its business is the generation of novel and elegant forms. The degree of "looseness" involved in this process would be entirely inappropriate in many contexts — e.g. in a social setting, while walking or driving, etc. But in the midst of the creative act, in the midst of interaction with the artistic medium, "anything goes." The limited scope of the perceptual-motor interface of a creative subself is essential.

It might seem an exaggeration to call the collection of creation-oriented procedures used by a creative individual a "subself." In some cases, perhaps this *is* an exaggeration. But in the case of the most strikingly creative individuals, it may be quite accurate. In many cases there is a surprising difference between the everyday personality of an individual, and their "creative personality." This is why so many people are surprised when they read their friends' books. The reaction is: "Good God! This is **you**? This is what's going on in *your* head? Why don't you ever tell any of this stuff to me?" The answer is that there are two different "me"'s involved. The book is written by a different subself, a different autopoietic system of patterns, than the subself who carries out conversations and goes about in the world. There is a relation between the everyday subself that the friend knows, and the creative subself that wrote the book — but not as much relation as our unified vision of personality leads us to expect.

A striking case of this kind of dissociation was Friedrich Nietzsche, who in real life was mild-mannered and friendly to everyone — but in his books was unsparing and ruthless, tearing his competitors to pieces and, in his own phrase, "philosophizing with a hammer." In his books he called Christianity a terrible curse and insulted Christians in the worst possible terms. In his life he was not only cordial but kind to many Christians, including his own sister and mother. Similarly, in his books he insulted the modern German race in the worst possible way — considering them the apex of egotistical, power-hungry, over-emotional conformist stupidity. But in his ordinary life he got on fine with his fellow Germans.

Not every human being has a creative subself. On the other hand, some rare individuals may have several. The conditions for the emergence of such a subself are not clear, but they seem to require that the human in question, during their youth:

1. develop a habit of carrying out some creative activity;
2. have an emotional or situational need to withdraw from the world into "their own universe."

These factors combine, in some cases, to allow a human mind to make a certain creative activity "its own universe." To an extent, the creative activity takes the place of the ordinary world, in which other people move. Just as different subselves normally develop to deal with different situations, a subself develops to deal with this situation, this particular subset of the world — which happens to consist of interaction with an artistic medium. And when this creative subself attains a certain level of coherence and autonomy, it gains a "life of its own" and begins to grow and develop like any other subself. It cannot flourish without access to the medium that is its world; thus it urges the other subselves to pursue the creative vocation that supports it. In some circumstances, the creative subself may be the only redeeming aspect of an otherwise execrable individual. In other cases, however, one might rightly view the creative subself as a kind of psychological parasite on an otherwise healthy organism. The creative vocation in question may have no value to the person in question; the passion to pursue this vocation may in fact destroy the person's life. The other subselves may not understand why they are unable to hold down a job, stay in a relationship, etc., when the answer is that the creative subself has gained a great deal of power, and is willing to sacrifice everything for steady access to what, in its view, is the only "real" part of the world.

In *From Complexity to Creativity*, five basic "principles of creative subself dynamics" are put forward:

1. Thorough integration of the creative subself into other subselves seems to contradict extremely productive and innovative creativity.
2. Often in creative activity, the ordinary subselves of the mind in question must play the role of a "critic," judging the results of the creative subself.

While all creativity involves some oscillation between creative and ordinary subselves, the *frequency* of the oscillation should correspond roughly to the amount of *constraint* involved in the creative medium. (In very unconstrained media such as modern poetry or abstract expressionist painting or jazz improvisation, the critic can come in only at the end, when it is necessary to decide on the value of the work as a whole. In very constrained media such as mathematics or portraiture, the critic must play an integral role, and become an integral part of the inventive process itself.)

3. Creativity requires continual adaptation of the creative/critical oscillation frequency, which can be most effectively achieved, in the long run, by I-You interactions between creative and ordinary (critical) subselves.
4. The feeling of "creative inspiration" is the feeling of emergent pattern viewed from the inside — i.e., the feeling of a perceptual-cognitive loop which encompasses only part of a broadly-based emergent pattern.
5. Which subselves will lead to more intense large-scale emergent patterns? The ones that are permitted a large but not too-large amount of creative disruption (memory reorganization and hierarchical-system crossover). Creative subselves tend to fall into this category; everyday subselves tend not to.

Principles 2 and 3 involve the "critic" role of ordinary subselves in the creative process. It is interesting to view this role in evolutionary biology terms. The frequency of critical intervention in the creative process may be understood as one component of the *harshness* of the environment facing the creative subself. In biology one may show that the maximum evolutionary innovation occurs in a moderately but not excessively harsh environment (this point is argued in *The Evolving Mind*). The same result appears to hold for creativity. A totally friendly, criticism-free environment places no constraints on creative output and is unlikely to ever lead to impressive creative works. Some constraint is necessary, or else all one has is a pseudo-random exploration of the space of all products of some collection of ideas. On the other hand, too much constraint, an overly harsh environment, is also unproductive, because it penalizes experimentation too severely. New ideas must be given some leeway, some extra time in order to bring themselves up to snuff. The exact degree of harshness which is optimal for a given situation is never the minimum or maximum, but rather lies somewhere inbetween, at a point which depends upon the amount of constraint inherent in the particular artistic medium, and also on more particular contextual factors.

Creative inspiration is the feeling of emergent patterns (personal or transpersonal) from the inside. Creative subselves are more likely to give rise to large-scale emergent patterns (the kinds that are large enough for a perceptual-cognitive loop to **fit** inside). Thus, the very looseness that characterizes creative subselves is essential to the experience of inspiration. And this looseness is due, in the end, to the "cushy" environment experienced by creative subselves, due to their relatively abstract and pliant "external world." Creative subselves are protected from the real world by the other subselves; this is the secret to their success, to their internal dynamics of emergence-yielding flexibility.

Of course, the presence of a creative subself has a major impact upon the remainder of the mind. After all, except in pathological cases, the different subselves of the same mind are not *entirely* different. Generally they will share many particular characteristics. Thus we have the known peculiarities of the "creative personality type."

Or, to look at it a different way, the Fundamental Principle of Personality Dynamics says that different subselves must relate to each other on a You-You basis, reacting to overall emergent patterns in each other's structure. But it would be very difficult indeed for a strongly *in*flexible ordinary subself to related to an highly creative subself on a You-You basis. Such a subself would not have the cognitive flexibility required to *understand* the emergent patterns making up the You of the creative subself. As a rule, then, one would expect that the ordinary subselves of a creative person would have to pick up a lot of flexibility from the creative subself.

Whatever the general merits of this speculative theory of creativity-psychology, it seems to be extremely useful as a guide for thinking about creativity in the Novamente context. The question it gives rise to is: What can we do to encourage Novamente to grow a highly active creative subself?

Of course, we can set up the preconditions for this very easily:

- Give it access to a creative medium, which is rich enough to engage all (or most) of the different parts of the mind;
- Ensure its initial motivation system is set up so that it can obtain pleasure from acting in this medium.

The "creative subself" theory suggests that, if:

- This creative medium is distinct enough from the domain of its everyday interactions with other minds;
- Its life is set up so that it spends a lot of time interacting with this creative medium.

then a creative subself will very likely form. Novamente will model itself interacting with the creative medium, and develop, to some extent, specialized theories of itself and its reality in the context of this medium.

Based on biographical studies, it appears that nearly all creative humans began to develop a "creative subself" in youth; and the reason for this seems clear. Once a strong set of interlocking subselves is set up, guiding all the mind's actions based on perceptions as related to goals, it can be very difficult for a fundamentally new structure to form and take away control from the existing subselves. This kind of rigidity should be less of a problem for Novamentes than for humans, as in Novamente it is always possible to make explicit modifications to system parameters to encourage new structures to form. On the other hand, some aspect of this phenomenon will likely still exist in Novamentes, hence we believe that, in the teaching and training of a Novamente, it makes sense to explicitly encourage the formation of a creative subself from an early stage.

Of course, from a practical AI engineering perspective – we do not yet have a completely working, fully implemented Novamente system – this sort of issue can seem slightly farfetched at times. However, we consider it important to have a relatively detailed game plan that will take us all the way from the initial coding of data structures and learning mechanisms, to the teaching and tutoring of a baby digital mind with roughly human-level intelligence. Certainly we will learn things along the way that necessitate modification of the game plan, but without a thorough plan, the odds of achieving the (very difficult) end goal would be extremely poor.

Chapter 19
Creative Intuition

In the minds of some highly creative people, intuitive insight melds with spiritual insight to yield powerful visions of the underlying nature of the universe. This phenomenon is relevant here from two perspectives: as a fascinating and revealing example of advanced mental functioning, and because the insights achieved by visionaries in this sort of mind-state have sometimes harmonized very nicely with the ideas of patternist philosophy, and give a valuable and different angle on patternist ideas.

In Vedantic terms, what I'm talking about is Intuition – by which is meant a kind of "higher intuition," which may be conceived as the movement of mind to adapt to input from the transpersonal realm, from the Overmind, *anandamaya*, the collective unconscious. Everyone experiences higher intuition, but some people experience it only occasionally, and only with slight intensity, and it plays a minor role in their lives. For others – including many of the best artists, writers, scientists and other "creative people" – higher intuition may be a dominant part of life.

Sometimes, in ordinary life, it is difficult to distinguish higher intuition from emotive thought. Any idea or inclination which goes against rational, logical thought is a candidate for the label "intuitive." In terms of the Vedantic hierarchy of being, however, the distinction between emotion and intuition is an extremely important one. If one has a certain craving — e.g. a craving borne of romantic love, or hunger, or frustration — and follows it, this is not higher intuition. It is a phenomenon fundamentally of the Body level, *pranamaya*. Sometimes intuitions may align themselves with emotions, circumventing the rational level; in other situations, intuition and emotion may oppose each other. Intuition is irrational, but in a different way from emotion: it takes instructions from somewhere else besides the body. It exists on the level of higher-order emergent patterns.

For some creative people, higher intuition exists in the service of rational and emotive thought. It is just a way of getting around obstacles that cannot be gotten around in any other way. For others, it is experienced primarily in the context of spiritual insight: it is a way of exploring the deeper realms of the universe, which may also happen to have practical consequences now and then. In these latter cases, higher intuition is melding into the next level up in the Vedantic hierarchy: *anandamaya*, Bliss. These are the cases that will preoccupy me the most in this chapter. These are the intuitions that pierce the sheaths of Being, that bring the personal and the universal together.

The most profound creative insights always seem to have something of the Realm of Bliss in them. They go beyond the purely personal level, and embody cultural or human universals. They show us nebulous forms that defy the range of the individual mind. In these cases, intuition is not acting in the service of rational/emotive thought; but precisely the opposite is happening. Reason and emotion are acting in the service of higher intuition.

In *anandamaya*, the Realm of Bliss, one experiences Being more directly than during intuitive insight. One experiences Being as a creator and destroyer of forms. One experiences a pattern-space, a universe of fuzzy, chancy forms, at once too disparate and too coherent to put into words or symbols. One experiences a vast intelligence, embodied in a boundless, flowing fabric of reality.

The only well-known image we have for the Realm of Bliss in our culture is Jung's "collective unconscious". The idea of the collective unconscious has never been accepted by scientific psychology, but it has survived anyway, because of its powerful resonance with peoples' subjective experience. The collective unconscious is a vast pool of partly-formed structures, which appear in the mind as definite symbolic images. Great creative discoveries work by drawing "universal" structures from the collective unconscious and elaborating them in individual, intelligent, appropriate ways.

By its very nature, the Realm of Bliss can never be scientifically studied; it is too unpredictable and ungraspable for that. Once a definite form emerges from *anandamaya* into the thinking, scientific mind, one is dealing with Intuition and not Bliss. Intuition is the highest level of the Vedantic hierarchy of being that is susceptible to any kind of scientific investigation.

In the hierarchical model of the universe, one sees a linear progression from Mind, through Intuition, to Bliss and then Being. But there is also another complementary point of view, which I call the "interpenetrative view", and which is based on the fundamental triad of Mind, Body and World. Mind, Body and World all create each other in a cyclic fashion; but all have Being within them. The realm of Quanta is obtained by injecting World with a dose of Being — by fluidifying the rigid structures of the world, breaking them down into more freely-flowing processes that interpenetrate and transform each other. The realm of Intuition is obtained in a similar way from the Mind: creative intuition is a process of loosening up thoughts from their habitual patterns, and allowing them to move around until they settle into a new, more inspired, pattern. And, similarly, though perhaps less obviously, the realm of Bliss is obtained by loosening up the rigid structures of the Body.

The Body is a physical entity, but it is also a metaphorical, metaphysical entity. A person's body-image is intricately tied up with their method of constructing concepts, of partitioning up the world. The boundary between Self and World is the same as the boundary demarcating each concept, the boundary between This-Idea and Remainder-of-Mind. By dissolving the boundaries between ideas, allowing all ideas to flow through each other in intricate, dazzling, beautiful patterns, the Realm of Bliss is also dissolving the body. In Intuition, one individual is receiving insights from above, pertinent to his or her particular life. In *anandamaya*, on the other hand, the individual body is vanished, or it is present in diminished form, as one among many fluctuating patterns. The individual self is gone. The Realm of Bliss is the Collective Unconscious, the Overmind — it is the transpersonal realm, in which the body, the granter of individuality, has been shed. This is the irony of practices such as Yoga, which teach one to control one's body, with the ultimate aim of transcending one's body.

In this chapter, I will explore Intuition and Bliss side by side, with a particular emphasis on experiences that play around the boundary of the two. As concrete human examples of Intuition and Bliss, I will discuss the creative/intuitive/ spiritual experiences of some of the writers whose works and lives I have studied. These writers illustrate the diverse ways that intuition melds into spiritual insight. In many cases, there is no separating the two: each one presents itself in the clothing of the other.

The use of writers — rather than, say, visual artists, scientists or musicians — to investigate Intuition is a somewhat arbitrary choice, but it does have one definite advantage: it allows one to deal with a direct way in issues of *language*. The nature of language is a crucial question for anyone concerned with the mind and universe. The intuitive experiences of writers provide insight into Intuition and language both.

Modern linguistics is largely founded on Noam Chomsky's concept of transformations that take a "deep structure," which is unconscious, to a "surface structure" which can be communicated to others. The details of Chomsky's linguistics are now in large part obsolete, but the basic concept of deep vs. surface structure remains. In the computational linguistics work I've done, the role of deep structure has been played by semantic representations at the border between logic and syntax. The logical predicate relationship:

```
eat(Ben, snake)
```

becomes the deep semantico-syntactic relationship-set:

```
subj(eat, Ben)
obj(eat, snake)
```

which may then generate multiple surface structures such as:

```
Ben eats a snake.
A snake is eaten by Ben.
```

There are linguistic rules mapping between logical relationships and deep semantico-syntactic relationships, and linguistic rules mapping between these deep structures and surface structures. These rules are very many in number, and are learned implicitly by humans in the course of hearing, reading and using language. The further one extends toward deep linguistic structure, the less personal and more transpersonal one gets. Syntactic surface structure is culture-dependent but deep syntactic and logical structure is universal and transcends the individual mind – moving up the Vedantic hierarchy.

In the view of linguistics, then, language consists of sequences of symbols, acted on by transformations. The vision of the Realm of Bliss in terms of language may, in this spirit, be interpreted as a vision of the Realm of Bliss in terms of abstract symbol-manipulation. In the end, I will argue, the moment of creative inspiration is always a moment of *perception of the world as virtual* – i.e. perceiving the world as a subset of pattern space, without any fundamental reality beyond the reality of patterns in patterns in patterns … patterns interpretable in various ways and mutable by the inspired creative mind. Inspired artists, at the moment of creation, have always lived in a virtual world – have always lived *explicitly* in pattern space. In the case of linguistic artists, the mode of pattern-space interaction has to do dominantly with symbol manipulation: representing patterns as symbols and using them within other patterns to build up recursively nested patterns evoking desired meanings. Other forms of creation such as music or abstract visual art are less fundamentally symbolic in nature, though still involving symbolism in an essential way.

Of course, the experiences of the writers I'll discuss here are not in any sense "characteristic." Visionary writers represent a particular type of human being, with their own peculiar way of experiencing material and spiritual reality. But as we move toward *anandamaya*, we have passed the point where any of the generalizations that I know can be useful. The unification of science and spirit is still meaningful, but in a less concrete way. Spiritual and scientific considerations can still inspire each other, through the discussion of these phenomena, but they cannot connect to each other as directly as on the mental and physical levels. Here things are too individual and shifting for scientific concepts to have any fixed relevance.

The experience of these writers is that *anandamaya* is made of language. This is a valuable insight, which gives us a new way of understanding the Realm of Bliss. It should not be taken as an "scientific truth" of any kind — it is just one way of experiencing things. But, as we are all users of language in so many ways, it is a way of experiencing things that is deeply relevant to all of us. This mode of experience gives us a new way of thinking about scientific linguistics — about the transformations involved in language. And it gives us a new way of thinking about our everyday, language-filled minds.

Many writers, I have found, shared a common vision of words and sentences and paragraphs as a way of goading the mind into recognizing a deeper reality — reality more fundamental than the one generally recognized as "real." This more fundamental plane is precisely *anandamaya*, the Overmind; and the state of mind that these writers wish their work to induce is largely equivalent to the recognition of the world as virtual. It is the recognition that the world is made of patterns, habits and structures; that everything is autonomous and free to act on anything else, subject only to the law that habits will tend to persist. The creative mind at the peak moment of creation lives not in objective nor subjective reality strictly conceived, but rather in pattern-space.

In the following chapter, I will (along with pursuing other themes) continue the same sort of theme but in a different context, turning from creativity to the meditative and psychedelic experiences. Meditative experience was the root of Vedanta, Buddhism and the other great Eastern mystical traditions. On the other hand, psychedelic drugs are probably the main route by which individuals in modern culture encounter the Realm of Bliss. They are an ancient technology, preceding the more difficult methods developed by the Oriental mystics. A brief consideration of the meditative psychedelic experiences gives a great deal of insight into the nature of the higher realms of being.

Inspiration and bliss are timeless things, intrinsic aspects of the cosmos — they have nothing to do with drugs, computers, writing, or any other technology. But yet these different technologies can serve as *portals* to the Realm of Bliss. The same cultural and physical constructions that prevent us from seeing the Realm of Bliss all around us, are ourtools in breaking through to underlying blissful reality. This is the supreme irony which one may call "the contradiction at the heart of wisdom." Science and literature and technology have emerged from, and solidified, a culture that exalts the individual and hence separates the individual from the cosmos as a whole. But they are leading us toward a world-view in which, once again, the individual stream of consciousness and the flow of time and form in the universe are explicitly perceived as a unity, via the perception of both as (dynamic, overlapping, emergence-spawning) subsets of pattern space.

Marcel Proust

At well over 3000 pages, Proust's *Remembrance of Things Past* (1982) is the longest continuous story in the history of "serious" literature. Yet the story which it tells spans only a few decades; indeed, around 80% of the book deals with a single decade of the narrator's life. And even within this short time scale, the narrative is focused on brief snippets. Hundreds of pages are devoted to a single dinner party; twenty pages to the experience of looking at a room of paintings, or listening to a piece of music. What is the purpose of this "magnifying glass" approach? Proust wanted to get at, not the broad contours of life that are captured in such things as "plots" and "characters," but the specific, situation-bound *feelings* that make up the concrete experience of living. The assumption is that these feelings are universal, and that they constitute a kind of reality which is deeper and more real than the particular collection of habits that is conventionally taken for absolute reality.

So-called "realistic" art, according to Proust, falsifies reality. The *truly* realistic work of art is not the one which produces a falsely frozen, "objectified" picture of the world, but the one which captures the fluidity of the world as perceived by actual human beings. As observed by the narrator of *Remembrance*, who toward the end of the book becomes more and more closely identified with the author himself,

> *Real life, life at last laid bare and illuminated — the only life in consequence which can be said to be really lived — is literature, and life thus defined is in a sense all the time immanent in ordinary men no less than in the artist. But most men do not see it because they do not seek to shed light upon it. And therefore their past is like a photograpic dark-room encumbered with innumerable negatives which remain useless because the intellect has not developed them.*

In the Proust-iverse, real life, the real world, is equated with literature. Each one of us, by living, writes his own book. *The Remembrance of Things Past* was Proust's attempt to create an *outer* book approximating his *inner* book — and thus, as a consequence, approximating *everybody*'s inner book:

> *I thought ... modestly of my book and it would be inaccurate even to say that I thought of those who would read it as "my" readers. For it seemed to me that they would not be "my" readers but the readers of their own selves, my book being merely a sort of magnifying glass like those which the optician at Combray used to offer his customers — it would bemy book, but with its help I would furnish them with the means of reading what lay inside themselves. So that I should not ask them to praise or to censure me, but simply to tell me whether "it really is like that," I should ask them whether the words that they read within themselves are the same as those which I have written (though a discrepancy in this respect need not always be the consequence of an error on my part, since the explanation could also be that the reader had eyes for which my book was not a suitable instrument).*

What is this "it" referred to in Proust's "'it really is like that'"? This "it" is, most obviously, the *true reality* of human experience — something which is not gotten across by ordinary novels, but which *Remembrance* makes at least an effort to transmit.

The true reality, in Proust's view, is somewhere different than in the solid world of objects studied by classical physics. First of all, it is more *psychological* than *physical*:

> *I had realized now that it is only a clumsy and erroneous form of perception which places everything in the object, when really everything is in the mind; I had lost my grandmother in reality many months after I had lost her in fact, and I had seen people present various aspects according to the idea that I or others possessed of them, a single individual being different people for different observers ... or even for the same observer at different periods over the years.*

But "everything is in the mind" does not imply that everything is in the *conscious* mind. Though writing before psychoanalysis became popular, Proust developed a very sophisticated understanding of the subterranean, inarticulate forces guiding behavior. And so, though he placed everything in the mind, he attached the greatest importance to those patterns emerging from **outside** the coherent, linguistic, thinking mind:

> *When an idea — an idea of any kind — is left in us by life, its material pattern, the outline of the impression that it made upon us, remains behind as the token of its necessary truth. The ideas formed by the pure intelligence have no more than a logical, a possible truth, they are arbitrarily chosen. The book whose hieroglyphs are patterns not traced by us is the only book that really belongs to us.*

The word "pattern" is a signpost for pragmatism: the first sentence of this quote makes the very Peircean observation that memory deals with habits, with patterns, with Thirds. Experiences, when we come across them for the first time, have a quality of immediacy that cannot be duplicated. When we visit them again via memory, we see only their patterns, their abstract structure. But the fact is that, even if one rejects the notion of an absolutely solid external reality, one must admit the existence of a source of patterns existing outside one's own consciousness, one's own "pure intelligence." This constant flow of unfamiliar patterns is the real crux of external reality: it is what makes the internal books of our lives interesting reading.

Books as Virtual Reality Machines

Proust wanted *Remembrance* to be a "virtual reality" in a very strong sense: it was supposed to create a simulated world in which the reader would experience *his own self* more intensely than in the real world. But this idea is not as crazy as it might sound. For, there is a sense in which the novel, as a technology, is a "proto virtual reality machine." It is obviously not a true virtual reality machine, but it is a precursor which has many of the same qualities as a true virtual reality machine.

Sitting down in front of a printed page sets the mind in a certain special state. Anyone who doubts this should reflect on the enormous subjective difference between reading words *printed* in a book and reading words *displayed* on a computer screen. The very permanence of the book imparts a certain reassuring feeling. And the feel of the spine or pages in one's hand also has a definite psychological effect. When I read a novel, I have, as the song goes, "the whole world in my hands." The fictional universe described by the book is demarcated as a part of *me*, as an element of my immediate physical world. And the fact that it is a part of *me* helps me to feel that I am a part of *it*.

The contrast between books and computer screens is somewhat similar to the contrast between movies and TV shows. When one stares at the TV, one is always aware that one is looking at a little box with pictures flashing on it. This effect can be counteracted somewhat by shutting out all the lights, but even then one does not obtain the same sense of totality that a movie screen automatically imparts. You can *enter into* the world of a movie precisely because the movie scenes identify themselves with your whole visual field. The movie is in *you*, so you can allow yourself to enter into *it*. Similarly, the book is *in your hand*, it is a part of your body, and hence you can thoroughly, unreservedly, attach your emotions to the world inside the book. This reflects the fundamentally self-referential nature of the universe: the world is inside the mind, which is inside the world. Every time we go to a movie, or read a book, we recapitulate this basic self-reference to a striking degree.

Perhaps someday, TV screens will be large enough and accurate enough that they will have the same effect as movies. And perhaps someday, the computer screen will seem so natural that it will promote the sensations of inclusion and reassurance currently associated with books. But today, there is a definite distinction between the two types of media. Books and movies are excellent proto virtual reality machines — we enter into them, we forget the world in which we "really" live. Word processing programs and TV shows are, except perhaps for a few unusual minds, *not* very strongas proto virtual reality machines. They may entertain us, but they never cause us to forget who and where we are.

Perhaps the biggest weakness of the book as a VR technology is its *one-dimensionality*. When reading a novel, it seems to us that events proceed in a fairly linear way. But in fact, is not *each* of the characters thinking and doing something at *each* point in time? When writing a third-person novel, the writer has no choice but to distort the time axis of his fictional world, in order to meet the demand of presenting different points of view *one after the other*.

Similarly, in a mathematics book, each definition, each theorem seems to follow very naturally from the last. This linear order does reflect the underlying deductive order of mathematics — but it also covers something up. The only really natural way to teach high school mathematics, for example, would be to teach algebra and geometry together. Doing algebra first, one inevitably loses something that would be obtained by doing geometry first; and doing geometry first, one inevitably loses something that would be obtained by doing algebra first.

Even in these relatively straightforward types of book, lineality is a minor problem. When writing a book such as *The Remembrance of Things Past*, whose central theme is precisely the importance of individual images and the absence of a unified overlying structure in the world, lineality becomes almost intolerably confusing and distorting. Thus Proust's novel is in a sense paradoxical. Its form is inevitably opposed to its content. It wants to transmit the contents of memory, but instead of taking on the shape of memory, it must take on the linear shape of a narrative. The one-dimensionality of time is just as illusory as every other aspect of the "shallow realism" that Proust derided.

There are books which struggle to go *beyond* the habit of lineality. The best novels of Philip K. Dick, to be discussed below, fall within this category by virtue of their constant shifts in space, time and authorial perspective. James Joyce's *Finnegan's Wake* (Joyce, 1999) is an even more valiant effort in this direction — it is like a system of equations, where one must understand *every* section in order to understand *any* section (of course, it is also nearly impossible to read). Octavio Paz wrote pairs of poems which can be read either separately, or in combination as a single poem.

But this sort of "nonlineality" can at best be partially successful. Reading, even reading radical writing such as Dick, Joyce and Paz, accustoms us to living in virtual worlds with a rigid *lineal* structure. Therefore it automatically predisposes us *against* accepting the multiversality of the world.

In fact, this is approximately what Marshall McLuhan (2005) meant when he wrote that TV and computers are moving us into a "global village," a new tribalism in which multiversality and multidimensionality replace objectivity and lineality as strategies for understanding the world. But the problem with this idea is that TV and computers are not such dramatically effective proto virtual reality machines. They are not easy enough to enter into. When we have real virtual reality machines, McLuhan's prediction should genuinely come true. We will no longer think in the lineal, objectivist way that printed matter tends to induce. We will think multidimensionally, fractally, hyperrealistically — not "tribally," not as they thought before printing was invented, but rather in a new way which overcomes the very issue of "lineality"....

Philip K. Dick

Another writer who was very explicitly concerned with issues of underlying reality was Philip K. Dick. Like Proust, Dick was concerned with "jolting" the reader out of his everyday mindset, impelling the reader's mind to a deeper understanding.

Dick might seem to be as far-removed from Proust as one could possibly get. Proust was highbrow; Dick was lowbrow. Proust wrote an endless literary novel, snobbishly obsessed with the rich and titled, full of fancy syntax and meandering sentences; Dick wrote a series of short science-fiction novels, with short, punchy sentences, a lower-middle-class sensibility, and a deep respect for the ordinary, undistinguished man. Beneath the surface however, there are several important commonalities. Both were concerned with getting at the essence of nature. Both were entranced with the human construction of reality. And both wanted to make an active difference in the realities of their readers.

Proust, born independently wealthy, spent the first half of his adult life in relative indolence, and devoted the second half to writing the "inner book" of these idle years. Philip K. Dick, on the other hand, spent much of his life scrounging for food and rent money (and unfortunately this was true even after he became a celebrated writer). He wrote science fiction for a living, and in order to sell books, he had to write stories with lively plots. His editors would not have allowed him to fill up his novels with Proustian philosophical observations. In order to tell his own "inner story," therefore, he had to be much more clever; he had to write in a kind of code, describing science-fictional "real-world" events which were actually inspired by events within his own mind.   As Dick reached the end of his life, however, a very strange thing happened: his creative process inverted itself, and he became convinced that his own inner life was affected by science-fictional things like alien artificial intelligences. In the conventional view of things, the only explanation for this strange occurence is that Dick "went insane." But the patternist perspective suggests a more detailed explanation (and one which is, as we shall see, supported by Dick's own autobiographical writings): as a part of his process of creative inspiration, Dick simply ceased to make a distinction between mind and reality. This did not render him insane in the sense of being unable to function in the world; it simply made his world-view *unusual*.

To understand Dick's peculiar form of "insanity" we must first of all discuss the actual contents of his books. Dick was one of the major science fiction writers of the century — and hence of all time — but, ironically, the science in Dick's novels is usually contrived and pseudoscientific, even by science fiction standards. What makes Dick's novels great has nothing to do with science, and only a little to do with science fiction. Dick once wrote that all his novels are concerned with two questions: *What is real?* and *What is human?*

Dick's concern for the essence of humanity is part of what makes his novels fun to read: although the science is hokey, the *characters* are real; they have real emotions, problems and concerns. And human morality is the central focus of one of Dick's most popular novels: *Do Androids Dream of Electric Sheep* (1996), which was the basis for the movie *Blade Runner*. As the title indicates, *Do Androids Dream* is concerned with the differences between humans and high-tech androids — the physiological differences, but more importantly the moral differences.

But it is Dick's preoccupation with the essence of reality that makes his novels unique in all world literature. Many have asked "What is real?" — but few have interpreted the question in so many different ways, and none have given so many interesting answers.

For example, *Man in the High Castle* (1992) is an excellent alternate-world novel, beginning from the barely science-fictional premise that the Axis, and not the Allies, won the Second World War. The hero of the book is a novelist, Abendsen, who has written a book called "The Grasshopper Lies Heavy" about an alternate world in which the Allies won World War II (for in the world of Dick's book, this is a fiction). Throughout his writing of the book, Abendsen has been using the *I Ching* to guide him — to tell him who should be elected president in his alternate world, and so on.

And then, at the end of *High Castle* — the very last page — Abendsen's wife asks the *I Ching* about "The Grasshopper Lies Heavy." The *I Ching* replies: True Reality.

This is what I call a "Phil Dick moment." Suddenly, the whole world, all the contents of your mind — *everything* is questioned. Abendsen's wife understands that, in reality, the Allies did win the war — than her world is only semi-real. Although *High Castle* is one of the best alternate-world novels ever written, it is the last page that sticks in your mind.

Another classic Phil Dick moment occurs in the movie *Total Recall*, and the story upon which it is loosely based, "We Can Remember It for You Wholesale" (1990). The lead character, played by Arnold Schwarzenegger, wants to go to Mars for vacation — but his wife wants to go to Titan. So he hears about a company which will implant in your mind the memory of a vacation. You strap yourself to their company computer, and in two hours when you wake up you feel as though you have gone to Mars — or Titan, or Venus, or wherever.

This is a brilliant twist on the idea that the best thing about a vacation is the remembrance of it. Our hero signs up for a simulated Mars trip, and the computer operators ask him what sort of vacation he would like. Would he like to experience Mars as an ordinary tourist, as a rich business executive, as a secret agent...? He chooses secret agent.

But the simulated trip goes awry. He experiences psychosis while connected to the machine, and actually destroys the machine as he struggles to escape the false reality. Then, on his way home to his apartment, he is ambushed by mysterious people who yell "You just couldn't keep quiet about Mars!" Only by the most outlandish luck does he manage to kill all his attackers and escape...

After a complicated sequence of events, our hero finds out the "truth" of the matter. He learns that he had previously been a secret agent, doing some kind of mysterious work on Mars, and that he had voluntarily had his memory blotted, so as to go undercover as effectively as possible...

Eventually he winds up on Mars, trying to get at the root of the phenomenon; and some of his old enemies try to arrest him. He pulls a gun and threatens them. They can't get him to surrender, so they get clever. They bring in a psychiatrist, and have him tell our hero that he is still attached to the simulated vacation machine — that he is not really on Mars at all, but in their office deep in psychosis, and that if he does not put down the gun he will sink even further into psychosis.

This is a Phil Dick moment, par excellence: Where am I? Who am I? What am I doing?....

But these are minor examples. Without a doubt two of the most impressive and profound of Dick's works are *Ubik* (1991) and *The Three Stigmata of Palmer Eldritch* (1991). Each of these novels presents the "What is real?" question in a clearer fashion than one would imagine possible.

Ubik presents a group of characters fleeing a mysterious attacker through a reality that keeps shifting. Specifically, their twenty-first century reality keeps getting earlier and earlier, until it reaches the 1930's. The only way out of the fake realities is to get ahold of a spray can of Ubik. Spray some Ubik around — and false realities disappear! Mysterious, right? But then the characters realize that they are not in physical reality at all — they were killed in an explosion, and they are stuck in cold-pac: a half-dead, half-alive state which allows the mind to wander through irrational dreams....

Three Stigmata, on the other hand, presents a somber future in which people are forced to leave Earth and live on Mars colonies, which consist of dingy, overcrowded hovels and parched, neglected farms. The only way to avoid being recruited for colonization is to become legally insane, which is accomplished by hiring a mechanical psychiatrist, "Dr. Smile"....

The Martians' only solace is a drug called Can-D, which works in conjunction with a "layout" — a physical model of an illusoryworld. In a particularly humorous touch, the specific illusory world in question is 1950's America — the layout consists of a doll-sized 1950's house, complete with a doll named Perky Pat and her husband doll, Walt. Taking Can-D gives the colonists the illusion of being Perky Pat and Walt, back in the (relatively) idyllic past.

But then eccentric billionaire Palmer Eldritch returns from Proxima Centauri with a new drug — something even better, called Chew-Z. You don't need a layout for it. "Be choosy, chew Chew-Z" proclaim the advertisements — and the colonists buy it. Chew-Z gives a far better trip: one experiences one's deepest fantasies, and different ones each time, instead of the same old Perky Pat. But there are three drawbacks....

One, a Chew-Z trip doesn't take any time. When you come down from your trip, even though subjectively hours or days have passed, it is exactly the same time as it was when the trip started.

Two, there is a certain evil presence hovering over you throughout the trip, playing nasty games on you and mocking you.

And three, you never really come down. The trip lasts forever.

As a symbol of this latter drawback, even after Palmer Eldritch is killed and Chew-Z is banished, everyone winds up walking around sporting the three most significant physical disabilities of Palmer Eldritch: a mechanical hand, a glass eye and a bad leg.

These are not the only profoundly ontological Dick novels. A complete treatment would have to include at very least *Eye in the Sky*, *Time Out of Joint*, *Martian Time-Slip*, *Flow My Tears, the Policemen Said*, *A Maze of Death* and *A Scanner Darkly*. But the most intriguing thing about Philip K. Dick is not what he wrote in his novels. It is, rather, the fact that, as he grew older, he came to believe his novels. He came to believe that the shifted, distorted realities of his novels represented a more fundamental truth than the spatiotemporal reality which we perceive around us.

Beginnings and ends of psychic events are always indeterminate. But for sake of simplicity, Dick always denoted the beginning of his spiritual "conversion" by the shorthand 2-3-74 — meaning February/March 1974.

What happened in 2-3-74 is hard to summarize. Dick's own diary tells the story better than I could:

> *March 16, 1974: It appeared — in vivid fire, with shining colors and balanced patterns — and released me from every thrall, inner and outer.*
> *March 18, 1974: It, from inside me, looked out and saw the world did not compute, that I — and it — had been liedto. It denied the reality, and power, and authenticity of the world, saying "This cannot exist; it cannot exist."*
> *March 20, 1974: It seized me entirely, lifting me from the limitations of the space-time matrix; it mastered me as, at the same instant, I knew that the world around me was cardboard, a fake. Through its power I saw suddenly the universe as it was; through its*

power and perception I saw what really existed, and through its power of no-thought
decision, I acted to free myself.

This is insanity, right? One cannot dispute such a diagnosis — the experience of being "possessed" by a superior, hyper-rational intelligence has a definitely loony ring to it! But, as Dick realized, even if 2-3-74 was insanity, it was also a spiritual revelation not apparently any different in kind from those associated with Buddha, Jesus, Zoroaster, William Blake, Nietzsche, Krishnamurti, and countless other mystical giants throughout history. 2-3-74 was an example of artistic intuition blossoming into something yet more profound — of Intuition melding into Bliss.

Just as ancient mystics integrated their experiences with their own cultures, so Dick conceptualized his experiences in the vernacular of the twentieth century. In his novel *VALIS*, Dick describes the events of 2-3-74 in some detail, and elaborates them into a fictional plot involving rock stars, psychedelic music, and the coming of the savior in the form of a two year old girl. The result is an outstandingly original novel, one of Dick's best. Robert Anton Wilson deemed the narrative originality of *VALIS* to be comparable with that of James Joyce's *Ulysses*.

The most honest and penetrating assessment of 2-3-74 and its psychological aftermath may be found in the eight thousand page "Exegesis," a journal in which Dick recorded his own personal reactions and speculations. Underwood-Miller has published some of the more interesting fragments of the Exegesis, under the title *In Pursuit of VALIS*. (VALIS, incidentally, stands for Vast Active Living Intelligent System — one of Dick's names for the "overmind" which contacted him in 2-3-74.)

Adding to the appeal of the Exegesis is the fact that, although he was emotionally overwhelmed by these visions, Dick never lost his skeptical side. He never lost sight of what he called the "minimum hypothesis," which was simply the possibility that he was totally insane and VALIS had no real existence. However, he was never completely successful at balancing his inherent skepticism with the passionate strength of 2-3-74. This constant tension is evident in the following passage from the Exegesis:

> *Here is the puzzle of [my novel] VALIS. In VALIS I say, I know a madman who*
> *imagines that he saw Christ; and I am that madman. But if I know that I am a madman I*
> *know that in fact I did not see Christ. Therefore I assert nothing about Christ. Or do I?*
> *Who can solve this puzzle? I say in fact only that I am mad. But if I say only that, then I*
> *have made no mad claim; I do not, then, say that I saw Christ. Therefore I am not mad.*
> *And the regress begins again again, and continues forever. The reader must know on his*
> *own what has really been said, what has actually been asserted. Something has been*
> *asserted, but what is it? Does it have to do with Christ, or only with myself?.... There is no*
> *answer to this puzzle. Or is there?*

In this passage, VALIS, the Vast Active Living Intelligent System, is described with the word "Christ." But this is not the key point. The main point is that Dick cannot accept that VALIS really exists, but yet he cannot reject it either. He is caught in a loop — if he accepts that VALIS exists, then he knows that this is a mad idea, so he must believe that he is mad, in which case VALIS does not really exist. But if he rejects the existence of VALIS, then this is a sane thing to do, so he is not mad (he has no other reason to consider himself mad), so therefore what he perceives must be taken at face value, and VALIS must be accepted to exist.   This regress of doubt, self-doubt, doubt of self-doubt, doubt of doubt of self-doubt, and so forth might seem to be the least productive thought process in the world. But no, in Dick's world things are exactly the opposite. The acceptance of this paradox becomes the paramount thing. Dick does not wish to overcome his skepticism, or to overcome his passionate belief. "The Sophists," he writes,

> *saw paradox as a way of conveying knowledge — paradox, in fact, as a way of arriving at conclusions. This is known, too, in Zen Buddhism. It sometimes causes a strange jolt or leap in someone's mind; something happens, an abrupt comprehension, as if out of nowhere, called satori. The paradox does not tell; it points. It is a sign, not the thing pointed to. That which is pointed to must arise ex nihilo in the mind of the person. The paradox, the koan, tells him nothing; it wakes him up. This only makes sense if you assume something very strange: we are asleep but do not know it. At least not until we wake up.*

According to this argument, the process of simultaneously doubting and believing in VALIS is analogous to continually repeating a Zen koan to oneself. The paradox "wakes us up."

"Waking up" is not a phrase which Dick takes lightly. In fact Dick's concept of "waking up" is inextricably tied to *Ubik*, in which people are kept half-asleep in cold-pac, while they experience realities that are only "semi-real." And what keeps us from waking up is not merely laziness, or cosmic chance; it, like Palmer Eldritch as he appears to Chew-Z chewers, is definitely evil:

> *The criminal virus controls by occluding (putting us in a sort of half sleep) so that we do not see the living quality of this world, but see it as inert. Man reduced to automaton. The occlusion is self-perpetuating; it makes us unaware of it...*
>
> *The process of thinking paradoxical thoughts is a pointer to something — a pointer to something which the "criminal virus" prefers us not to see. And this something to which the paradoxical process points, is itself paradoxical:*
>
> *Something ("Y") is recognized as its own antithesis ("~Y"). This sounds like Zen or Taoist thinking. But this is oxymoron thinking. ("A thing is either A or ~A" what could be more obvious? How can A=~A? There is no such category of thought; literally, it cannot be thought; it can be recognized about reality, however, as I did in 3-20-74).*

In other words, the true underlying world — which Dick calls by the Greek name macrometasomakosmos, but he might as well have called *anandamaya*, the Overmind; or in patternist-philosophy terms, "pattern space" — is paradoxical. The true world is built on a foundation of paradox; and by presenting our minds with irresolvable paradoxes, such as the existence of VALIS, the true world alerts us to its own existence.

By this logic, Dick's own novels are part of the process by which the true underlying reality makes itself known:

> *My writing deals with hallucinated worlds, intoxicating & deluding drugs, & psychosis. But my writing acts as an antidote, a detoxifying — not intoxicating — antidote... My writing deals with that which it lessens or dispels by — raising those topics to our conscious attention.*

The goal of Dick's novels is therefore not to get across any particular message or content, but just to wake us up.

Overall, one may summarize Dick's peculiar metaphysics in three key points:

- The everyday spatiotemporal world is at most semi-real.
- There is another world underlying it which is truly real.
- The thing which prevents us from seeing this true world is evil.

This is, as he points out, a combination of Gnostic and Platonic philosophical doctrines — 1) and 2) are from Plato, and the idea of evil is from the Gnostics. But on a more essential level, it is a product of his own personal experience.

As a philosopher, Dick is a strange mix of Plato, Kant and Nietzsche. He is admittedly confused, and from one novel to the next he drifts from one view to the next. In *Time Out of Joint* the real world is finally attained through perseverance and intelligence; in *Three Stigmata* it is snapped away by evil, and the question of its recapture is left ambiguous; and in *VALIS* the whole idea of contacting true reality is made to seem faintly ridiculous.

Dick's final view, however, as we shall see a little later, veers remarkably close to Peircean pragmatism and patternist philosophy: this mysterious "true world" is revealed to be nothing more than a world of pattern, of structure. The "evil" of which Dick speaks is thus nothing other than the mental block that prevents us from seeing reality as virtual, and seeing that what's "really" there is a collection of intercreating patterns. This evil is, in Buddhistic terms (and to anticipate the ideas of the following chapter), the "mental knots" obstructing the smooth flow of the stream of consciousness. Dick, through his artistic intuition, saw through the surface structure world, through to another world beneath. But except in his most inspired moments, he was "asleep" like the rest of us — his vision was occluded by mental knots, and he could not see through to the truth of things.

One of Dick's favorite ways of explaining his mystical experience was by reference to Plato's notion of learning as anamnesis (anamnesis = loss of amnesia, i.e. recovery of memory; remembrance). According to this idea, the most ignorant schoolboy actually "knows" every theorem of advanced mathematics, so that one does not actually need to teach anyone anything; one only needs to make them remember what they have "forgotten."

The Platonic theory of learning is based on Plato's vision of the Realm of Ideas — an *anandamaya*-like cosmos where abstractforms exist in and of themselves, ready to be instantiated in the realm of spacetime. Plainly, Whorf's *Arupa* is the same as Plato's Realm of Ideas — only, Whorf's emphasis was on language, whereas Plato's was on ideas in general.

Aristotle very convincingly argued against this Platonic idea of learning, in favor of the more modern theory of learning by induction: once we see that some rule holds in a number of cases, we make the leap of inference and decide that it always holds. Because the sun has risen every day for the last ten thousand days, we assume that it will rise again tomorrow.

But of course, two millenia after Aristotle, Hume came along and devastated the concept of learning by induction. And then, a few hundred years later, the theory of probability finished the job, demonstrating beyond all possible doubt that there is no real logical justification for assuming that the sun will rise tomorrow, just because it has risen every day for the last ten thousand days.   The Humean argument, made more rigorous by probability theory, is as follows: when we reason by induction, we must have some rule telling us how many cases we must see before we accept something as a general law. How many times must we see the sun rise before we decide that it will rise every day? Once, twice, five times, a hundred? But this rule for guiding inductions — how do we arrive at this? If we arrive at it by induction, then we face the same question again — we enter into an infinite regress of inductions. But if we just pull some rule for guiding inductions out of the blue, then what validity do our inductions have?

Hume's final answer is that, in reality, the regress bottoms out after a finite number of levels — and at this bottom level the decision is made by "human nature". At first this may seem to be a disappointing conclusion for such a virtuosically logical train of thought. But we will see a little later that there is more in Hume's answer than the first glance reveals.

So the Aristotelian theory of learning has its flaws too. Dick is proposing to reconsider the Platonic theory. Or, more precisely, he is proposing to combine the two:

> *This meta-abstracting due to anamnesis is equal to the following. A child learns that one apple plus one apple equals two apples. He then learns that one table plus one table equals two tables.... Then a day comes when he abstracts; it is no longer one apple plus one apple nor one table plus one table; it is: one and one equal two. This is an enormous leap in abstracting; it is a quantum leap in brain function.... But I say, Another leap exists, beyond this; another quantum leap. And this next leap does not occur to everyone; in fact it only occurs to a few.... It is truly dependent on anamnesis, whereas the above, as Aristotle rightly pointed out, does not depend on anamnesis. The child does not in fact remember or recollect that one and one equals two; he extrapolates from the concrete examples of apples and tables. Plato knew that another and higher leap existed, based on anamnesis, and it meant a leap from thespatiotemporal world into another world entirely...*
>
> *I am saying "One plus one equals two" to people who are saying, "One apple plus one apple equals two apples. One plus table plus one table equals two tables." It's not their fault. I'm sorry, but the difference between my meta-abstraction as a brain-function and their abstracting, their brain-function, is that great. I'm lucky... my blocked memory of my prenatal life was disinhibited. After making the initial leap into meta-abstracting my brain drew conclusion after conclusion, day after day; and I saw world more and more in terms of conceptual or morphological arrangement and less and less in terms of the spatiotemporal; I continued to abstract reality more and more, based on the hierarchy of realms (each higher one possessing more unity and ontology than the lower) that Plotinus describes.*

This is an intriguing passage. Dick accepts both Aristotelian induction and Platonic intuition, but assigns them different roles. He feels that ordinary thought is based on induction, but higher thought is based on intuition. In essence, this is just a distinction between Thought and Intuition. He says that he is truly "remembering" things, seeing them in their essences, whereas others are merely inducing — meaning simply that he is more Intuitive than the others around him.

What Dick means by meta-abstracting, by seeing things as they are, is more or less seeing pattern as pattern. To look at something, and see that it is nothing but a pattern in other entities, an ordered arrangement of other entities — that it is just a pattern and has no substance in itself. This is precisely the type of realization that Dick was writing about! This is nothing more or less than the recognition that everything is virtual. One sees through the particular instantiations of abstract, *anandamaya* forms, and sees the underlying abstract patterns, with their more fundamental reality, closer to the realm of Pure Being.

Further insight is given by the following passage:

> *The issue is not reality or ontology but consciousness — the possibility of pure, absolute consciousness occurring. In terms of which material things (objects) become language or information, conveying or recording or expressing meaning or ideas or thoughts; mind using reality as a carrier for information, as a lp groove is used to carry information: to record, store and play it back. This is the essential issue: this use of material reality by mind as a carrier for information by which information is processed — & this is what I saw that I called Valis...*

This is about as Whorfian as you can get. Language as the ultimate reality! Everything, at bottom, is made of "language or information" — i.e., pattern.

And there's more. Recall that, in the the psynet model, the structure of an entity is the set of all patterns in that entity. And recall also that, in this view of mind, patterns are active. Similarly, according to Dick:

> *The agent of creation ... is at the same time the abstract structure of creation. Although normally unavailable to our cognition and perception, this structure — and hence the agent of creation — can be known...*

And pattern space is the fundamental reality:

> *[T]his insubstantial abstract structure is reality properly conceived. But it is not a God. Here, multiplicity gives way to unity, to what perhaps can be called a field. The field is self-perturbing; it initiates its own cause internally; it is not acted on from outside.... [I]t is not physical ... it is known intelligibly, by what Plato called Noesis, which involves a certain ultimate higher-order meta-abstracting...*
>
> *I posit ontological primacy to the insubstantial abstract structure, and, moreover, I believe that it fully controls the physical spatiotemporal universe.*

In the end, a careful textual analysis shows that what Dick's crazy mystical vision comes down is simply a vivid, visceral realization that the world is made of pattern. Physical objects, and everything which appears real to us, are just epiphenomenons of the underlying reality of pattern. And that this underlying reality can be seen — it is possible to see an object as a pattern as well as an object.

"I posit ontological primacy to the insubstantial abstract structure..." — there is no better way to summarize the crux of computational pragmatism, the enigma of fundamental pattern-space, the primacy of *anandamaya*, the intrinsic virtuality of the world? Dick's "Platonic reality" is not a reality of abstract, perfect Ideas, but a reality of informational relations, of connections between things, of communications. It is the web of relation announced in the Vedas; the universe of interrelating dynamic quanta proclaimed in *The Will to Power*.

In this context, we can understand exactly what Dick meant when he wrote about "meta-abstracting" the fact that $1+1=2$. He meant nothing other than recognizing "$1 + 1 = 2$, and this is a pattern which has its own independent existence; it is not inherently tied to any specific 'substrate' reality." But when one recognizes this, it does not come parceled up into clauses — "this and this and this" ... it comes all at once, in a single intuitive flash. It comes, in short, as an emergent pattern. Dick's and Whorf's visions blend naturally into the hierarchical network of the psynet model of mind — which itself blends naturally into the Vedantic hierarchy, and Buddhist psychology. There is only one reality here, only one type of experience, being expressed in many different languages.

The hilarious and disturbing reality games of Dick's novels, when seen in this context, make perfect sense. Although the supposedly solid "underlying realities" of the novels are constantly shifting and disintegrating, the novels are anything but absolute chaos. The personalities are constant, and the interactions between people are the same. The basic human situations are invariant, even when the decade spontaneously changes, or Mars becomes Earth, or everyone grows a mechanical hand. The basic structures of life remain the same, because it is after all structure which is basic. The moral of Dick's novels, considered as one long interconnected story, is that human consensus reality is more basic and powerful than physical reality. This is an incredibly important lesson to learn.

Dick asked "What is human?" and "What is real?", and his corpus points inexorably to the conclusion that the two questions have the same answer. There is no absolute reality; but for us, in the "region" of pattern space in which we live, the closest thing to a constant reality is the human consensus reality. Ergo, humanity is reality; and reality is humanity. QED.

And this brings us right back to Hume's *Treatise on Human Nature*. At the conclusion of his devastating critique of Aristotelian logic, Hume's conclusion is that the infinite regress of inductions bottoms out with human nature. What this means is that there is no absolute justification for anything, because there is no absolute reality; but at bottom, human inferences are justified by reference to the consensus subjective reality in which we — approximately — live. They are justified by their survival value — by their affiliation with networks of patterns that are able to survive multiversal dynamics, able to survive the recurring "ontological primacy" of structure.

The bottom line of Dick's crazy visions is a very simple one. Reality is virtual. Fundamentally and essentially, we are living in pattern-space, not in any kind of objective spatiotemporal reality. More specifically, we are living in a region of pattern-space that is designed by, for and of human nature. According to Phil Dick's brand of informational mysticism, the apparently real world is an illusion; the only truly real world is information, language, structure, and pattern.

Arthur Rimbaud: The Systematic Disorganization of All the Senses

The French Symbolist poet, Arthur Rimbaud, is another excellent example of an artist who used language to see through the world. Let us begin with the poem *Drunken Morning*[71] (1967):

> *O my Good! O my Beautiful! Appalling fanfare where I simply cannot stumble.*
> *Enchanted rack! Hurrah for the wonderful work and for the marvelous body, for the first*
> *time! In the midst of children's laughter it began, and with their laughter will it end. This*
> *poison will remain in all our veins even when, the fanfare turning, we shall be restored*

---

[71] Translated by Ben Goertzel and Gwen Goertzel, sometime in the late 1980's. The translation may not be perfect but we consulted a lot of other translations and didn't find any of them so great either, though we were amused at their wildly disparate choices in many places. The translation process becomes annoying right from the start where there is no way to gracefully carry over to English the French alliteration "O mon Bien! O mon Beau!" (we thought of "O my Good! O my Gorgeous!" but it doesn't work; gorgeousness doesn't carry enough of the deep polysemous overtones of beauty…).

the old disharmony. May we, so worthy of these tortures, now take up fervently that superhuman promise made to our created body and soul: that promise, that madness! Elegance, science, violence! They promised to bury in darkness the tree of good and evil, to deport tyrannic respectability so that we might bring here our love so very pure. It began with a certain disgust — and it ends, — unable instantly to grab this eternity, — it ends in a riot of perfumes.

This poem, from the collection entitled *Illuminations*, is a remarkably potent concoction — it reminds one of a confusing but moving Picasso painting, all full of distorted women and staring eyes; or a rambling John Coltrane saxophone solo, soulful blues mixed up with fancy Arabian scales mixed up with plain old discord.... Not at all bad, especially considering the author was a drug-addled, sexually confused teenage delinquent.

Prose poetry of this nature is more shocking in French, in which the rules of prose composition are stricter. But even in English, if *Drunken Morning* is interpreted by the rules of proper descriptive prose, it is absolute nonsense. Yet *Drunken Morning* is certainly not structureless, meaningless gobbledygook. It describes a certain experience, a certain feeling, and it does so in a uniquely vivid style.

Part of the beauty of the poem lies in the way abstract ideas and states of mind are represented in terms of familiar images. Rimbaud didn't write "to get rid of rules" or "to banish morality" — he wrote "to bury in darkness the tree of good and evil," giving a similar meaning along with intense visual imagery. He didn't say the experience ended in a potpourri or a miasma or a mixture of perfumes, but rather a riot of perfumes. One would not normally speak of a riot of perfumes: the juxtaposition of words gives an image of extraordinary olfactory assault, and it brings the mind back to the passage "Elegance, science, violence!" a few lines up.

But it's not only the words. Part of the beauty is the tone, the hurried, frenzied, enthusiastic rhythm of the sentences. This forward-going rhythmic thrust is broken up only by the dashes at the end, which surround the description of the end of the experience. As the breaking-up of the experience is described, the verbal rhythm breaks up also.

So, in order to get across certain experiences, Rimbaud employed unusual combinations of words and an unusual tone. But this is not the full story. As the following oft-quoted letter indicates, not only was Rimbaud's style a conscious decision, it was the product of a furious quest (Rimbaud, 1967):

> *One must, I say, be a visionary, make oneself a visionary.*
>
> *The poet makes himself a visionary through a long, a prodigous and rational disordering of all the senses. Every form of love, of suffering, of madness; he searches himself, he consumes all the poisons in him, keeping only their quintessences. Ineffable torture in which he will need all his faith and superhuman strength, the great criminal, the great sickman, the accursed, — and the supreme Savant! For he arrives at the unknown! Since he has cultivated his soul — richer to begin with than any other! He arrives at the unknown: and even if, half-crazed, in the end, he loses the understanding of his visions, he has seen them! Let him be destroyed in his leap by those unnameable, unutterable and innumerable things: there will come other horrible workers: they will begin at the horizons where he has succumbed.*
>
> *So then, the poet is truly a thief of fire.*

> *Humanity is his responsibility, even the animals; he must see to it that his inventions can be smelled, felt, heard. If what he brings back from beyond has form, he gives it form, if it is formless, he gives it formlessness. A language must be found...*
>
> *This eternal art will have its functions since poets are citizens. Poetry will no longer accompany action but will lead it.*
>
> *These poets are going to exist!*

Rimbaud, like Phil Dick and Friedrich Nietzsche, took the "quantum leap," and "arrived at the unknown," albeit at least "half-crazed!"

Knowing all we do about the biology and psychology of perception, we must take Rimbaud seriously when he speaks of "the prodigous and rational disordering of all the senses." During his years of poetic activity, Rimbaud was perpetually on one drug or another — opium, hashish, alcohol, and probably others as well. And he behaved so outrageously that he even offended other young poets. We know that a great deal of the construction of a person's subjective world takes place in the sense processing centers oftheir brain. And we know that drugs can modify the operation of various levels of sensory processing, as can "natural" insanity.

Most people who take drugs and act crazy are, it would seem, trying to force themselves to perceive the world differently. But Rimbaud's case was unusual, in that his motivation was not merely hedonistic but artistic as well. He wanted to stretch his mind to the limits, in the service of art. He wanted to get rid of the preconceptions by which we perceive the world: to penetrate to a deeper, more magical level, and then to bring some of the magic back to the rest of humanity.

Prometheus stole fire from the gods and gave it to humanity; for this he was punished with an eternity of torture. When Rimbaud describes the poet as a thief of fire, he means that the poet is a martyr, sacrificing himself to give humanity magical power. Somehow, he seems to imply, the reader of the poetry can experience a bit of divine magic, without the anguish of stealing it from the land of the gods. To put it more concretely, the poetry can loosen the bonds of language and perception, of consensus reality, on the mind. The poetry is not a static artifact but an active, dynamic process; and once it enters the mind it sets to work at changing it.

The way I read "Drunken Morning" is a description of precisely the state of sensory disorganization which Rimbaud sought to achieve. What is this "poison," which elicits opposite emotions simultaneously? It brings tremendous exuberance ("O my Good! O my Gorgeous!"); and it brings an "appalling fanfare" every time he fails to stumble. What is this "poison," which will continue to infect him even as his present state fades into "the old disharmony"... which makes him a "superhuman promise" of going beyond morality, beyond the strictures of "tyrannic respectability", of "burying in darkness the tree of good and evil?" This poison, which promises to transcend all these things that restrain "our love so very pure?" Certainly, the poison is alcohol or some other drug. But it is also the "other world," the world of the supernatural, the mad, fierce, fiery world from which inspiration springs. The other world promises deliverance from the restrictions of ordinary life, it promises the release of pure love; but it always fades and then only the poison remains. The task of the visionary poet is to grab something of this superhuman promise, of this transcendent life, and bring it back to this world.

But in fact, is this "other world" not the same as the macrometasomakosmos of Philip K. Dick — or the *anandamaya* of the Vedas? Is Rimbaud's unnamed "poison" all that different from Palmer Eldritch's Chew-Z? Is Rimbaud, in his famous letter, not essentially saying the same thing that Dick said in his Exegesis?

The disorganization of the senses is one way to disrupt the ordinary pattern recognition routines. Normally we don't realize that the world in front of us is made of pattern — we're so accustomed to seeing it there that we assume it to be absolutely real. The basic pattern of pattern – Gregory Bateson's "metapattern: that it is pattern which connects" is hidden from us. Disrupting the senses, whether by drugs or by other means, forces one to face the relativity of existence; it is an ontological challenge. It destroys the false division between world and mind, transforming the strange into the ordinary, the ordinary into the strange, and the real into the hyperreal.

Once the normal assumptions were cast aside, Rimbaud's native intelligence was able to recognize new and different patterns in the world, patterns that were obscured by the process of ordinary perception. The whole universe of *anandamaya* was open to him. Of course, anyone is capable of disrupting their senses — but not everyone is able to take advantage of the freedom thus obtained to recognize and create striking new forms. Not everyone is able to transform sudden flashes of insight into *anandamaya* into serious, sustained bouts of Intuition.

The pain involved comes — as discussed above — from the re-adjustment to the normal mode of perception. After seeing amazing, exciting patterns that contradict the patterns of consensus reality, one comes down from the "trip" and must accept the patterns of consensus reality as being in some sense "real." This can be a horrifying experience, it can give one the feeling that all the tremendous visions one has had are completely worthless. But the poet, in order to be consistently brilliant, must learn to live with this feeling. Rimbaud did not learn to live with it; he gave up poetry before his twentieth birthday. The mental knots took over. He never learned to untie them on a permanent basis: he only learned to take brief vacations into non-ordinary states of consciousness, and this was not enough.

Octavio Paz: Through Us the Universe Talks to Itself

Rimbaud, Dick and Whorf were loners. But there have also been cultural movements, involving large numbers of people, concerned with language as a tool for exploring the deepest levels of reality. The best example is the Surrealist movement.

Probably the best-known Surrealist is the painter Salvador Dali. But in fact, the original Surrealist movement was focused on literary and political endeavors much more than on visual art. Andre Breton, Paul Eluard, Phillipe Soupault and most of the other early French Surrealists were writers, with a very different philosophical agenda from Dali (who eventually left the Surrealists to pursue his own vastly less influential, but much more entertaining, "paranoiac-critical method").

What united all the Surrealists, however, was a love for intellectual game-playing, and a commitment to revealing the underlying truth of mind and world, which is ordinarily hidden by consensus categories of thought. The Surrealist movement was temporarily allied with the Communist Party — but soon enough the Surrealists recognized that the Marxists were more concerned with control than with liberation, and they broke the relationship off.

For example, a standard Surrealist literary game was taking an ordinary sentence, and replacing each word with the word that appears eleven words before it in the dictionary. Or: taking an article about a gathering of famous politicians, and re-writing it to refer to a meeting of famous murderers. The point was to jar the mind, to reveal hidden structures in things; to make one think in unusual, nonconforming ways; to associate things that would not ordinarily be associated. Dali, using his own idiosyncratic vocabulary of tricks, accomplished this same goal in the realm of visual art, with admirable elegance and wit.

The surrealist artist whom I wish to discuss here is the Mexican poet Octavio Paz. Paz was not one of the original Surrealists; and his artistic approach differed significantly from that of the founders of the Surrealist movement. For one thing, he was not French — not even European. And, for another thing, whereas the literary experiments of the French surrealists were often somewhat academic and contrived, Paz's writing flows straight from the heart as well as the mind. Like Dali, he is intensely personal and emotional, as well as intellectually and transpersonally surrealistic.

In their most celebrated work *The Immaculate Conception* (1992), Breton and Eluard run through a list of mental disorders, and seek to give each one an appropriate prose-poetic expression. And in *The Magnetic Fields* (see Breton and Eluard, 1992) Phillipe Soupault writes randomly, incomprehensibly. These works are entertaining, and they serve the Surrealist purpose of making the reader throw off his or her mental categories — so as to see the hidden meanings inside ordinary things. But they do not do what Paz's writing does — that is, hint at another world beyond the world of immediate appearances, a truer reality. And this is precisely because, whereas the surrealism of Breton, Eluard and Soupalt was based largely on game-playing, Paz's surrealism was based on mystical, emotional intuition.

Let's take a simple, out-of-context example. Paz (1991) speaks of:

> *...Images buried*
> *in the eye of the dog of the dead*
> > *fallen*
> *in the overgrown well of origins*
> > *whirlwinds of reflections*
> *in the stone theater of memory*
> > *images*
> *whirling in the circus of the empty eye*

This passage may be interpreted in several ways. Let's just consider the first line. "Images buried in the eye of the dog of the dead" may be taken as a literal image, in which case it makes little sense but leads to an interesting picture. Or, by some stretch of the imagination, it might be taken as a metaphor. Or, finally, it may be taken as a meta-linguistic comment, as "Images buried in 'the eye of the dog of the dead'". All these meaningsare equally valid; and Paz probably intended them all.

Paz's poetry unearths the images buried in "the eye of the dog of the dead" and other magical, half-nonsensical, intuitively potent juxtapositions of words. He rearranges the language we use to describe reality and thereby draws out the hidden structure of reality — the inner world. When he wrote, in another poem, "The amphitheater of the genital sun is a dungheap," he was not just stringing words together at random, he was trying to use unusual juxtapositions of words to connect parts of the mind that are not usually connected. Without knowing the psychological specifics, he realized that thought depends upon associative memory, and he was trying to form new connections in the memories of his readers, so as to free their thinking from its conventions. Reading Paz's poetry is like having the experience of brainstorming imposed on you — this is the essence of Surrealism, and in Paz's work it is realized to an unparalleled extent.

Does "the genital sun" imply that the genitals are fundamental life-giving sources of energy, or does it imply that the physical sun is in some sense a sexual symbol? Is "the ampitheatre of the genital sun" a colorful way of referring to the world? Shakespeare said "all the world's a stage," so why not an amphitheatre? Does Paz mean to say that our world, particularly insofar as it is illuminated and empowered by sexuality, is a dungheap? The point is not any particular interpretation, it is the process of shifting symbols and subliminal connections.

And just as he saw images hidden in linguistic forms, Paz saw language hidden in everything:

> *Are there messengers? Yes,*
> *space is a body tattooed with signs, the air*
> *an invisible web of calls and answers*
> *Animals and things make languages,*
> *through us the universe talks with itself.*
> *We are a fragment —*
> *accomplished in our unaccomplishment —*
> *of its discourse. A coherent*
> *and empty solipsism:*
> *since the beginning of the beginning*
> *what does it say? It says that it says us.*
> *It says it to itself. Oh madness of discourse,*
> *that cause sets up with and against itself!*

We are, Paz says, a fragment of discourse. The universe is a discourse, a coherent talking-to-itself which ultimately says nothing. And what does it talk to itself about? It talks to itself about the fact that we are one of the things it talks about, and about the madness of the fact that it is talking! It talks to itself about talking! Just like Dick's "VALIS" voice — which was the "abstract structure of the universe," and which spoke to Dick specifically to tell him that it existed. Just like Rimbaud's "inner world," full of fire, this filled him with inspiration and impelled him to write poems about — precisely this inner world. There might seem to be something bogus about a revelatory experience that gives you revelations only about itself. But the important thing is that this process of self-revelation automatically awakens the mind to the fact that everything is composed of this same process of self-revelation. X speaks, to spread the word that X exists, but the very existence of X is synonymous with its omnipresence. Not merely "I am that I am," but "I am that I tell you I am" — and, furthermore, "I am that I tell you I am, and so is everything else!"

Language speaks, to reveal the transcendental existence and fundamental power of language, and in the process it reveals that language underlies everything. The circle of creation connecting language, reality and mind was at the center of Paz's poetry. He used surreal language to jar the mind into perceiving alternatives to ordinary reality, and to depict the language inherent in ordinary objects, in the universe, in space, in our bodies.

As a final example, let consider the following fragment, an excerpt from a poem written as a reaction to the music of John Cage:

> *I am*
> *an architecture of sounds*
> *instantaneous*
> *on*
> *a space that disintegrates itself.*
> *(Everything*
> *we come across is to the point.)*
> *Music*
> *invents silence,*

architecture
invents space.
Factories of air
Silence
is the space of music:
an unextended
space:
there is no silence
save in the mind.
Silence is an idea,
the idee fixe of music.
Music is not an idea:
it is movement,
sounds walking over silence
(Not one sound fears the silence
That extinguishes.)
Silence is music,
music is not silence.
Nirvana is Samsara,
Samsara is not Nirvana.
Knowing is not knowing...

John Cage was one of the most radical composers of the century. He composed music for the lid of a piano, and for pianos prepared by hanging paper clips and other objects from the strings. Sometimes he composed by playing a normal song, recording it, then cutting up the tape and reassembling it in a random order. Once he composed a piece consisting of four minutes or so of silence — the idea being that there is no such thing as silence, except for a deaf person; that whatever sounds you hear during that four minute interval of silence are just as valid a piece of music as Beethoven's Ninth. He was fond of observing that even in an anechoic chamber, one hears two sounds, a low one and a high one. One's heart and one's nervous system.

Paz connects Cage's analysis of silence and music to the Buddhist concepts of Samsara and Nirvana. Yes, Nirvana is Samsara. God is everywhere. As the Gnostic version of Jesus said, split a stick and I am there. But no, Samsara is not Nirvana. There is a difference between seeing a rock and seeing a rock with Buddha in it. And yet there is no difference. Under a certain interpretation, the distinction between silence and music is a particular version of the distinction between void and form. Remember, the Buddhists often pointed out that Void is never truly Void.

Here Paz is subtly juxtaposing concepts generally considered unrelated, with the goal of transmitting the mystical aspects of the experience of listening Cage's music. Unlike Rimbaud, Paz does not feel a deep need for the prodigious and rational disordering of all the senses. He concentrates more on disordering linguistic forms — on illustrating the linguistic forms underlying everyday physical and mental reality, so as to make vivid the Whorfian idea that everything is at bottom linguistic.

For after all, what does Paz mean when he writes "I am an architecture of sounds?" An architecture is a form, a way of arranging space. "An architecture of sounds" is simply a piece of music. Does he mean that he is, literally, music? To describe music in spatial terms is to emphasize the transcendent nature of pattern, of form. An arrangement of sounds, he is saying, is essentially the same as an arrangement of physical space — what is essential is the arrangement.

He is an arrangement, a pattern, which may be expressed in words, in sounds, or in pictures — in Rupa or in Nama, to use Whorf's interpretation of the Sanskrit terms.

Samuel Beckett: No Body in No Place

The Surrealists were known for excess. As a contrast, it is worth considering an artist who revered simplification above all else, who used his deepest intuitions to bring back visions of asimpler, more elementary realm than the ordinary world. This person is Samuel Beckett.

Beckett is best known for his plays, *Waiting for Godot* (1997) and *Endgame* (1970) which present a unique combination of surreal dialogue, stream-of-consciousness monologue, and physical comedy. He also wrote a number of outstanding conventional novels. But what I am interested in here are the prose poems which Beckett produced with increasing frequency toward the end of his life.   To start off, let us take a look at a fragment from the first page of *How It Is* (1964), written in 1964. It has been remarked that whereas Shakespeare only said "Life is a tale told by an idiot," in this book Beckett demonstrated it:

> *past moments old dreams back again or fresh like those that pass or things things always and memories I say them as I hear them murmur them in the mud*
> *in me that were without when the panting stops scraps of an ancient voice in me not mine*
> *my life last state last version ill-said ill-heard ill-recaptured ill-murmured in the mud brief movements of the lower face losses everywhere*
> *recorded none the less it's preferable somehow somewhere as it stands as it comes my life my moments not the millionth part all lost nearly all someone listening another noting or the same.*

This is a far cry from Paz's florid linguistic excesses, or Rimbaud's tightly woven networks of sensations and images. It is, rather, language stripped bare to the bone. Language with all the excesses removed. It is primal expression, as simple as expression can possibly be. Is there any way to compactify a phrase like "my life last state last version ill-said ill-heard ill-recaptured ill-murmured in the mud"? There is no metaphor here, or almost none. Whether or not the character is really in the mud is hardly relevant. This is "my life my moments" recorded "as it stands as it comes," and this mode of expression is "preferable somehow somewhere" even though it is "not the millionth part," even though it is "all lost nearly all."

Instead of trying to remake language in the image of the inner world, Beckett sought to strip language of everything but the simplest core, the idea being that this core corresponds to the inner world. In this view, fancy language — perhaps all language — only distracts us from ultimate reality. Listen to the final page of *Company* (1996):

> *...Till finally you hear how words are coming to an end. With every inane word a little nearer to the last. And how the fable too. The fable of one with you in the dark. The fable of one fabling of one with you in the dark. And how better inthe end labour lost and silence. And you as you always were.*
> *Alone.*

Here language is taken as a metaphor for external reality. The entire book *Company* recounts the experiences of someone lying on his or her back in the dark, listening to a voice, which may or may not be coming from someone beside him/herself. The person talks to the voice, but s/he is never sure if the voice is really responding or just going on of its own accord. Most of the book details either bodily sensations or communication with the voice. Language is portrayed as the only point of contact with the outside world: indeed, the only evidence of the existence of the outside world. Language is the outside world, in the universe of *Company*. And each word is inane. "How much better in the end labor lost and silence, and you as you always were, alone." Beckett saw words as a tool for perpetrating illusion. His contradiction was that he nonetheless wrote. But he attempted to resolve this contradiction by stripping his language of as much illusion as possible, by being maximally direct.

In *Worstward Ho* (1993), Beckett took this approach to its ultimate extreme. The subject of the book is nothing less than the creation of the universe; the emergence of pattern from nothingness. Here is how it begins:

> *On. Say on. Be said on. Somehow on. Till nohow on. Said nohow on.*
>
> *Say for said. Missaid. From now say for be missaid.*
>
> *Say a body. Where none. No mind. Where none. That at least. A place. Where none. For the body. To be in. Move in. Out of. Back into. No. No out. No back. Only in. Stay in. On in. Still.*
>
> *All of old. Nothing else ever. Ever tried. Ever failed. No matter. Try again. Fail again. Fail better.*
>
> *First the body. No. First the place. No. First both. Now either. Now the other. Sick of the either try the other. Sick of it back sick of the either. So on. Somehow on. Till sick of both. Throw up and go. Where neither. Till sick of there. Throw up and back. The body again. Where none. The place again. Where none. Try again. Fail again. Better again. Or better worse. Fail worse again. Still worse again. Till sick for good. Throw up for good. Go for good. Where neither for good. Good and all.*
>
> *It stands. What? Yes. Say it stands. Had to up in the end and stand. Say bones. No bones but say bones. Say ground. No ground but say ground. So as to say pain. No mind and pain? Say yes that the bones may pain till no choice but stand. Somehow up and stand. Or better worse remains. Say remains of mind where none to permit of pain. Pain of bones till no choice but up and stand. Somehow up. Somehow stand. Remains of mind where none for the sake of pain. Here of bones. Other examples if needs must. Of pain. Relief from. Change of.*
>
> *All of old. Nothing else ever. But never so failed. Worse failed. With care never worse failed.*
>
> *Dim light source unknown. Know minimum. Know nothing no. Too much to hope. At most mere minimum. Meremost minimum.*

This astoundingly simple passage has more to say more about language, mind and reality than a hundred academic papers in linguistics, psychology and philosophy journals. In Beckett's world, saying is equivalent to creation. The world is formed through language. First comes "On," and immediately afterwards comes "Say on." The first act is existence, and the second is speech acknowledging existence. "Say bones. No bones but say bones. Say ground. No ground but say ground." Bones and ground do not exist, but in saying them one grants them a degree of being. They may be illusory, but they are viscerally present.

"Other examples if needs must," he wrote. "All of old. Nothing else ever." To Beckett diversity of form is meaningless. It's all just nonexistence made real through speech. All we ever see is "dim light source unknown," about which we can only know the "meremost minimum." But we create from nothingness, from the dim light, mind and pain. "Say remains of mind where none to permit of pain." We speak mind into existence, in order to make pain possible.

Beckett is, in essence, giving us a prescription for constructing virtual reality — not from the engineering point of view but from the phenomenological point of view. First the body. No, first the place. Bones. No bones, but say bones. Somehow the bones may stand. He is telling us, from the standpoint of basic human experience, what the basic ingredients of a world are. And he is expressing a persistent disgust with the process. Why create all this stuff, when there's nothing really there. Say remains of mind where none to permit of pain. Why create a mind when it will only feel pain?

Say yes that the bones may pain till no choice but stand. Why would the bones want to stand? — only because it hurts not to. The world creates itself because it hurts not to exist, but then it hurts to exist as well. It's a lose-lose situation.

Language, Beckett says, creates reality. The world is virtual, and the programming language in which it is written is plain old English (or maybe French, in which Beckett alsowrote...). For Beckett the crucial point is not the ultimate reality of the world, but rather the unpleasantness of the world. Just as for Dick the crucial point was not ultimate

Goethe

Now, near the end of our journey through the world of literary/philosophical/spiritual intuition, it's time to change course a little.

Many of the writers mentioned above were unhappy men, who lived tragic, unfulfilled lives. One should not think from this, however, that deep pattern space voyaging is necessarily cruel, that artistic hyperspace always exacts a harsh toll on the human mind. Rather, such men as Philip K. Dick, Rimbaud and Samuel Beckett were unhappy before their journeys into the deeper realms of pattern space — it was, in part, their unhappiness which spurred them on to transcend conventional modes of perception. If ordinary waking consciousness had been more pleasurable for them, they probably would not have sought out alternatives. But it is also perfectly possible to be spurred into a creative use of pattern-space out of an energetic, restless dissatisfaction rather than a brooding, depressed dissatisfaction. Probably the best case in point is the German writer/philosopher/ scientist, Goethe.

Goethe journeyed deep into the realm of pattern space, and came back with such jewels as *Faust* — one of the most brilliant parables of pattern space ever created. And all the while he carried out a remarkably various and fulfilling life in the "real" world. Goethe's life, as well as being intriguing in its own right, is valuable as a demonstration. Goethe shows us that success in artistic contact with pattern space is not mutually contradictory with success in ordinary life.

Goethe was blessed with a keen, wide-ranging intelligence and a huge amount of energy. He was constantly bounding from one project to another and from one type of endeavor to another: painting, drama, prose, poetry, biology, physics, political administration ... After barely squeaking through law school, he won fame, and helped start the Romantic Movement in literature, with his novel *The Sorrows of Young Werther*. He wrote some poems and popular dramas, and then devoted himself for a decade to the administration of the small duchy of Weimar. Finally he tired of politics and left for Italy, telling no one of his departure and traveling under a false name. In Italy he painted, and lived the life of an artist. His energy and intelligence were unable to compensate for his basically mediocre artistic talent — but the trip rejuvenated his energies. "All the dreams of my youth," he said, "I now see living before me. Everywhere I go I find an oldfamiliar face; everything is just what I thought it, and yet everything is new. It is the same with ideas. I have gained no new idea, but the old ones have become so definite, living and connected with each other, that they may pass as new." He paid notably little attention to the artistic and historic wonders of Rome and Florence, spending his time instead painting, thinking, and studying plants, observing that "The book of Nature is after all the only one which has in every page important meanings."

Clearly Goethe's trip to Italy was an experience of deep spiritual insight and personal growth. The direct perception of deep interconnections, the increased vividity and vivaciousness of the everday — all this speaks plainly of hyperrealistic experience, of openness to underlying pattern space. After this he could not go back to politics — fortunately, the Duke of Weimar saw this, and paid Goethe his high salary as a kind of court genius rather than as an administrator. Unlike most creative artists, Goethe, for most of his life, was paid very highly just to think, theorize and create: he didn't have to worry about the marketability of his creations, nor about working a "side job."

In addition to *Werther* and *Faust*, Goethe wrote two great novels, *The Travels of Wilhelm Meister* and *Elective Affinities*, a wonderful autobiography *Poetry and Fiction*, popular plays like *Gotz* and *Hermann und Dorothea*, and a number of classicist dramatic poems — *Tasso, Egmont, Iphigenie* ... He was not a consistent writer; these profound artistic successes were just a fraction of his total literary output, some of which was embarrassingly weak. But his vast and various output is an indicator of his immense creative energy.

Furthermore, a large proportion of his time was spent on nonliterary pursuits. After Italy, he gave up painting, but he took up biological and physical science with an increased passion. To him all these different pursuits were as one: all part of an attempt to understand and participate in the wholeness of the universe.

Goethe's theory of the morphology of plants was revolutionary and in essence correct. He viewed different parts of plants, such as leaves, flowers and stalks, as coming from the same underlying form. Time, he said, caused the fundamental forms to develop in different ways, creating different overt forms. He drew attention to the structural and geometric similarities of different plants, and different parts of the same plants. All these ideas seem fairly obvious today, in the context of evolutionary and genetic theory, but at the time they contradicted the established dogmas. It took a great deal of courage to publish and promote them in the face of almost universal rejection from the scientific establishment.

His theory of colors was also reviled by the scientific establishment, but in this case justly: he was incorrect. Rejecting Newton's concept of the spectrum, he viewed all colors as made up from the two different principles of Darkness and Light. This fundamental opposition was expressed in different ways depending on different circumstances, thus yielding different tcolors. He argued vehemently for his view of things, and collected masses of interesting optical data. But still he could not explain the disturbing fact of the rainbow; and, due to his disinterest in mathematics, he never really understood the subtlety of the Newtonian view. Had the scientists been more receptive to his botanical insights, he might have accepted their views on optics more readily; but as it was, he had evolved a very bitter attitude toward scientists, and especially toward their use of mathematics. In the end it seems that most of the energy Goethe spent on optical research was wasted — but anyhow, he of all people had energy to spare!

Throughout his life Goethe had been interested in writing a version of the *Faust* story. Dr. Faust, so the legend went, was a kind of traveling alchemist and quack doctor, traveling the world in search of arcane and occult knowlege. Finally he sold his soul to the devil in exchange for knowledge and understanding. Goethe wrote a few fragments of the first part of his *Faust* in his thirties, but there was then a twenty-year gap before he finished what is now known at *Faust, Part One*. Finally, nearing the age of eighty, he returned to his vision and wrote *Faust, Part Two*, quite different from its predecessor, but yet embodying the same themes, and carrying them to a higher level.

God and the Devil, Mephistopheles, wager on whether the Devil will be able to corrupt Faust. Then Mephisto makes Faust, knowledge-seeker, a bargain: he will show Faust all there is in the world, give him an endless and universal feast of information and experience, if Faust will only promise not to become *absorbed* in this parade, if Faust will retain his detachment and never become fully satisfied with the world. Once he has run the gamut of earthly pleasure, orgies and all, and he has run the gamut of intellectual pleasure, had his fill of science and philosophy, Faust is very nearly tempted by deep love relationship with a woman, Gretchen. But he remains aloof, cruelly abandoning Gretchen to continue his quest.

Faust, Part One is an incredibly diverse and entertaining dramatic poem, full of colorful scenes and characters — but this entertaining intricacy only lends irony to the fundamental point, which is the *frustrating nothingness of existence* – the lack of any absolute meaning. Even the passion Faust feels for Gretchen, vehement as it is, is feverish and transitory. He seeks the Absolute, which can never be found.

In Part One of Faust, it has often been said, the problem of existence is stated but not solved. What Goethe has produced here is a profoundly existentialist document. It poses the anguish, the angst of existence, without more than hinting at a way out.

Part One is artistically beautiful, but in a deep sense, it is not complete. Had it been written by a depressed, angst-ridden individual, it would be at least psychologically complete. But it was not: it was written by Goethe, by an energetic, excited man, in love with the world in spite of its fundamental ungroundedness. *Faust, Part One* does not even come close to fully expressing Goethe's spiritual insights.

Goethe had no affection for organized religion. But neither did he advocate atheism. Each individual, he believed, must cultivate their own religion. True spirituality was to be found, not in nihilistic freedom and denial of the Supreme Being, but rather in the everyday recognition of the Supreme Being within oneself:

> *'I believe in God' is a beautiful and praiseworthy phrase; but to recognize God in all his manifestations, that is true holiness on earth...*
>
> *[In the Four Gospels] there is a reflection of a greatness which was emanated from the person of Jesus, and which was of as divine a kind as was ever seen upon earth. If I am asked whether it is in my nature to pay Him devout reverence, I say — certainly! ... If I am asked whether it is in my nature to reverence the sun, I again say — certainly! For he is likewise a manifestation of the Highest Being. I adore in him the light and the productive power of God, by which we all live, move and have our being. But if I am asked whether I am inclined to bow before a thumb bone of the apostle Peter or Paul, I say — away with your absurdities! ...*
>
> *Let mental culture go on advancing, let science go on gaining in depth and breadth, and the human intellect expand as it may.... [A]s soon as the pure doctrine and love of Christ are comprehended in their true nature, and have become a living principle, we shall feel ourselves great and free as human beings, and not attach special importance to a degree more or less to the outward forms of religion.*

To Goethe, God was to be found in mind and nature, not in the rituals of organized religion. And it is this attitude which, finally, gains its highest expression in *Faust, Part Two*.

The First Part of *Faust* is concrete, full of vivid portrayals of everyday events. It is not a drama to be performed, but still, when read it evokes genuine dramatic scenes in the reader's mind. The Second Part is different: it exists on a higher, more abstract plane. It has left the trappings of the real world behind, and has ascended into the realm of pattern space. It is not so entertaining as the First Part — though there are some funny scenes, such as the one where an impish homunculus in a bottle, created by Wagner who has taken Faust's place, undertakes to instruct the devil in classical aesthetics. But the obscure passages are more than redeemed by the immeasurably beautiful ending.

Faust, finally, finds something with which he is satisfied, and he agrees to deliver himself over to the Devil. But it is not any worldly pleasure, sensual or intellectual, to which he sacrifices himself — it is a vision. It is a deep vision of the world: a vision, in my language, of pattern space as pattern space! Having seen to the essence of the world, having seen the underlying unity amidst the dizzying wonderful diversity, having finally come to see God as manifested in concrete processes of creation, Faust is contented. The brilliance of the moment makes his future fate irrelevant. He is interconnected with the world, and thus his personal future is no more important than the future of the rest of the world. Existentialism is overcome as the symptom of a false attitude. Visionary ecstasy is plucked out of the jaws of absolute despair. The unity of individual consciousness and cosmic consciousness is made manifest. Patterns and forms coalesce together, in the manner of *anandamaya*, and thus the universality of structures of consciousness ascends above the apparently divisive forms created by the lower realms of Being.

It is crucial that Faust's blissful vision occurs at an extremely unlikely moment, at the conclusion of a nightmarish vision in which he is visited by a vision of four grey women, Want, Guilt, Necessity and Care. Faust argues mightily against the four demon hags, celebrating the spiritual beauty of the everyday:

I only through the world have flown
Each appetite I seized as by the hair;
What not sufficed me, forth I let it fare,
And what escaped me, I let go.
I've only craved, accomplished my delight,
Then wished a second time, and with might
Stormed through my life: at first 'twas grand, completely
But now it moves most wisely and discretely
The sphere of Earth is known enough to me
The view beyond is barred immutably;
A fool, who there his blinking eyes directeth
And o'er his clouds of peers a place expecteth!
Firm let him stand, and look around him well!
This World means something to the Capable.
Why needs he through Eternity to wend?
He here acquires what he can apprehend.
Thus let him wander down his earthly day;

When spirits haunt, go quietly his way;
In marching onward, bliss and torment find,
Through, every moment, with unsated mind!

Faust affirms the power of transformation, development, motion. But the battle is a difficult one, and he is blinded by the demon hag Care –

Throughout their whole existence, men are blind

So Faust, be thou like them at last!

Faust conceives a vast construction project:

The night seems deeper now to press around me
But in my inmost spirit all is light
I rest not till the finished work hath crowned me

The master's Word alone bestows the night
Up from your couches, vassals, man by man!
Make grandly visible my daring plan!
Seize now your tools, with spade and shovel press!
The work traced out must be a swift success.
Quick diligence, severest ordering
The most superb reward shall bring
And, that the mighty work completed stands,
One mind suffices for a thousand hands

he proposes to drain a swamp and fill it with soil on which millions can live. And in this activity, this creation, he finally sees a meaning in life. It is the action, the creativity, that fills him with joy. The process of making, of giving birth — this not the abstract Christian God but more like a de-superstition-ized version of an erotic pagan mother-Goddess. This is the Perceptual-Cognitive Loop, the abstract form of creation, but enhanced with a higher awareness, an awareness of absolute Being — it is the self-creating feedback loop of spiritual awareness. Just envisioning this vast project of his, this tremendous instance of physical creativity, fills Faust with such joy that he utters the magic words, the words which give Mephistopheles permission to take possession of his soul — "I now enjoy the highest moment":

Below the hills a marshy plain
Infects what I so long have been retrieving
This stagnant pool likewise to drain
Were now my latest and my best achieving
To many millions let me furnish soil,
Though not secure, yet free to active toil;
Green, fertile fields, where men and herds go forth
At once, with comfort, on the newest Earth,

And swiftly settled on the hill's firm base,
Created by the bold, industrious race.
A land like Paradise here, round about:
Up to the brink the tide may roar without,
And though it gnaw, to burst with force the limit,
By common impulse all unite to hem it.
Yes! to this thought I hold with firm persistence;
The last result of wisdom thinks it true:
He only earns his freedom and existence
Who daily conquers them anew.
Thus here, by dangers girt, shall glide away
Of childhood, manhood, age, the vigorous day:
And such a throng I fain would see,
Stand on free soil among a people free!
Then dared I hail the moment fleeting!
"Ah, still delay — thou art so fair!"
The traces cannot, of mine earthly being.
In eons perish, they are there!
In proud fore-feeling of such lofty bliss,
I now enjoy the highest Moment — this!

With these words, Faust delivers his soul to the devil, Mephistopheles — having understood the power of creation, the infinite significance of the single moment, and the interconnectedness of his life with the rest of the world. "The traces cannot, of mine earthly being, in eons perish" — because his earthly being will continue in the people he has helped, and, ultimately, in all things! The Devil, lacking a deep spiritual understanding, is baffled:

No joy could sate him, and suffice no bliss
To catch but shifting shapes was his endeavor;
The latest, poorest, emptiest Moment — this —
He wished to hold it last forever.

How can an empty moment in fact interdepend with other, fuller moments: moments in the past and moments to come? How can there be an order outside of time, which makes life worthwhile, which is an inner light even when the eyes are blind? The Devil will never understand this, which is precisely what makes him the Devil.

Inevitably, under these circumstances, the Devil's grasp on Faust is called into question. God has the power to forgive — and what better situation in which to exercise this ability? In the end Faust's vision, of happiness through good works and divine interconnection, must redeem him. And, in a typical Goethe masterstroke, Faust's arrival redeems Heaven as well. In the Prologue to *Faust*, Heaven is a rather lifeless, antiseptically blissful sort of place. But in the conclusion it is viewed as a realm full of life, development, action. It is exciting rather than sleepily euphoric. No devout Christian, Goethe put the standard Christian mythology into *Faust* only for sake of concreteness and ease of communication. In the end, he created a Devil more mischievous than evil, and a Heaven eminently suited to Faust, or Goethe.

Faust's voyage, obviously, parallels Goethe's own. Goethe had no lack of body-level pleasures: he took many lovers, and finally married a coarse, sensual working-class woman. He lived the life of the Court, was an inveterate prankster, and experimented with other lifestyles as well, as during his incognito trip to Italy. He threw himself into science, art, literature and politics with a passion. In the end, nothing finally contented him: he kept moving from writing to science, from science to writing, from fiction to poetry, biology, optics, politics, and so forth. He had a serious fear of romantic commitment; even once married, he had numerous affairs, and frequently made long journeys alone. He experimented extensively in all domains, but nothing wholly satisfied him until finally, in his last decades, he found contentment, not in any new passion, but in an understanding of the whole. He understood that the passion, not the object, is the crucial thing. He exalted process, development, growth — creation! — as the manifestation of God in the world. God, he saw, is morphological development, creative process. Love for corporeal women was beautiful but not fundamental; the essence was the Goddess, the universal process of giving birth, or, in his famous coinage, the eternal-feminine.... Feeling unity with the Goddess, he distanced himself from the everyday world, assuming what has frequently been called an "Olympian detachment."

In the end, for all his human flaws — his egotism, his occasional dogmatism, his strange relations with women — Goethe was a very rare thing in human history: not only a great achiever but a truly great person. He is remarkable not only for what he did but for what he was. He saw deeply into pattern space, and in a thoroughly Western way, through diverse and impassioned activity. He died reconciled to death, understanding his own life cycle as part of the big picture of universal creative and destructive force. His life unfolds like a symphony, reaching a dramatic conclusion in which, finally, the divine is perceived in its natural place, in the everyday world.

Too many creative visionaries remain stuck in the world of *Faust, Part One*. Dazzled by the variety of forms emanating from *anandamaya*, but realizing their ultimate insubstantiality, they remain in a state of confusion. The ecstatic act of intuitive creation fails to induce a lasting mental transformation. But Goethe's story indicates the possibility of going beyond this stage — without submitting to Oriental routines of meditation, sleep deprivation, and so forth. Goethe indicates a kind of spiritual "enlightenment" achieved through systematic and diverse creative activity.

It is for this reason, essentially, that Nietzsche took Goethe as a model for his *Ubermensch*, Superman.... Goethe was not superhuman, but he is an extraordinary lesson as to the tremendous potential of a single human life. Goethe lived the kind of life that many of us would create for ourselves if we had the ability to structure our own virtual world. Each of us has different flaws, different talents, and a different trajectory of development. But we should not forget the concreteness and possibility of the universal goal — dynamic, flowing, creative harmony; harmony inclusive of creative conflict; deep perception of pattern space and the interconnectedness of the world. Or, as Goethe put it, in *Faust*'s closing verse,

> *All things transitory*
> *Are but as symbols sent;*
> *Earth's insufficiency*
> *Grows to Event;*
> *The Indescribable,*
> *Here it is done:*

The Eternal-Feminine
Draws us on

Nietzsche, with his half-serious, half-ironic spite toward all things feminine, considered these closing lines Goethe's biggest joke. Above all, I suppose, it was a statement about deep structure versus surface structure, essence versus appearance. The "eternal-feminine" of Goethe's lines had little to do with the foolishness and finery of the average female in Goethe's or Nietzsche's time, but had to do rather with the power of birth and creation – which Goethe viewed as the eternal and essential part of femininity. Nietzsche insisted fervently that there are no essences, no noumena – the apparent, phenomenal world is all. Yet for one who chooses to perceive abstract patterns – "essences" – directly and form them vividly in their mind, these essences are not noumena but phenomena: they are experience realities, Firsts. It is the abstract pattern of creation that draws us onward: without this pattern, pattern-space would be static and there would be no such things as experience or thought.

The Language of Structure, the Structure of Language

I have lightly skimmed over the work of a number of linguistic visionaries — men who, by dint of their personal, intuitive experiences with language, arrived at penetrating insights into the nature of mind and world. Each of these men had his own peculiar preoccupations, his own talents and shortcomings. But yet, among their very different analyses of linguistic reality, one finds a remarkable amount of common ground. There is an unvarying vision of language as indicative of another world deeper than the spatiotemporal one — an underlying realm of pattern, information, structure. In short, we may say that *anandamaya* presented itself to these men as a realm of abstract linguistic forms. In the psyche of such a creative genius in the moment of creative flow, intuition adopts the role of an agent transforming these abstract linguistic forms into sentences, paragraphs, words, story ideas. And so in such minds it is language that brings together the personal and the universal, putting the timelessness of the present moment ahead of the sheaths that cloak Being.

The linguistic bias of the above is obviously a consequence of the fact that we are dealing with writers here, rather than (say) visual artists, priests or musicians. But it is also indicative of a deep truth, which is that language plays a significant, perhaps dominant role in maintaining the lower levels of being, especially the "central triad" of ordinary waking consciousness, Mind, Body and World. This single statement packs so much information that thousands of pages of detailed analysis would be required to adequately explore it and document it. But, at bottom, the phenomenon that it identifies should not be at all surprising.

Language is a large part of what causes different minds to judge things similarly. Individualized thought that does not follow standard linguistic patterns is less likely to be common among various minds. It follows from this that, to a certain extent, it is language which maintains consensus reality: the world is built out of language. One may arrive at this conclusion by logical or empirical analysis; or, like our linguistic visionaries, one may apprehend it directly. One may apprehend it directly by seeing patterns as patterns — and thus "remembering" what consensus reality causes us to "forget," which is that we are all just interconnected subsets of pattern space.

Our linguistic mystics, our creative literary visionaries, have found their muse in the notion that everything is language and information — that reality is a "pure simulation," a trick with linguistic and algebraic mirrors, a construction of self-deceptive logic. This realization loosens the bonds of consensus reality, and thus encourages the mind to explore and create patterns outside the bounds of consensus.

All creative visionaries experience, on some level, a similar view of the world (world-as-pattern, world-as-information). Some artists devote substantial time to understanding their own creative process; others proceed on a more implicit basis. But, one way or the other, in many different cases it is *seeing everything as pattern* that opens up the mind to consider new and innovative patterns, patterns that contradict the patterns hitherto assumed absolute and irrefutable. Intellectually realizing that everything is pattern is not enough; the realization must be integrated into one's perceptual and cognitive systems. In order to derive truly deep and fantastic creative accomplishments, one must move toward the farthest edges of Intuition, toward the Realm of Bliss, where definite inspirations and ideas meld into general, transpersonal patterns — where forms play, multiply and subdivide, and the goals and biases of humans are just particular forms dancing among many, many others.

Chapter 20
Mindfulness and Evil[72]

"Evil" is a somewhat comical word to me, since I don't believe in any absolutist religious or moral framework. I'm a long-time fan of Nietzsche-style analysis of morality, which observes that in many cases in human history, "evil" has been defined as "whatever's bad for some particular social class in some particular society at some particular point in time."

My favorite story about evil involves a conversation I once had with a friend who founded (and eventually sold, at a large profit) a successful software consulting firm oriented toward serving the US military intelligence community. He had a second-in-command who bore a significant physical resemblance to Superman (with a dash of Elvis thrown in), and I was needling my friend about the superhero theme, in the vein of "If your second-in-command is Superman, then which superhero are you? How do you top Superman?" and so forth.

His reply was "In truth, I've always viewed myself as less of a superhero and more of the dark overlord type." With a shaved head, a small, well-sculpted beard and moustache, and a propensity for wearing long dark coats, he was definitely right – he looked the part of a Darth Vader of military IT.

But I protested: "Wait! Isn't there a conflict? How can an evil overlord have an assistant who's a superhero on the good side?"

My friend laughed. "I didn't know you were so naïve," he said. "You're still viewing good versus evil as a contradiction."

"Sure," I said, "how do you view it?"

"I view good versus evil," he said, pausing for effect, "... well, let's just say it's a *mutually rewarding business arrangement*."

This of course is the early-21'st-century-America update of Nietzsche. The elites of the US and Arab world each define themselves as good and their opponents as evil, and they each use this to help maintain power structures providing themselves with benefit.[73]

In spite of this cynical and pragmatic aspect, however, I think there is still some deep truth underlying the concept of "evil" – and I will continue to use the word "evil" here because I don't know a better one, even though the concept of evil as I mean it is not exactly the same as the concept of evil in Christianity or any other religion.

---

[72] The ideas in this chapter were partially crafted via conversations and ongoing dialogues with two people: Allan Combs and Gwen Goertzel; and were refined somewhat via conversations with Izabela Freire Goertzel.

[73] Yes, of course, this is a very incomplete analysis of the current geopolitical situation, and of course I ultimately have more empathy for the US than for the Arab world, but still, this aspect is there and it would be foolish to deny or ignore it.

In fact the approach I will take here is more nearly Buddhist than Christian, though it's not exactly in line with any particular school of Buddhism either. I will draw on the philosophical thinking of David Bohm, a quantum physicist who was also a good friend and sympathizer of Krishnamurti, the Indian mystical philosopher. Buddhism teaches that "All existence is suffering," meaning not that everything is pure suffering, but rather that all experiences have some aspect of suffering to them (an insight Nietzsche expressed in *Thus Spake Zarathustra* with his lines "Have you experienced one joy? O my friends, then you have experienced all woe as well! All things are enchained, all things are entwined, all things are in love!"). In these terms, the problem of evil may be most simply cast as: "Why does so much pain exist?" Why so much suffering?

The existence of so much pain in the world doesn't imply that any human or other organism is necessarily being evil in a moral sense – but it does seem to suggest that the universe itself is evil in a moral sense. One thinks here of Leibniz's much-mocked philosophy that "this is the best of all possible worlds" (see Leibniz, 1985). Is it really? Or can't one conceive of a world that has the good things of this one in it, but without nearly so much suffering?

From a physiological, objective-reality-oriented perspective, the answer to the "problem of evil" thus formulated is obviously that pain is an organism's way of registering threats to its survival. These signals have now been subverted beyond their original purpose, so that we can feel psychological pain when we fail to prove a math theorem or when a lover insults us, even when these things are no threat to our short or long term survival or the survival of our DNA. But the evolutionary origin of suffering is obvious: what we experience as pain and torment is the quale-level reflection of body-level "avoid this!" messages.

It follows from this that, in principle, an all-powerful organism would not need to experience any pain – since there would be no threats to its survival. This is an obvious observation but it turns out to be an interesting one, when one combines it with other observations made from a more subjectivist point of view – as we'll see below.

The analysis of emotion given above also needs to be pulled in here. Human emotional dynamics have the property of taking elemental *pain* and building it up into deeper, more systematic *unhappiness*. This has to do with human cognition and its ability to build complex systems out of memories of pain and anticipations of pain. Pain affects us through the unconscious portions of our nervous systems, thus impacting us in the powerful way of all our emotions; but memories and anticipations of pain can enact the unconscious portions of our nervous systems and affect us in the same sort of way.

In the rest of this chapter I will continue in this general vein, but moving in the direction of subjectivity. What is the property of subjective worlds, of minds, I will ask, that necessitates so much suffering? This is a question that Eastern philosophy turns out to say a lot about — and what I'll conclude here is that the wisdom of the East, when cast into the philosophical language I've been using in this book, implies a very crisp formulation: *the root of all evil is limited computational capacity.* Ultimately, in other words, suffering is caused by not being smart enough.

In a human sense, some suffering is caused by us not using our given resources well enough. But some of our suffering goes deeper than that – it is caused by the resource-limitations that are innate to our existence as humans, and without which we wouldn't be ourselves.

Of course, a phrase like "suffering is caused by not being smart enough" is ripe for misinterpretation – I don't mean to imply that more intelligent humans are systematically happier; in fact this seems not to be the case. Some of the happiest people I know are mentally retarded; and tormented geniuses are so common as to be almost cliché'. What I mean is rather that human organisms are wired for suffering in various ways, and that this wiring evolved as an adaptation to the limited memory and processing power of our brains. Furthermore, any other intelligent system (including an AI) that has severely limited processing power is also going to necessarily be wired for suffering in some way.

However, the amount of intelligence or computational resources of a system do certainly not determine the amount of suffering experienced by the system in any simple way: the architecture of the intelligent system makes a huge difference. Humans are definitely not wired for suffering-minimization. I like to think about this in quasi-mathematical terms, even though there is nowhere near enough detail or data here to approach doing actual calculations. For any particular amount of intelligence I and computational resources R, one may posit a minimum amount of suffering $s(I,R)$. One of my claims is that, if I is fixed, then $s(I,R)$ decreases as R increases. On the other hand, if R is fixed, then the dependency on I is subtler. One way (not the only way) to think about this dependency is in pragmatic terms related to organism-environment interactions. In some situations it may be that $s(I,R)$ increases with I: the stupider the happier, if survival is simple enough, because greater intelligence just brings greater opportunities for frustration. On the other hand, if the situation is more difficult then greater intelligence may reduce suffering.

But why does stupidity induce suffering, in this sense? This question leads us to the concept of "mindfulness," which will occupy most of the rest of the chapter. The basic idea I'll pursue is that, from a subjective and experiential perspective, suffering is primarily due to the inability of the individual mind to understand its own internal actions in detail – in Buddhist lingo, to a lack of "mindfulness." This inability, in humans, stems in part from cultural and psychological issues – but it also has a basis in fundamental computational limitations. There is no way for a system with severely limited computational power to both do complex things and possess full mindfulness while doing them. So to the extent that unmindfulness is the root of suffering, the conclusion is the one I've given above: suffering is the inevitable conclusion of the quest for significant intelligence within limited computational resources.

Proprioception of Thought

To initiate my in-depth analysis of the relation between suffering and computational capacity, I now turn to the philosophical thought of David Bohm, the late quantum-physicist-turned-philosopher. One of the last books Bohm wrote before his death, *Thought as a System* (Bohm, 1994), is wonderfully relevant to these issues.

Bohm's views on mind are substantially in sympathy with my own. He pictures thought as a system of reflexes — habits, patterns — acquired from interacting with the world and analyzing the world. He understands the self-reinforcing, self-producing nature of this system of reflexes — the emergence of autopoietic subsystems. And he diagnoses our thought-systems as being infected by a certain malady, a malady which he calls *the absence of proprioception of though*t. Though Bohm's language is unfamiliar, it turns out that what he is talking about is very familiar indeed.

Proprioceptors are the nerve cells by which the body determines what it is doing — by which the mind knows what the body is doing. To understand the limits of your proprioceptors, stand up on the ball of one foot, stretch your arms out to your sides, and close your eyes. How long can you retain your balance? Your balance depends on proprioception, on awareness of what you are doing. Eventually the uncertainty builds up and you fall down. People with damage to their proprioceptive system can't stay up as long as as the rest of us.

According to Bohm,

> *... [T]hought is a movement — every reflex is a movement really. It moves from one thing to another. It may move the body or the chemistry or just simply the image or something else. So when 'A' happens 'B' follows. It's a movement.*
>
> *All these reflexes are interconnected in one system, and the suggestion is that they are not in fact all that different. The intellectual part of thought is more subtle, but actually all the reflexes are basically similar in structure. Hence, we should think of thought as a part of the bodily movement, at least explore that possibility, because our culture has led us to believe that thought and bodily movement are really two totally different spheres which are no basically connected. But maybe they are not different. The evidence is that thought is intimately connected with the whole system.*
>
> *If we say that thought is a reflex like any othermuscular reflex — just a lot more subtle and more complex and changeable — then we ought to be able to be proprioceptive with thought. Thought should be able to perceive its own movement. In the process of thought there should be awareness of that movement, of the intention to think and of the result which that thinking produces. By being more attentive, we can be aware of how thought produces a result outside ourselves. And then maybe we could also be attentive to the results it produces within ourselves. Perhaps we could even be immediately aware of how it affects perception. It has to be immediate, or else we will never get it clear. If you took time to be aware of this, you would be bringing in the reflexes again. So is such proprioception possible? I'm raising the question.*

"Proprioception of thought" is a fancy phrase, a weird concept, a brain-stretcher. But a very similar idea has been proposed within the Zen Buddhist religion, under the much simpler name of "mindfulness." In the words of Zen Master Thich Nhat Hanh (1999).

> *[T]he seed of mindfulness — when manifested, has the capacity of being aware of what is happening in the present moment. If we take one peaceful, happy step and we know that we are taking a peaceful, happy step, mindfulness is present. Mindfulness is an important agent for our transformation and healing, but our seed of mindfulness has been buried under many layers of forgetfulness and pain for a long time. We are rarely aware that we have eyes that see clearly, a heart and a liver that function well, and a non-toothache. We live in forgetfulness, ignoring and crushing the precious elements of happiness that are already in us and around us. If we breathe in and out and see that the tree is there, alive and beautiful, the seed of our mindfulness will be watered, and it will grow stronger...*
>
> *Mindfulness makes things like our eyes, our heart, our non-toothache, the beautiful moon and the trees deeper and more beautiful. If we touch these wonderful things with mindfulness, they will reveal their full splendor. When we touch our pain with mindfulness, we will begin to transform it. ...*
>
> *Mindfulness is something we can believe in. It is our capacity of being aware of what is going on in the present moment. To believe in mindfulness is safe, and not at all abstract. When we drink a glass of water, and know that we are drinking a glass of water, mindfulness is there.*

This is just a different way of formulating familiar ideas from Zen Buddhist philosophy. But it is an interesting reformulation indeed.

Mindfulness is the mind acting, and knowing exactly what it is doing as it is acting. In other words, and without stretching things at all, mindfulness is proprioception of thought. Bohm's is a scientist's formulation. It begins with the behaviorist view of the mind as a collection of reflex-arcs, the system-theorist's conception of thought-systems as self-producing, and the physiological fact of proprioception — and it arrives at the same conclusion as Thich Nhat Hanh did, by pure intuition and experience, with ultimate simplicity. The conclusion is that, if mind were immediately aware of what it was doing, we would be a lot better off.

According to Thich Nhat Hanh, mind cannot become aware of itself by logical analysis, or by emotion. Mindfulness is a matter of directing attention toward the lower realms: *anandamaya, pranamaya, vignanamaya*. But this "attention" being directed is coming down from the upper realms, from above *manomaya*. In this way mindfulness is in fact a form of deep contemplation, a form of intuition — *vignanamaya*. Mindfulness is a bridge between the upper and lower levels, a shaft of light piercing the mental knots that bind up *manomaya* and *pranamaya*. Applying deep intuition to itself, the mind becomes aware that its internal systems are just held together by autopoiesis, without any absolute solidity or reality. The mind becomes aware that, in reality, its forms and patterns are just floating in *anandamaya*, the realm of Bliss.

Zen is not the only wisdom tradition to arrive at the same conclusion as Bohm. Yoga exercises are largely exercises in proprioception. They teach control of breathing and heartbeat, and in their slow methodical body motions they teach body awareness, oneness with the body — enhanced overall proprioception. The concept is that, through improved proprioception of the body, proprioception of thought will follow.

The reader versed in Western philosophy will note that mindfulness and proprioception are also very Schopenhauerian notions. True reality, according to Schopenhauer, was chiefly perceived through willing, through the feeling of the body responding to one's commands — in short, through proprioception. Creation and appreciation of art, particularly music, was said to give a similar feeling of immediate awareness, implying that thought is most proprioceptive when it is most creative, a conclusion that makes eminent sense.

Thich Nhat Hanh, in the Zen tradition, promotes meditation (rather than, say, yoga exercises or mushrooms or artistic creation) as a path toward mindfulness. He places particular emphasis on mindfulness of *breathing*: "Breathing in, flower; Breathing out, fresh".... From mindfulness of basic body processescomes, gradually, heightened mindfulness of the abstract processes of thought. He focuses on mindfulness of body processes because these are simplest to understand. But he also talks about *samyojama*, mental knots — which, we have seen, are basically nothing other than self-supporting thought-systems:

When someone says something unkind to us, for example, if we do not understand why he said it and we become irritated, a knot will be tied in us. The lack of understanding is the basis for every internal knot. If we practice mindfulness, we can learn the skill of recognizing a knot the moment it is tied in us and finding ways to untie it. Internal formations need our full attention as soon as they form, while they are still loosely tied, so that the work of untying them will be easy.

Autopoietic thought systems, systems of emotional reflexes, guide our behaviors in all sorts of ways. In his various writings, Thich Nhat Hanh deals with many specific examples, from marriage woes to warfare. In all cases, he suggests, simple sustained awareness of one's own actions and thought processes — simple mindfulness — will "untie the knots," and free one from the bundled, self-supporting systems of thought/feeling/behavior.

Psychedelic Psychotherapy and Cognitive Compression

Mindfulness meditation a la Thich Nhat Hanh is one powerful way of getting beyond the control of our more idiotic habits and seeing ourselves more directly as what we are – networks of patterns interpenetrating with the other pattern-networks in pattern space. The exalted trancelike creative state reported by Nietzsche, Dick and others is another approach to a similar end. Probably the most popular path beyond ordinary consciousness in the modern Western world, however, is the chemical path. Psychedelic drugs provide mind-liberation and immersion in pattern-space in a much faster and more immediately shocking way than meditation or artistic creation. Like all ways of opening the individual mind more fully to pattern-space, they have their strengths and weaknesses, and can be used in ways that are ultimately destructive as well as ways that are ultimately beneficial. But they deserve far more serious attention than they receive in modern Western society today, where they are lumped together foolishly with psychologically-shallow mind-altering drugs like heroin, marijuana and cocaine.

In this vein, I now turn from Bohm and Hanh to another stellar modern thinker, the transpersonal psychologist and psychedelic psychotherapist Stanislaw Grof. Among all the key thinkers of the discipline of transpersonal psychology, it is, in my view, Grof who has come closest to the essence of psychological dynamics.

One of the most common criticisms of psychedelic-drug-induced insights is that they are short-lived. More often than not, they start fading when the drug leaves one's body … and after a few weeks they are often a dim memory, enriching one's world-view but definitely not filling one with the same immediate light and knowledge as was the case during the peak of the experience. Now, one must not make the mistake of thinking that ephemerality dilutes the power of a mystical experience. In a very powerful sense, insight is eternal: once you are in hyperspace, the temporal continuum doesn't matter. Time is a creature of the physical, mental and body worlds: there is no time in *anandamaya*, and even in *quantum-maya* time only has a limited validity.

But even so, there is something very satisfying about integrating mystical insights into everyday life. This is important with all mystical insights, but it is especially crucial with drug experiences, due to their intensity and short lifespan. This kind of integration is the aim of Stan Grof's *psychedelic psychotherapy* (2001). In this therapy, instead of merely talking about their problems, psychological patients use LSD as a tool for seeing to the core of their problems. The therapist guides their trips and helps them, between trips, to gracefully meld their psychedelic insights with their daily lives. As Grof says,

> *The main objective of psychedelic therapy is to create optimal conditions for the subject to experience the ego death and the subsequent transcendence into the so-called psychedelic peak experience. It is an ecstatic state, characterized by the loss of boundaries between the subject and the objective world, with ensuing feelings of unity with other people, nature, the entire universe, and God. In most instances this experience is contentless and is accompanied by visions of brilliant white or golden light, rainbow spectra or elaborate designs resembling peacock feathers. It can, however, be associated with archetypal figurative visions of deities or divine personages from various cultural frameworks. LSD subjects give various descriptions of this conditions, based on their educational background and intellectual orientation. They speak about cosmic unity,* **unio mystica, mysterium tremendum,** *cosmic consciousness, union with God, Atman-Brahman union, Samadhi, satori, moksha, or the harmony of the spheres.*

*This peak experience is intended to help the patient **transcend** their psychopathology — a radical shift in emphasis from conventional psychotherapy, with its focus on the verbal exploration of the roots of pathology. In LSD therapy the objective is to untie the knots directly, via immediate experience of their non-absolute, self-producing nature, rather than indirectly via talk.*

Grof speaks, not of mental knots, but rather of "COEX systems" — systems of compressed experience. A COEX system is a collection of memories and fantasies, from different times and places, bound together by the self-supporting process dynamics of the mind. Elements of a COEX system are often joined by similar physical elements, or at least similar emotional themes. An activated COEX system determines a specific mode of perceiving and acting in the world. A COEX system is an attractor of mental process dynamics, a self-supporting subnetwork of the mental process network, and, in Buddhist terms, a samyojama or knot. LSD therapy unties these knots, weakens the grip of these COEX systems. The therapist is there to assist the patient's mental processes, previously involved in the negative COEX system, in reorganizing themselves into a new and more productive configuration. This therapeutic process has been dramatically successful on many occasions, and has enjoyed especial success with alcoholic patients. The thought-systems causing alcohol addiction are, it seems, particularly easily "dissolved" by the ecstasy of the psychedelic experience.

Grof's emphasis on the *compression* of thoughts and experiences is both interesting and important. For as I noted above, if one takes a computational view, pattern itself is just a different way of looking at compression. According to the algorithmic information approach to pattern theory, a pattern in some entity is just some process that allows one to produce that entity in a particularly simple way. Linguistic systems are patterns in the world in precisely this sense — they simplify the world for easy memory and comprehension. Grof's concept of compression brings us back once again to the pattern philosophy, the linguistic nature of the world, and the psynet model.

To recognize a pattern in something is to compress it into something simpler — a representation, a skeleton form. It is inevitable that we compress our experiences into what Grof calls COEX's. This is the function of the hierarchical network that structures each individual human mind: to come up with routines, procedures, that will function adequately in a wide variety of circumstances. We can never know exactly why we do what we do when we lift up our arm to pick up a glass of water, when we bend over to get a drink, when we produce a complex sentence like this one, when we solve an equation or seduce a woman. We do not need to know what we do: the neural network adaptation going on in our brain figures things out for us. It compresses vast varieties of situations into simple, multipurpose hierarchical brain structures.

But having compressed, we no longer have access to what we originally experienced, only to the compressed form. We have lost some information. This is the ultimate reason for what Bohm calls the absence of proprioception of thought. It is the reason why mindfulness is so difficult. Thought does not know what it is doing because thought can do what it does more easily without knowing. Proceeding blindly, without mindfulness, thought can wrap up complex aggregates in simple packages and proceed to treat the simple packages as if they were whole, fundamental, real. This is the key to abstract symbolic thought, to language, to music, mathematics, art. But it is also the root of human problems – the root of suffering.

Limited Computational Resources as the Root of All Evil

Intelligence, itself, rests on the lack of mindfulness. It rests on compression: on the substitution of packages for complex aggregates, on the substitution of tokens for diverse communities of experiences. This is the essence of the cognitive equation presented above: yesterday's emergent pattern is tomorrow's atomic token from which patterns emerge. It requires us to forget the roots of our thoughts and feelings, in order that we may use them as raw materials for building new thoughts and feelings. But this forgetfulness, after it has helped us, then turns around and stabs us in the back. It works against us as well as for us. Intelligence is a mixed bag.

To put it a little differently, proprioception of thought is closely related to *reversibility.* If the mind remembers everything it did to get to its current state, then in principle it at least has a reasonable chance of restoring itself to its previous state, if it wants to. But it's not possible to maintain this kind of proprioception and reversibility in the context of complex decisions involving complex knowledge-structures. In a subjective-reality sense, the loss of cognitive proprioception is closely connected to the introduction of irreversible psychological time. Physically, it is tied to the inevitability of decoherence in complex chaotic systems.

Compression is inevitable if one wants complex structures and dynamics within limited resources. But compression is lack of mindfulness which leads to suffering. This, quite clearly and crisply, is the sense in which lack of computational resources is the root of all "evil." Eastern philosophy and related philosophy like that of Bohm and Grof tells us that the root of all evil is that mind knows not what it does — but cognitive science tells us that the reason mind knows not what it does is that full cognitive proprioception would exhaust its computational resources in a wholly implausible way.

Bohm and Thich Nhat Hanh are optimists, in the sense that they believe proprioception can be indefinitely extended. This is an admirable point of view, but in the end, it is only correct if one considers the possibility of essentially infinite computational power and the total dissolution of the integrity and boundaries of human selves. Cognitive proprioception cannot be indefinitely extended within the constraints of the human form – though certainly it can be extended much further than we humans typically do in our everyday lives.

Thich Nhat Hanh goes along with many other commentators in blaming modern society for the rampant lack of mindfulness he observes around him. Alcohol, TV, junk food and the commuter culture are pinpointed as the culprits. But surely this misses the main point. The ancient Orientals were not such a universally enlightened people – and nor were they, historically, prior to their involvement with the West. China's bloody and tumultuous history is evidence of this, as is the repressive nature of traditional Oriental culture. Thich Nhat Hanh's own country, Vietnam, is hardly a paradigm case of peace, balance and harmony, for reasons that have roots in Oriental culture as much as in Western influence. The truth is that, while aspects of modern culture certainly work against the quest for deep insight, the real problem lies in the nature of mind itself. Modern culture is not responsible for (to revert to Vedantic lingo for a moment) the sheaths obscuring Pure Being from itself. Even primitive tribal cultures lack mindfulness to some degree — though they are closer to Thich Nhat Hanh's ideal than Oriental or Occidental culture. None of the key structures and dynamics of mind escape the fundamentally irreversibility of compression, the non-proprioceptiveness of the unconscious. They merely express this compression in different sorts of ways.

This said, however, it must be admitted that there may exist differences of degree. Mindfulness may be easier in a tribal setting because life is simpler. We, in our culture, rely on intelligence for nearly everything. Twelve or more years of school are required in order to teach basic cultural competence. This reliance on the symbolic and abstract seems to carry with it a systematic lack of mindfulness. Essentially, the more complex the tasks one has to carry out, the more difficult it is to be mindful of one's actions. It's not so hard to be mindful while walking, or picking strawberries, or washing dishes. It is much more difficult when, say, writing a book like this, or solving an equation, or manipulating columns of numbers in a spreadsheet. In these instances one has to think fast, and strain one's mind to encompass more and more ideas, operations, transformations. One's mind is strained to the limit already, without the additional task of monitoring itself. In a sense, it seems, cultures based on pushing the mind to the limit naturally work against proprioception of thought.

Although, of course, there is also a sense in which mindfulness is necessary for a complex culture like ours. TV, books, cars and movies are all the products of intensely creative minds. And the creative process relies essentially on deep awareness of interior mindspace. In order to carry out deep creative innovations, the mind must be self-proprioceptive, to whatever degree of approximation it can muster: it must be intensely and probingly aware of what it is doing. Mindfulness is not useful for rote brainwork, but it is crucial for creative brainwork.

Very noteworthy here is the extreme simplification of lifestyle that goes along with the monastic lifestyle. If you're doing nothing but sitting, meditating and walking outdoors and praying, your mind is not being occupied with much complexity and it's relatively easy for it to monitor what it's doing. But this is basically making a tradeoff between

- Devoting one's resources to complex things like math, science, business or art — but then not having resources left over to monitor one's own processing;
- Spending time doing things that are very undemanding of computational resources (like emptying one's mind, walking, or raking leaves), so that there are ample resources left over for self-monitoring).

The role of limited computational capacity becomes extremely glaring here. There is a conflict between the goals of peace and creation/complexity. In practice, the choice of the Eastern masters is peace and cognitive proprioception – "stilling the mind" and giving up the illusion of complex, systematic thought. On the other hand I have made a different personal choice – I aim to maximize my own mindfulness and self-understanding, but, not at the cost of reducing the complexity and subtlety of the structures and dynamics in my mind. Because of this choice I am surely a lot less enlightened than I could be (if I had, say, given up complex cognition for meditation 15 years ago), but I've had a lot of interesting thoughts....

Generally speaking, if one's goal is to increase mindfulness (thereby reducing evil) while retaining intelligence, then there are two approaches to take. One is to work to increase the mindfulness of humans – bearing in mind however that one is always working with the complexity vs. proprioception tradeoff mentioned in the previous paragraph. Another is to work to expand the limitations of the human – increasing the computational capacity devoted to mind, which thus decreases the need for suffering. Spiritual work and transhumanist work, from this perspective, may be seen as two different quests with significantly overlapping goals. In this view, however, spiritual work is viewed as more of a "local optimization" approach, whereas transhumanist work is a "global optimization" approach – seeking solutions that are a bit further away from our current condition, but with the capability for dramatically superior properties.

The parable of Goethe's Faust is apt here. Faust sold his soul to the Devil, in exchange for an understanding of the world — i.e., in exchange for the development and flowering of his own mind, for greater interior structural and dynamic complexity. And – waxing poetic for a moment –

it's not too outrageous to metaphorically observe that *this is exactly what the universe itself has done*. The universe has bought its immense beauty and intricacy, at the cost of selling off fragments of its soul, so to speak. It has created forms using autopoiesis — and these forms then stick in its throat, prevent its parts from seeing the whole, from seeing their true nature. But fortunately, there is still some soul left! There are still enlightened minds, tremendous experiences, perfect moments. There is still the experience of the world as hyperreal — the infusion of *anandamaya* into the mind, loosening up forms and creating a smooth two-way flow of information throughout all the levels of being. Slowly, and in a way we can hardly understand, we may be able to help the cosmos to heal its self-inflicted wounds — bringing the universe to a point where the structural identity of individual and cosmic awareness becomes more dominant, and the sheaths that create forms and cloak being melt a little further into the background. But a complete "healing" of the "evil" of irreversible computation and compression might well mean the annihilation of all complexity and intelligence, at least in any form that the latter are comprehensible or knowable by human minds.

Philip K. Dick expressed this same feeling by saying that the universe, which was perfect and healthy, also had a sick twin universe — a universe with a flaw. We, he said, live in the sick twin universe, Hyperuniverse II. But the divine entity has invaded our universe, and is attempting to cure things. This theme is explored brilliantly in his novels *The Divine Invasion* and *Valis*. Although the details of his visions are odd and often amusing, I sympathize entirely with the core concepts and feelings underlying them. I would add only that *we are all this divine invader*. We all have the power to put more self-awareness back into the world. As Thich Nhat Hanh says, "A bodhisattva doesn't have to be perfect. Anyone who is aware of what is happening and who tries to wake other people up is a bodhisattva. We are all bodhisattvas, doing our best." We are all mental knots, blocking the free flow of information — and we are all untying these knots, enabling and surfing on this free flow. This is the wonder and the paradox of being. As the transhuman age begins, this fundamental dynamic and paradox will not change – but the computational constraints will lessen, which means – if things go well — more complexity and beauty and intelligence and less suffering.

Chapter 21
Immortality[74]

      Ever since I understood the idea of death – at age three or four or so – my heart and mind have rebelled against it. And I'm clearly not the only human with the deep-seated feeling that it would be good to live forever – or at least to live a very long time, without the fear of involuntary death, until such (hypothetical, perhaps never-arriving) point as I decide freely that continued life is no longer of interest.

      In this chapter I probe a little deeper into this feeling than is usually done. What does it really mean to want to live forever? Delving into the meaning of "live" and "forever" isn't that fascinating to me in this context, though it does give rise to some significant issues (to what extent is our sense of self tied to our physical body? what are the odds that the physical universe containing us will end in a Big Crunch or some similar calamity, making literal immortality impossible?). What intrigues me more is digging into the "I" who wants to live forever. What exactly is it that's being perpetuated, in the hypothesis of eternal life? How can we tell whether, in a particular scenario, this "I" is really being perpetuated or not?   This ties back into the psychology of "self" presented in previous chapters.

      Immortality-wise, I take a lot for granted before I even begin this conceptual investigation.   I assume that human immortality is physically possible, sociologically, economically and psychologically practical, and morally acceptable. I realize that all of these points have been and still are disputed by a variety of intelligent and thoughtful individuals, but the arguments on both sides have been repeated many times and there would be no point in my recounting them.   It seems very clear to me that pharmacology or nanomedicine will eventually enable human physiological immortality, and that uploading will eventually allow humans to copy their minds into computers of some sort. It seems very clear to me that socioeconomic problems ensuing from widespread human life extension would be solvable via deployment of appropriate technologies – e.g. overcrowding of the planet could be solved at first via digging underground and building huge skyscrapers, then later by colonizing other planets, building huge spaceships and/or uploading into virtual realities. And although I agree that much of the meaning of current human life ensues indirectly from the inevitability of death, it seems to me that humans freed from the fear of death would find new sources of meaning.   All these points, though currently controversial from a mainstream point of view, seem obvious and not worth debating. But the debates over these trivial points tend to obscure the deeper issues that make the issue of immortality really interesting.

---

[74] The ideas in this chapter were partially crafted via conversations and ongoing dialogues with Bruce Klein, Martine Rothblatt and Izabela Goertzel.

To illustrate the deeper issues I want to raise, I'll first outline a few thought-experiments regarding hypothetical scenarios that may be enabled by future technology. The general points I want to make are not tied to these particular scenarios; the scenarios are just illustrative.

FutureBen and FutureBush

The first scenario involves a future version of myself – FutureBen, let's call him – who is allowed to grow, learn and change freely as he wishes. Suppose FutureBen lives ten billion years and increases his intelligence by a factor of ninety-seven quintillion. His human body was shed after a few thousand years of life – and he's placed the episodic memories of his first century of life (the part that took place in humanly-embodied form) in a very-rarely-accessed portion of his memory, since it's really not very interesting compared to some of the things that have happened to him since.

Now let's suppose President George W. Bush has also spawned a future version of himself – FutureBush. And suppose that FutureBen and FutureBush made friends in their eight billionth year and decided to link their lives together in a kind of posthuman mind-partnership. I submit that, after a billion years of collaborative, mutually-coupled growth and change, it may be rather difficult to distinguish FutureBen from FutureBush. Perhaps they have even exchanged ancient episodic memories, so that each of them has complete first-person memories of the other one's life. They may be using the same black holes in the same galactic cores as their wormhole-coupled quantum-gravity cognitive processors. FutureBen and FutureBush may be an awful lot like each other — and very, very little like Ben Goertzel or George W. Bush from 1985, 2005 or 2035.

The question is: What difference does it make that FutureBen happened to evolve out of Ben Goertzel instead of George W. Bush, PeeWee Herman, or for that matter, one of 2005-Ben's pet guinea pigs Coffee or Tea? It surely doesn't matter much to FutureBen. If I, 2005 Ben Goertzel, am simply going to serve as the initial state for a completely different sort of being, then in what sense am "I" really becoming immortal? Why not just let myself die, and then let other sorts of beings comparable to FutureBen and FutureBush take my place? Or why not, for example, die and let my children take my place, and let their children take their place, eventually after N generations resulting in completely different sorts of beings?

One possible answer is that this is not really immortality for Ben Goertzel of 2005. The difficulty posed by this answer, however, is: Where do you draw the line? 2005-Ben is very different in many ways from 1985-Ben, and more so from 1970-Ben (who was 3 years old). Is it the continuity of the physical body that's critical? It's easy to make counterarguments to that, as my next hypothetical scenario will illustrate. Is there some critical threshold of change, beyond which "selfness" is not preserved? Is there some special emergent "self-pattern" which exists in all the human Bens mentioned above (emerging in different ways from the differently mature Bens, but still maintaining its own integrity), but is lost in FutureBen? I think the latter is fairly close to the truth – there is an emergent self-pattern spanning all these human Bens which is not there in the hypothesized far-future FutureBen. The question then is: What is this emergent self-pattern? How real is it? Why is it important? These are critical questions.

Another possible answer to the FutureBen question is that this scenario is real immortality for 2005-Ben, and that there is some sort of value in the continuity of consciousness between 2005-Ben and FutureBen. In this view the immortality lies in the process of continuous awareness, even if this includes continuous radical growth and change. Continuity of awareness is posited as a primary value, right along with awareness and life itself. This is a philosophically respectable view, but it gives rise to further subtle questions. What is this continuity-of-consciousness and why it is important?

The Philosophical Significance of Uploading

Following up on the prior thought-experiment, I will now introduce some additional hypothetical scenarios, intended to highlight the philosophical subtleties associated with various futuristic "mind-uploading" techniques. These scenarios get directly at the issue of continuity of consciousness raised at the end of the prior section. I'll define a number of different NearFutureBens (NFBens), and pose the question of which ones have a greater claim to Ben-ness.

NFBen1 is produced via recording complete information about the physical parameters of all particles inside Ben Goertzel at a particular point in time, storing this information in a database, then completely annihilating Ben Goertzel – and then, 666 hours later, re-creating an exact copy of the previously annihilated Ben Goertzel from the database records.

NFBen2 is produced via making the same kind of recording and re-creation, but doing the re-creation 10 microseconds after the annihilation, rather than 666 years.

NFBen3 is produced the same way but with a delay of 3 femtoseconds.

NFBen4 thru NFBen7 are produced like NFBen1 thru NFBen3, but the reconstruction takes place inside a computer rather than in physical reality. These Bens are software, though they may feel like they have physical bodies, due to having the option to cruise around in Ben-like bodies in an Earth-like virtual reality. When embedded in the virtual reality, they feel like physical Bens, but they know they're "just software."

NFBen 8 thru NFBen10 are produced like NFBen1 thru NFBen3, but the reconstruction takes place differently: Ben's physical body is re-created correctly particle by particle, but Ben's brain is replaced by an Asimov-style positronic brain or a digital computer that contains all Ben's memories and realizes Ben's thoughts and feelings exactly in spite of having a different physical substrate.

Next, imagine variant NFBen11, who is an exact particle by particle replica of regular old Ben, but with one difference: the particles from the original Ben are gradually transplanted into him. First NFBen11 is created by the same method as NFBen3, then one by one the actual cells from the original Ben are exchanged with the cells in NFBen11, until eventually all the original Ben's cells are in NFBen11, and all the original NFBen11's cells are in what used to be the original Ben.

All of these NearFutureBens will feel like they are Ben – they will have Ben's memories and Ben's self and will feel like their consciousness is continuous with Ben's. They will have as much right to feel like Ben as I, writing these words, have to feel that I am the same Ben who went to sleep last night. I went to sleep, disappeared, and then woke up in the morning – a new person, with a feeling of having the same self as the guy who went to sleep in my bed last night, and a feeling of having a sort of temporarily interrupted continuity of consciousness with that guy.

One can argue that all these NFBens really are Ben. Or one can argue that some of them are just new people who feel like they're Ben. My sympathy is strongly with the former position. The distinction between the Bens who look and feel and think and act exactly like Ben, and the Bens who actually are Ben, feels to me like an over-mystical distinction-without-a-difference. If this is accepted, the implication is obvious: Ben *is* uniquely identifiable via a certain a pattern of arrangement. Ben is not uniquely identifiable via a particular collection of particles, nor even necessarily by a pattern of arrangement of particles – he is uniquely identifiable via a pattern of arrangement that may emerge from particles or bits or potentially anything else.

I'm not committing to a completely reductionist view of the mind here. I tend to think there are certain aspects of subjective experience that aren't well-captured by the scientific, reductionist perspective. I wouldn't want to quite say that "Ben is a pattern of arrangement." But I would say that from the perspective of looking at various physical systems and deciding which ones are Ben or not, the pattern of arrangement is all that matters. The subjective view, from inside Ben, is at least in part another story – but we've already posited that, in these scenarios, all these NearFutureBens subjectively feel just like Ben (they have the feeling of "being Ben").

Self, Continuity of Consciousness, and Other Illusions

So what does immortality mean, really?

One case is absolutely clear: If a person maintains their human body forever, and doesn't alter their ways of thinking and feeling too much, then they will "live forever" in the same sense that a person now lives 60 or 75 or 90 years or whatever. This doesn't solve any of the philosophical puzzles, however, it just defers them to commonsense.

It is worthwhile to question the meaning of the commonsensical sense in which a person now lives 60 or 75 or 90 years or whatever. We change immensely over our lifetime, and we disappear nearly every night and reappear in the morning – with what justification do we say that it's the same person?

I do believe there is some continuity of structure there – I have many traits in common with my 1970 self, and there are in particular many aspects of my *self-model* that are still there after all these years. At a high level of organization, Ben is still Ben, even though the particular memories and ideas out of which this high-level organization emerges have changed a lot. And this high-level organization is linked in with a bunch of physical and emotional peculiarities that haven't changed over the years. There are many continuously existing patterns there, unlike the Ben vs. FutureBen contrast made above.

There is a lot of neuropsychological research showing that the "self" is in a strong sense an illusion – much like its sister illusion, "free will." Thomas Metzinger's recent book Being No One (2003) makes this point in an excellently detailed way. The human mind's image of itself – what Metzinger calls the "phenomenal self" – is a construct that the human mind creates in order to better understand and control itself, it's not a "real thing." Various neuropsychological disorders may lead to bizarre dysfunctions in self-image and self-understanding. And there are valid reasons to speculate that a superhuman mind – be it an AI or a human with tremendously augmented intelligence – might not possess this same illusion. Rather than needing to construct for itself a story of a unified "self entity" controlling it, a more intelligent and introspective mind might simply perceive itself as the largely heterogenous collection of patterns and subsystems that it is. In this sense, individuality might not survive the transcendence of minds beyond the human condition.

But, illusory or not, the patterns of my human self-model have largely persisted in me since early childhood – and in that sense "I" do have some persistent existence, even if this "I" is really an illusory phenomenal self a la Metzinger.

Next, in addition to the continuity of structure, there is a perceived continuity of consciousness. I feel like I'm participating in a largely continuous stream of thought and feeling from one minute to the next, and when I wake up in the morning I often (not always) feel like I'm resuming the stream of thought and feeling from the night before. Sometime this continuity is quite vivid – as when I fell asleep thinking about some problem, and I'm still thinking about it when I wake up. Other times the perceived continuity is hardly there at all, as when I'm traveling and I wake up in a hotel room and have a hard time remembering where I am or how I got there, until I fully re-emerge into consciousness.

My contention is that, just like the phenomenal self, the continuity of awareness is also a psychologically-constructed illusion. I have a memory of my prior thoughts and feelings, and so I construct within myself a story of continuous flowing from these prior thoughts and feelings to my current ones. But this process is not unlike how I construct within myself a story of a "whole self" rather than a disparate and multi-faceted population of mental and emotional processes; and it's not that different from how I construct within myself a story about "free will", telling myself that I consciously and rationally control my actions when in fact it's usually the case that my conscious rationalizations follow the unconsciously-determined decisions the rest of my brain has made.

If a person wants to preserve their continuity of consciousness, and their internal self-model, that's fine – it's a valid value judgment. But they should make this value judgment based on the understanding that these things are not "real" – they're psychologically constructed illusions, that our brains have come to make through some combination of evolutionary utility-seeking and self-organizational pattern formation. It may be that an individual decides these illusions, like the illusion of free will, lie at the essence of humanity, and are worthy of preservation as a fundamental core value.

On the other hand, it's equally valid to judge that the fundamental value lies in overcoming these illusions of self, will and continuity – and seeing that they don't have any true reality. Perhaps the overcoming of these illusions is the right path toward discovering a better way to exist – the human condition having many well-known and well-documented flaws. This is the direction in which Vedanta, Zen Buddhism and all the other mystical traditions point. Science has discovered the illusory nature of free will, self and continuity of consciousness only recently, but in the domain of "wisdom traditions," all this is quite old news.

Complex systems theory also has something to add to this discussion. I have argued above that the self may be thought of as an "attractor" of complex brain-mind dynamics. This suggests that perhaps there's some critical threshold, so that if one hasn't saved *enough* of a person to meet the threshold, then that person's self-attractor will not be able to form from the saved traces. As a related point, I know a woman with severe brain damage from a car crash, and she says she knows she is not herself anymore: some aspects of her brain-patterns constituting her feeling of integral "self-ness" are gone. She lost more than just memories and cognitive abilities, she lost a significant part of her self-attractor.

But even if this intuition about thresholds and self-attractors is correct, it doesn't tell us how much is enough! Where is the critical threshold? My intuitive feeling is that, most likely:

- A bunch of questionnaire answers recording a person's views is well below the threshold.
- Copying the complete brain state of a person at the neuron and neurotransmitter level is above the threshold.

I am undecided about avenues like:

- Less complete brain scans,

- Detailed video/audio recordings of substantial portions of a person's life, to be postprocessed later by an advanced AI that tries to solve the "inverse problem" of reconstructing the person from these records.

AGI systems like Novamente will provide one medium for experimenting with issues like these. Presumably once we have a really advanced Novamente system we will be able to see its "self-attractor" form and observe the conditions for its maintenance.

The Future of Mind

One can perceive the preservation unto eternity of the human illusions of free will, self and continuity of consciousness as a good thing – or one can view it as a burden, like the preservation unto eternity of stomachaches and bad tempers and pimples. An equally valid, alternate perspective holds that human-style individual minds, ridden with illusions as they are, are merely an intermediary phase on the way to the development of really interesting cognitive dynamics.

Among humans, illusions like will, self and consciousness-continuity are just about inevitably tied in with intelligence. Highly rigorous long-term routines like Zen meditation practice are able to whittle away the illusions, but as noted above, they seem to have other costs – I don't know of any Zen masters who make interesting contributions to science or mathematics, for example. Among humans, the reduction of these illusions on a practical day-to-day basis seems to require so much effort as to absorb almost the entire organism to the exclusion of all else. Yet the same will not necessarily be the case for superhuman AI's, or enhanced human uploads, or posthuman humans with radical brain improvements. These minds may be able to carry out advanced intellectual activity without adopting the illusions that are built into the human mind courtesy of our evolved brains.

A mind without the illusions of self, free will or continuity of consciousness might not look much like a "mind" as we currently conceive it – it would be more of a "complex, creative, dynamical system of inter-creating patterns." FutureBen and FutureBush, as envisioned above, are actually fairly unadventurous as prognostications of the future of mind – as described above, they're still individuals, with individual identities and histories; but it's not at all clear that this is what the future holds in store. If one's value system favors general values like freedom, growth and joy, rather than primarily valuing humanity as such, such a posthuman relatively-illusion-free mind may be considered superior to human minds … and the prospect of immortality in human form may appear like a kind of second-rate "booby prize."

Why Immortality?

All these issues center around one key philosophical point: What is the goal of immortality? What is the goal of avoiding involuntary death? Is it to keep human life as we know it around forever? That is a valid, respectable, non-idiotic goal. Or is it to keep the process of growth alive and flourishing beyond the scope painfully and arbitrarily imposed on it by the end of the human life?

Human life as it exists now is not a constant, it's an ongoing growth process; and for those who want it to be, human life beyond the current maximum lifespan and beyond the traditional scope of humanity will still be a process of growth, change and learning. Fear of death will largely be replaced by more interesting issues like the merit of individuality and consciousness in its various forms — and other issues we can't come close to foreseeing yet.

It may be that, when some of us live long enough and become smart enough, we decide that maintaining individuality and the other human illusions unto eternity isn't interesting, and it's better to merge into a larger posthuman intelligent dynamical-pattern-system. And it may be that others of us find that individuality still seems interesting forever. Resource wars between superhuman post-individuals and human individuals can't be ruled out, but nor can they be confidently forecast — since there will likely so *many* resources available at the posthuman stage, and diversity may still seem like an interesting value to superhuman post-individuals (so why not let the retro human immortal individuals stick around and mind their own business?).

These issues are fairly hard to "feel out" right now, stuck as we are in this human form with its limited capacity for experience, intelligence and communication. For me, the quest for radical life extension is largely about staying around long enough, and growing enough, to find out more about intriguing (philosophically, scientifically and personally fundamental) issues like these.

Chapter 22
Compassion and Ethics

Most of the ethical principles that people live by (or pretend to live by) are merely culturally-specific conventions of behavior. But it's tempting to look for an ethical core that goes beyond any particular culture or belief system. In a transhumanist context, one even wants to find an ethical core that goes beyond any particular species, or any particular type of mind or organism. Ethics is a separate issue from philosophy of mind, but it's closely related (ethics are rules created by minds for guiding minds), and it's worth asking what patternist philosophy has to say about these issues.

One strategy for seeking an "objective" grounding for ethics has been the evolutionary approach. Researchers in this vein try to explain why apparently-altruistic behavior would emerge in organisms evolving via a "selfish-gene" dynamic. This approach has led to a host of interesting insights, many of which are reported in Matt Ridley's excellent book *The Origins of Virtue* (1998). Ultimately, however, I think the evolutionary approach fails to get at the essence of ethical philosophy. I do think evolution is relevant — but I think that, to get at the crux of the matter, one has to look at evolution and ethics as intertwined manifestations of the same deeper phenomena.

I'll consider ethics here from two perspectives: first, using patternist ideas to define ethics that I personally think are valuable; and then, using patternist ideas to explore the possibility of "universal ethics" – ethical principles implicit in the universe as a whole. The personal ethics I'll argue for are "joy, growth and choice", and the universal ethic I'll identify is a form of "patternist compassion." Joy, growth and choice are related to compassion but are not exactly the same thing – they can be viewed as particular ways of expressing compassion; or, using some technical vocabulary to be introduced below, as particular ways of directing the intrinsically *pattern-compassionate* nature of the universe.

Joy, Growth and Choice

In *The Path to Posthumanity* (Goertzel and Bugaj, 2005), musing about what principles I'd like to see guide the long-term future of the universe, I proposed an ethical principle of "Voluntary Joyous Growth" – which holds basically that a good ethical system is one that tries to balance the three factors of choice, joy and growth. The first task I'll take on in this chapter is to defining these three concepts in a patternist way. This task is not so easy – they're all a bit slippery, and each one conceals a vast mass of ambiguity, subtlety and human history.

I note that, in order to impart an ethic to another mind (AI, human, whatever), explicit definitions are of limited use. Examples are more important; and, most important of all is the collective exploration of scenarios – living and working through ethics in "real life." But nevertheless, explicit definitions can be of some value in guiding these more practical and essential aspects of ethics.

Growth is the simplest of the three basic principles. Most abstractly, one may conceive growth as the creation of more and more pattern. Recall that, if one defines a simplicity measure, one can mathematically define the "intensity" of patterns, thus quantifying what it means to have more and more pattern. Furthermore, if one accepts the theory of consciousness I've presented above, then more intense patterns correspond to more intense qualia, so that *more patterns means more experience*. So growth means that the universe becomes more aware.

Choice is a little more complicated: choice, as I've analyzed it above, has to do with the maintenance within intelligent systems of a "virtual multiverse model" of the universe, in which multiple potential future universes are studied, and this study is used to guide system dynamics. Valuing choice means valuing systems that guide their actions using internal virtual multiverse models. In essence, then, choice means valuing the explicit embodiment of the multiverse within the universe. Valuing choice means valuing *virtual multiverse creation*.

In brief: More pattern is good, more experience is good, and multiplicity of (anticipated and sometimes realized) possibility is good.

Choice helps with growth, because systems that embody virtual multiverses tend to be good at generating new patterns and experiences. Growth helps with choice, because it takes fairly complex patterned systems to spawn internal virtual multiverses.

Joy – as treated in my above discussion of emotions – is what happens when a system finds itself overwhelmed with system-wide response patterns that occur in reaction to the successful achievement of its goals. More specifically, "spiritual joy" may be conceived as what happens when a system is overwhelmed with a complex dynamical pattern that embodies harmony between the inside of the system and the world in which the system is embedded.

Valuing joy means valuing systems setting goals and achieving them. Valuing spiritual joy means valuing the quest for harmony between self and universe. Ergo – speaking loosely and poetically — according to "Joy, Growth and Choice," achieving goals is good, and harmony with the universe is good.

Successful goal-achievement, in combination with the values of choice and growth, means that the values of choice and growth are propagated through the various systems of the universe via their goal-systems.

Spiritual joy is a reflection of the basic dynamic by which the universe seeks to overcome the paradox of one-versus-many (for a poetic exploration of this theme, see Goertzel, 1998). All is one, yet each thing is separate and distinct – this paradox lies at the heart of being, and one view of our ultimate purpose is to overcome this distinction and completely manifest both our unique separateness and our oneness at all times. Seeking harmony between the interior and exterior worlds is the way to fulfill this purpose.

One point to note is that, in the patternist perspective, all of these three values are defined in terms of the concept of "pattern" – but "pattern" is not an objective concept. The mathematical theory of pattern defines pattern in terms of a more elemental concept of "simplicity." Thus, judgments of the amount of choice, growth or joy in the world are ultimately dependent on the simplicity measure implicit in the judging mind. This leads to an important point regarding the future creation of powerful nonhuman AI's valuing growth, choice and joy. If these AI's measure simplicity very differently from humans, they will perceive different things as growing, joyful and free. Of course, it's not viable to restrict a superhuman AI to have the same simplicity measure as humans – but at very least, one can ensure that one's AI has a simplicity measure that's *inclusive* of human assessments of simplicity, in the sense that whatever humans perceive as simple, so does it. This means that the AI will see choice, joy and growth where humans do – but may also see it in other places.

Of course, I don't claim to have fully explicated any of my three basic values – joy, growth or choice – but I hope to have elucidated a little bit of what lies inside them. There is much more to discover.

Universal Ethics versus Particular Ethics

Ethics, by and large, is normative rather than descriptive. It doesn't tell you how the world it, it, it tells you how you should like the world to be. Linguistically, it consists most centrally of commands rather than statements.

However, it's interesting to explore the boundary between normative ethics and descriptive philosophy. There are some properties of the universe, I believe, that can loosely but justly be described as "ethical." So, one can speak about the existence of a kind of "universal ethics" – though this ethics is not fully satisfying from a human perspective, due to its somewhat "cold and impersonal" nature.

I'll describe here a general property of the universe called "continuous pattern-sympathy," and I'll present an argument that, because of this property, an ethic of compassion (under a certain definition) is a natural and inevitable part of the universe. This argument not only explains the existence of compassion but it also explains some of the particular properties of compassion as we observe it among humans. I'll then explore the relation between ethics and evolution by analyzing natural selection itself as a manifestation of continuous pattern-sympathy on the genotypic and phenotypic levels. Compassion as we humans experience it in our lives thus emerges as a particular example of continuous pattern-sympathy, intertwined with the continuous pattern-sympathy embodied in the evolution that gave rise to us.

What this all adds up to is an argument that compassion is a kind of "universal ethic" — in that its existence follows naturally from a very abstract property of the universe. This argument doesn't imply that we should all be maximally compassionate – it just argues that some degree of compassion is natural and inevitable. Freedom, joy and growth as analyzed above then emerge as closely related to universal compassion – as particular manifestations that universal compassion may take. The conclusion is that the universe does have its own ethic, intrinsic to its nature as a self-organizing pattern-dynamical system – but that this ethic doesn't give us everything we, as humans, would like to see as ethical principles. This is hardly surprising because the universe's implicit ethic is oriented toward maintaining and growing the universe as a whole, whereas our own human ethics are oriented toward maintaining and growing human systems (a distinction that is reflected e.g. in the orientation of human ethical systems toward choice and freedom, which are not really universe-level properties but rather properties of particular kinds of pattern-systems such as human minds).

Connecting universal and particular ethics, one may ask, for instance: What is the optimal degree of compassion that should exist, in order to maximize the amount of freedom, growth and joy in the universe? And one may ask: What is the optimal degree of compassion that I myself should manifest, in order to maximize this goal? The answer, pretty clearly, is that the optimal degree of compassion is neither zero nor maximal: rather, an intermediate level of compassion is almost certainly going to be optimal. Of course, quantifying this kind of conclusion in any useful way is far beyond the scope of contemporary science and mathematics. But even the qualitative conclusion seems worthwhile.

My own feeling is that the level of compassion displayed by the average human is somewhat below the optimal level for maximizing my pet goals of freedom, growth and joy. I believe that if humans were a bit more compassionate, we'd move toward these goals faster. But of course I can't prove this – and to mount a careful argument for the point would require a detailed analysis of human history and human affairs, which I'm not going to undertake here.

At one time, I puzzled at great length over the paradoxical peculiarity of Buddhism, which both asserts that the world doesn't exist, and asserts that we should be compassionate to other beings. If the beings aren't real, I wondered, then why the heck should be bother being compassionate to them? I answered this question for myself in a personal sense a long time ago, but the ideas in this chapter seem to form a crisper answer than I've come up with before. The crux is that "the world" as such doesn't exist, but patterns exist, and patterns persist and grow gradually over time ("continuous pattern-sympathy"), and compassion arises naturally from this process.

Continuous Pattern-Sympathy, Evolution, and Compassion

Compassion, the root of all ethics, appears to be a trivial and natural consequence of the property of continuous pattern-sympathy, introduced in Chapter 11 above. Suppose we have a population of minds, each one considered as a bundle of patterns. And suppose that the dynamic by which these minds change over time obeys the principle of pattern-sympathy. This means that each pattern in each one of the minds "wants" to perpetuate itself. Then it follows that a mind will want to ensure the survival and productive activity of other minds that share a lot of its patterns. In other words, it's not that "I" am compassionate toward "you" – it's that a large number of the patterns in my mind are compassionate toward their clone or near-clone patterns in your mind. This explains very neatly why compassion, in practical human life, is roughly proportional to similarity.

Continuous pattern-sympathy means: of all the patterns that exist right now, the ones that are going to persist into the long-term future are generally going to be the ones that persist into the short-term future. And this means that survivor patterns are going to be ones that gradually increase their intensity as patterns.

In the case of patterns that achieve their intensity via repeatedly occurring in several different substrates – for instance patterns that occur in many different organisms, or many different minds – then clearly the long-term persistence of a pattern is going to be achieved via, in the short term, maximizing the number of organisms or minds containing the pattern. Ergo, in a dynamical system displaying continuous pattern-sympathy and patterns that achieve their intensity via repeated instances, one will see pattern-bundles (minds, organisms, etc.) that seek the continuation and flourishing of similar pattern-bundles.

But similarity is not the only driver of compassion. We humans are also generally compassionate, for example, toward beings with whom we've been substantially involved, whether or not we are similar to them. This is because our involvement with the other being has led to the emergence of patterns between that being and ourselves, and these emergent patterns also want to persist.

And what does this tell us about ethics and compassion in humans and other evolved organisms? Ridley and other classical evolutionary-ethicists focus on the genetic roots of human ethics. They argue that apparent altruism in humans and other organisms is a consequence of the phenomenon in which a gene or gene-combination, existing in multiple organisms, maximizes its short and long term survival potential by inducing cooperation among the various organisms that host it. This is a valid observation and is doubtless a critical part of the origin and development of human and animal ethics, but I think these authors tend to underestimate the importance of purely phenotypic pattern-sympathy induced via co-adaptation.

The co-adaptive aspect of compassion is simply that, once a community of compassionate entities exists, there is a strong bias for the evolution of new compassionate entities. In a community of nasty bastards, patterns that are stuck in an organism displaying compassion will likely get screwed. In a community of compassionate individuals, patterns in an organism displaying compassion will get rewarded, because they'll get to enter into a community-wide exponential pattern-spreading dynamic. Thus, compassion itself – independent of any genotypic basis for compassion – will tend to propagate itself. Compassion itself, as a trait of organisms, is a particular pattern that spreads exponentially according to the logic of continuous pattern-sympathy. And it spreads faster than would be predicted via looking at gene-level dynamics alone.

Normative and Transhumanist Implications

I've argued for the ethic of compassion as a natural consequence of the fundamental nature of the universe. Not as a consequence of the particular nature of our physical universe, but as a consequence of the nature of basically any sensible universe – any universe displaying the property of continuous pattern-sympathy. I've also argued that compassion, as a property of humans and other biological organisms, emerges naturally from natural selection, which is itself a manifestation of continuous pattern-sympathy.

Of course, these ethical generalities don't give any kind of particular ethical guideline. OK, compassion is good – but are some kinds of compassion better than others? How bad is eating meat, as opposed to failing to send your extra dollars to Darfur and spending it on a night at the movies instead? How important is it to ensure that humans survive the next million years, as opposed to promoting the ongoing development of more and more advanced mind- and life-forms? Unfortunately or fortunately, I don't think that abstract philosophical or scientific analysis will ever be able to answer this sort of question.

I've described the kind of ethics that seems to exist in the universe, and explained why it's here in terms of other, non-explicitly-ethical aspects of the universe. To turn these ideas into a prescriptive ethics one needs to articulate some particular value system. As I noted above, my own value system centers on the notions of freedom, growth and joy. Given this particular value system or any other, one may ask how much compassion, or what kinds of compassion, are likely to provide maximum value. And in the context of my own value system, it seems clear to me that *if we assume the universe can only contain a finite number of patterns*, then there is some optimal degree of compassion which is greater than zero but less than the maximum.

Growth, in a finite universe, is clearly not maximized by maximal compassion: maximal compassion breeds stasis, because it means that patterns that exist will very strongly reinforce themselves, preventing new patterns from forming. As Nietzsche observed, in a finite universe, progress requires some degree of *hardness*: it requires letting old patterns die so that new and better ones may form.

Freedom also seems incompatible with maximal compassion: in order to provide choice, we must allow minds to choose to be evil if they want to, at least to an extent. At the very least, we must allow minds to be evil to *themselves* — otherwise we're not really allowing any kind of choice at all.

Of my three basic values, it seems that only joy is compatible with maximal compassion. If joy is the only goal, then everything in the universe can be maximally compassionate to everything else, no matter whether the universe is finite or not – there's no growth and no freedom, just a pulsing field of radiant bliss!

We thus return the conclusion of Chapter 18, that the root of "evil" – if evil is defined as incomplete compassion – is either the finite capacity of the universe, or the ethic of growth that seems to be embodied in the universe (alongside the ethic for real but partial compassion). Only in an infinite universe can we have growth and compassion side by side without contradiction.

This line of thinking has interesting consequences for the puzzle of "Friendly AI". Put briefly, this puzzle asks how one might create a superhuman artificial intelligence, with the ability to modify itself freely and to assimilate all the matter in the universe if it so wishes – yet create this AI in such a way that the survival of the human race isn't threatened. The good news is that any AI one creates is reasonably likely to be compassionate in a sense, since compassionate is intrinsic to the universe. The bad news is that compassion, as embodied in the universe, is a highly abstract thing. The fact that patterns tend to compassionately cause like patterns to persist isn't much consolation if you're one of the patterns that happen not to persist. There seems to me little hope of creating an AI that, once it becomes unboundedly intelligent and powerful, will care more than a little bit about persisting the particular patterns that are human beings. So the key to Friendly AI may well be *engineering a situation in which caring a little bit is enough*. In other words: make sure the AI has a really big universe to play in, so that it doesn't need to annihilate our patterns in order to make room for its. In this case just a little bit of specially-focused compassion on us humans will be enough to keep us around.

The issue of Friendly AI highlights the weakness of any approach to ethical philosophy as general as the one I've given here. To answer any particular ethical question – even one as seemingly general as whether it's ethically desirable for humans to survive the next millennium – requires one to get a lot more fine-grained than seems possible to do in the language of "universal ethics." But even so, it seems valuable conceptually to observe the ways in which the particular ethics by which we guide our behavior have a general root in the dynamics of the universe. Particular ethical rules and systems are cultural and psychological manifestations that fill in the blanks left by the powerful but abstract ethics of compassion that is an intrinsic consequence of the basic nature of the universe's dynamics.

Conclusion

The end of this book occurs, inevitably, at a relatively arbitrary point. Since the main common thread running through the diversity of the various chapters is the theme "everything is pattern," it stands to reason that there could be a chapter on basically anything. And this isn't an empty general observation: the patternist perspective does in fact have concrete insights to give on a lot of topics not covered here. I've considered writing a book on patternist music theory, for example. And sociology has barely been touched. My current theoretical research focuses on the implications of patternism for developmental psychology in AI's and humans.

If I were to become immortal, this book could go on forever. Perhaps one day I'll create a Ben-clone to do nothing but write this book on and on, chapter after chapter, until finally at the end of time every single phenomenon in the multiverse has been exhaustively discussed from the patternist perspective.

If there was a reason for stopping where I did, aside from the simple desire to spend time on other things, it was that I felt enough phenomena had been discussed to make the basic point: there really is significant insight – both broad and deep — to be gained via taking the "everything is pattern underneath" perspective.

A goodly number of philosophers have written books reducing everything in the universe (or, rather, in practice, some hopefully-representative subset thereof) to some one or two or few "elemental" categories. How does patternist philosophy differ? One unique aspect, I suggest, is the neat way in which patternist thinking spans the scientific and subjective worlds. Another is the way it spans all the branches of hard and soft science, intersecting nontrivially with many different scientific languages. Patternism ties in with computational mathematics, with brain science, with evolutionary biology, with quantum physics, with complex systems science – and also with Zen and Vedanta and the phenomenology of creativity and psychedelic experience. It provides a foundational perspective that spans different views on the world that are typically viewed as radically separate and opposed, and says interesting, detailed things about various particular phenomena viewed as important within these different views.

What's the next step along the patternist path? There are many, of course – it's more a "garden of forking paths" than an inevitable linear progression.

Nearly every chapter in this book would merit substantial elaboration. Most chapters could become books on their own, with a bit of attention.

Beyond that, one path I'm eager to follow (when I get the time) is to work out more of the details of the probabilistic formulation of pattern theory described in Appendices A and B of this book, and try to use this to push pattern theory further in a scientific direction, using it to formulate precise hypotheses in the spirit of the psychological and complex systems principles discussed above. It may be possible, with a lot of concerted effort, to turn pattern philosophy into a practical tool for science, which could then (thinking ambitiously!) have the effect of morphing various branches of science into explicitly "patternist science." I have a feeling that this would be a very good approach to working out a detailed theory of the dynamics of cognition, for example – something that will be badly needed once brain scans get more refined and AI's more advanced.

Another inviting direction is to essentially re-do the traditional Indian and Chinese phenomenologies of consciousness-states using pattern theory instead of the less precisely formulated, traditional mystical notions. I'm not thinking about religion per se, more about the mapping of mind-states and their interrelationships. Modern culture doesn't really have a language for this kind of mapping; I think pattern theory may be able to provide one.

Another direction, less well-formulated, is something I've been thinking of privately as a "general philosophy/science of the development of form." Look at three examples: embryogenesis, the birth of the universe from the Big Bang, and the development of the baby's mind. In each case we have the emergence of complex, evolving, autopoietic networks of interlocking patterns – out of nothing. How does it happen? Are there general pattern-dynamical laws describing all of these phenomena, and more?

For me, as I said in the introduction, developing the ideas presented in this book has been a personal as much as an intellectual quest. I have struggled, over the last 20-odd years since the truth of patternism hit me, to really truly *see the universe as pattern* – not just to understand the physical universe as pattern, but to directly experience my own subjective world as pattern, because in a strong sense, that's what it really is. Seeing the universe around and within me as a collection of patterns — or more accurately a network of interlocking, interemerging patterns – is a different experience from how I look at the world when in a more "ordinary, everyday" state of consciousness. It's a more objective perspective than the everyday-human-consciousness one (in which certain patterns are accepted as "absolutely real" and others dismissed as just "imaginary"), in terms of agreement with modern physics and modern analytical philosophy; and it's also a more spiritual perspective, displaying great harmony with the Perennial Philosophy underlying the mystical aspects of all world religions. Of course, seeing a piece of paper as a pattern doesn't mean you can't also see it as a piece of paper – hidden patterns may be hidden and revealed at the same time, that's part of their beauty.

But, OK, that's enough. The pattern that is Ben Goertzel has spilled out enough word-patterns on the theme of patterns hidden and explicit – for now.

Appendix A
Toward A Mathematical
Theory of Pattern

Introduction

This Appendix and the next pursue a line of conceptual development closely related to that of the main body of the book, but with a different focus. Here the stress is on formalization of core concepts. There is not any really deep mathematics; what is accomplished is rather to give mathematical and semi-mathematical formalizations for the key notions underlying patternist philosophy, such as pattern, emergence, complexity, intelligence, mind, self-modification, and so forth. This is a body of ideas that I have, in prior writings, collectively called "pattern theory". This Appendix deals with basic notions regarding pattern, emergence and so forth; then the following one builds on these basic pattern-theoretic concepts, and explicitly presents a formal theory of intelligence and mind. This material draws on earlier published previous works, but contains various new theoretical and mathematical developments as well.

As an example of the application of this sort of work, these pattern-theoretic formal ideas also allow us to precisely define the "derived hypergraph" associated with a complex system, a formalization of the concept of "Novamente maps" loosely introduced above. Furthermore, the pattern-theoretic definitions given here are explicitly used at a few key points in the Novamente AGI design:

- The system studies its own Atomspace and recognizes patterns (using the definition given here to assess pattern intensity), embodying the most intense patterns as encapsulated Nodes.
- One of the ingredients used in assessing the strengths of CausalLinks is the pattern-intensity of a "posited causal mechanism" (represented as a SchemaNode)
- Maximizing the amount of pattern present in the system (the "structural complexity" of the system) is one of the high-level goals that the system strives to achieve, in choosing its actions and in optimizing its parameters.
- The definition of an IntensionalInheritanceLink, in PLN, is given in terms of sets of patterns, based on the definition of pattern given here.

A lot of work has gone into the development of mathematical pattern theory, over a period of more than a decade. Even so, however, the current state of the theory is not particularly advanced. This is mainly because I have been the sole developer of the theory, and I haven't devoted much time to pattern theory as such, the intellectual portion of my life having been mostly occupied with other things, such as working out the conceptual ideas of patternist philosophy and designing and building and applying Novamente and earning a living – and secondarily because are very thorny concepts, from the point of view of conventional mathematics and computer science.

Although we will formalize notions like emergence and cooperativity here, even after doing so, we will be nowhere close to being able to demonstrate formally what sorts of emergence and cooperativity and so forth will exist in any particular complex intelligent systems. This kind of formal analysis of the systems' emergent structure and dynamics may well be possible, but it will be large task, perhaps equal to or greater than the task of actually implementing the system.

In the same vein, one thing the mathematically inclined reader will notice is that the formal definitions given here are "messy" in a way different from traditional mathematics. Many seemingly arbitrary choices will be made, on the path from conceptual ideas to rigorous formulations. Thus, the ideas presented here are presented as plausible initial formalizations of the concepts involved, not as final and crisp immutable mathematical structures. There is nothing here as clean and elegant as, say, the definition of a derivative in calculus, or of a group in abstract algebra.

I will present two different formalizations of the key notions of pattern theory – one based on algorithmic information theory and the other based on probability theory. Mathematically, the two are not equivalent but are closely related. The probabilistic formulation is more recent and is the one that I currently prefer, for one thing because of its utility in the PLN theory of probabilistic inference, which holds a key position within Novamente. I suspect that the probabilistic formulation will be more tractable, in terms of both practical applications and formal mathematical development.

Defining Pattern

The root concept, from which all the other formal definitions given in this chapter will follows, is that of "pattern." In my previous works I have given a very simple mathematical definition of "pattern," and used it to model numerous psychological and biological processes. Much of the material given here is closely modeled on this previous work, and in particular on the treatment in *From Complexity to Creativity*. However, a number of new details have been added here, based on the working-out of concrete examples in the Novamente context. And, as noted above, the final section presents a mathematically different but conceptually similar approach to the same issues, based on probability theory.

Pattern is closely related to computation, and I will use this connection here avidly. I will frequently invoke ideas inspired by algorithmic information theory (Chaitin, 1987), representing pattern using the theory of universal Turing machines. However, although grounding pattern in computation is convenient, I feel the concept of pattern is arguably *conceptually* more basic than the theory of universal computation.

Informally, the approach taken here is to define a pattern as "a representation as something simpler." In symbols, this means, roughly speaking, that a process p is a *pattern* in an entity e if:

- The result r_p of p is a good approximation of e,
- p is simpler than e.

More formally, suppose we have a space E, which we'll call the space of "entities," and a space P, which we'll call the space of "processes." Assume is a mapping from $\mathsf{P} \rightarrow \mathsf{E}$, called "production", denoting the idea that each process produces a certain entity. The entity in E produced by a process p, we will call r_p.

To say formally what it means for a process p to be a pattern in an entity e, we first need to define two concepts: *basic complexity* and *distance*. We need to define basic complexity for both entities and processes; and then we need to define distance between entities. (Distances between processes will be needed later for another purpose, and will be defined later.)

After using these notions to define the concept of pattern rigorously, I will extend the basic definition in a few ways, to obtain several related notions such as relative pattern, dynamic pattern, emergence, complexity and cooperativity.

Bit Strings and Novamente Subgraphs

The concepts of entity and process used here are pretty general, and could be applied in many different mathematical settings, e.g. within quantum theory, or a theory founded in uncomputable mathematical processes, etc. For simplicity's sake, however, in giving concrete examples of pattern-theoretic ideas here, I will invariably give computational examples.

Specifically, in this Appendix, we will make use of two cases:

1. Entities are bit strings and processes are computer programs represented as bit strings;
2. Entities are subgraphs of the Novamente Atomspace, and processes are Novamente schema or MindAgents.

In theoretical terms, Case 2 can be reduced to Case 1. First, a MindAgent or schema is a computer program, which can be represented as a series of characters. For instance, in the current Novamente implementation, MindAgents and schema are both C++ objects, and C++ objects have an obvious translation into bit strings (the one worked out by the C compiler). Next, a set of Novamente Atoms can be converted into bit strings using one of many possible simple encodings. Ultimately an Atom just consists of: lists of numbers, and lists of pointers to other Atoms. Each Atom has an object called a Handle, by which it may addressed; and Handles may be expressed as lists of numbers. To encode a Novamente Atom as a bit string, one merely has to encode lists of numbers as bit strings. Then, to encode a set of Atoms as a bit string, one merely has to encode a list of lists of lists of numbers as a bit string. This is very standard stuff.

In practice, though, it is awkward to explicitly convert Novamente entities into bit strings, and so when computing patterns, complexities and so forth using Novamente entities, it's better to stay in the Novamente domain. However, it's important to observe that theoretically, the choice of representation doesn't make much difference. The conversion from Novamente entities to bit strings and back again is a relatively low-complexity operation.

Basic complexity

A "basic complexity" measure is a function:

$$c \in [C \rightarrow R^+]$$

$$C \subseteq E \cup P$$

It is called a basic complexity function because it will be used within the definition of pattern, which will then be used to define *another* notion of complexity, one that may be thought of as "emergent complexity." A basic complexity measure gives rise to a measure of pattern, which gives rise to an emergent complexity measure.

The most typical way to define the basic complexity of an entity is as its size. In the case of a bit string, the basic complexity can be taken as string length. In the case of a Novamente subgraph (a set of Nodes and Links) it can be taken as the number of nodes and links in the set, or using some conceptually analogous but formally more sophisticated measure.

The case of processes is a little subtler. Here, it seems, there is no one "intuitively correct" basic complexity measure. The most practical course is to think about an integrative basic complexity measure, which is a weighted combination of several different measures.

Assume for simplicity that our space E of entities is represented as a space of binary strings. In this case, "size" is just bit string length. And let us consider processes p to also be represented as binary strings[75], which are interpreted computer programs P running on some given reference computer (which must be a Universal Turing Machine, like all modern general-purpose computers)[76]. Note that for now we are just considering programs that have no free input variables; we're just looking at a program P that runs and computes an output value r_p. A little later we will mention the case where the program P has a single fixed input ("assumed background knowledge"), but this does not change things significantly.

This grounding of processes in terms of computation is not the only way to develop pattern theory, but it is a very convenient path. One way to visualize this computational grounding of processes is to consider a fixed universal Turing machine which takes *two tapes* instead of the usual one. Tape 1 contains the program, Tape 2 contains the output of the program.

There are three major factors to consider in assessing the basic complexity of a computer program:

- Size (program length),
- Runtime (Koppel, 1987),
- "Crypticity" – which means, loosely, the difficulty of finding P, given r_p and the desire to find a program with small basic complexity (Bennett, 1990).

Size and runtime depend only on the reference computer assumed. Crypticity depends upon who is doing the finding! This is a major dependency, and yet one can make reasonable arguments for including crypticity here. What good, one may ask, is a program computing some desired entity, if it's impossible for any reasonably intelligent learning algorithm to find? Shouldn't we consider an alternative program less complex if it's 1000 times easier to conceive of, even if it is a little longer and slower?

We have been talking about bit string entities and programs, but if we have Novamente subgraphs for entities and Novamente schemata for programs, the story is essentially the same. In theoretical terms, we can simply encode these as bit strings using a small constant-size encoding

---

[75] For the mathematics of algorithmic information theory to apply properly, one must assume that processes p are represented as self-delimiting programs on the reference computer in question; i.e., programs that know their own lengths. The reasons for this are technical and are given e.g. in (Chaitin, 1987).

[76] If one is given a situation in which either entities or processes are not represented as binary strings, it's easy enough to convert any other representation into bit string form, to make the current theoretical constructs apply. By assumption, we are dealing only with finite sets, so no subtleties are involved.

program. In practice, we can simply define size, runtime and crypticity directly on these subgraphs, similarly to how one does with bit strings.

Similarly, the same concepts apply if one's entities are scenes in a simulation world and one's processes are programs computing images. The beauty of computing theory is that it is essentially representation-independent. The same concepts apply to bit strings, Novamente subgraphs, and images, and many other domains as well. (Of course, many pragmatic aspects of Novamente would become a lot more cumbersome if the bit string representation of entities and processes were used directly, but this is just a matter of "constant overhead", not fundamental complexity).

Measuring distance

Next, before formally defining pattern, we must define a metric d on the space of entities. There is a little subtlety involved in doing this properly.

First of all, the metric d must be scaled reasonably in regard to the basic complexity function c. One approach here is to specify that:

$$\frac{mean_{e,f \in E}d(e, f)}{mean_{e \in E}C(e)} = W$$

where W is a given constant. The choice of W is a weighting of accuracy of simulation (of the process p approximately producing the entity e) versus compactness (of the entity). If one has a base metric D(e,f), one can set $d(e,f) = c_1 D(e,f)$, and set c_1 to achieve the desired W. In real applications it is rarely possible to compute the means involved in the above equation precisely, but one can approximate them via random sampling.

There are many plausible choices for the base metric D, in any practical situation. Below we will define a metric that applies in the particular case where one has entities that are decomposable into parts, so that, e.g. an entity e can be written as an ordered list $(e_1,...,e_n)$ where the e_i themselves lie in a space with the metric d_c, and where the e_i and f_i (the i'th entries of two entities e and f) are in a sense "comparable."

Suppose the space E of entities is taken to be the space of possible subsets of pixels of a computer-screen "canvas" (i.e. the set of possible black & white pixilated pictures at the given level of granularity, defined as the number of possible "subsets colored black"). Then the basic complexity of an entity may be defined as the number of pixels in the entity. The base metric D(e,f) may be defined as the number of pixels in the symmetric difference of the entities e and f. Processes may be understood as computer programs that output subsets of pixels (and formally defined relative to some given Universal Turing Machine, for instance relative to the UTM defined implicitly by Novamente schemata). Leaving crypticity aside, the basic complexity of a process may be defined as a weighted average of *program length* and *program runtime*.

If one wishes to include crypticity in this example calculation, it should be defined in terms of a particular Novamente system, or set of Novamente systems, of interest. It may be defined as the *expected* time that it takes the Novamente system(s) under consideration, given the entity $e = r_P$, to search the space of programs and find that indeed P is a program so that $r_P=e$.

Representing Novamente Subgraphs as Vectors

For instance, a Novamente subgraph can be written as an ordered list of Atoms. Suppose one has a fixed Novamente instance, and e and f are each either *subgraphs* of this Novamente, or *approximations* to subgraphs of this Novamente.   In this case, if we want e_i and f_i to be comparable, we can create vector representations of e and f so that e.g.

- e_i = the Atom with handle i, if this Atom is in e
- e_i = 0 (the "empty Atom") if the Atom with handle i is not in e

For instance, if e consists of the Atoms with handles 1, 3 and 5, but f consists of the Atoms with handles 2 and 3, then the componentized versions are:

e = (Atom w/handle 1, 0, Atom w/handle 3, 0, Atom w/handle 5)
f = (0, Atom w/handle 1, Atom w/handle 3, 0, 0)

If e and f are exact subgraphs and have nonzero elements in entry i, then they have identical elements in entry i, because a given Novamente instance only contains one Atom with handle i. But if one of e or f is a just an approximation to a Novamente subgraph, then it may contain an Atom in position i which is not the same as the actual Atom with handle i in the Novamente instance in question.

To handle this case, we need first to define a metric d*(A, B) that applies when A and B are Atoms. Various ways of metrizing Atom space are discussed in *Probabilistic Logic Networks*. The simplest approach is to use vector space distance; i.e. to consider an Atom as a vector of components $A = (A_1 \ldots A_n)$, and define:

$$D(A, B) \;=\; \sum_{i=1}^{n} w_i \; d_{ci}(A_i \; B_i)$$

The metrics d_i compare particular Atom components to each other.

For instance, if Atom components are ordered such that A_1 is the truth value of Atom A and A_2 is the importance of Atom A, then $d_1(A_1, B_1)$ is the difference between Atom truth values and $d_2(A_2, B_2)$ is the difference between Atom importances. This simple metric works for all purely numerical components of an Atom, including strengths of links to other Atoms.

On the other hand, in the case of a NumberNode or CharacterNode, there will be A_i that represent numbers or characters. For instance, the distance between two characters should be 0 if the characters are the same, 1 if they are different. The distance between two numbers may be taken as the Euclidean difference, or as some kind of scaled version of this.

Processes, in Novamente, may be understood as computer programs that produce subgraphs; specifically, they may be understood as Novamente schema that produce Novamente subgraphs, and the basic complexity of such processes may be defined as above, as a weighted average of program size and runtime complexity, and also the difficulty of *Novamente itself* finding the process. Here the crypticity has a definite meaning, since we know who is doing the finding: Novamente.

Pattern Intensity

Given all these preliminaries, we may now define the notion of *pattern intensity*. The intensity with which *p* is a pattern in *e* is given by the formula:

$$IN(p, e) = \left(\left[1 - \frac{d(r_p, e)}{c(e)} \right] \star \frac{c(e) - c(p)}{c(e)} \right)^{+}$$

The term

$$\frac{c(e) - c(p)}{c(e)} = 1 - \frac{c(p)}{c(e)}$$

gauges the amount of simplification or "compression" provided by using p to represent e. If p provides no compression, this yields 0; in the limit where p is entirely simple *(c(p)=0)*, the term yields its maximum value of 1 (indicating 100% compression). If, say, $c(p) = .5c(e)$ then one has 50% compression.

The term

$$1 - \frac{d(r_p, e)}{c(e)}$$

allows for approximate matching or "lossy compression": it has a maximum of 1 when r_p, the result of carrying out process p, is exactly the same as e. The maximum intensity of a pattern, according to this formula, will be 1; and anything with an intensity greater than zero may be considered to be a pattern. This term can give a negative result in some cases, which is the reason for the superscripted + outside the whole algebraic expression, which means that a negative value results in a zero pattern intensity.

Relative Pattern

Pattern intensity, as defined above, measures the extent to which a process, considered as an entity operating alone ("in a vacuum"), can represent an entity as something simpler. In real life, however, processes often do not operate in a vacuum. They often operate in the context of large amounts of background knowledge. For instance, a Novamente schema does what it does using, potentially, all the knowledge implicit in the Atomspace of the Novamente instance it is embedded within.

To deal with this important phenomenon, the definition of pattern may be modified to give a definition of pattern *relative to a given base of knowledge*. This modified definition is very useful for the analysis of certain aspects of Novamente in terms of pattern theory.

In the relative approach, instead of looking at the basic complexity $c(e)$, one assumes some knowledge base K and looks at the basic complexity $c(e|K)$. This means, intuitively: how simple is e if one assumes that complete knowledge of K is given?

Formally, one way to represent relative pattern is to introduce a fixed universal Turing machine which takes *three tapes* instead of the usual one. Tape 1 contains the program, Tape 2 contains the output of the program, and Tape 3 contains the knowledge base. In working with relative basic complexities, one looks at programs P that are allowed to use the contents of the data tape K in doing their processing.

Using relative simplicities throughout the definition of intensity, one obtains a definition *IN(p,e|K)*. Similarly one may define a structure *St(f|K)*. In general, any quantity defined in terms of pattern theory may be relativized in this way. In Novamente it is often valuable to consider patterns where the implicit knowledge base *K* is the *set of Atoms in a Novamente system at a given time*.

Pattern Theory and Algorithmic Information Theory

As I've already noted, both formally and conceptually, pattern theory is closely related to algorithmic information theory (Chaitin, 1987). In fact, pattern theory might fairly be described as building on the basic concepts of algorithmic information theory, using similar ideas to produce a more flexible and general mathematical framework, more suitable to serve as an "information theory" of complex cognitive systems.

Given a bit string e, the algorithmic information I(e), defined relative to a Turing machine with a program tape and a data tape, is defined as the length of the shortest (self-delimiting) program that causes e to appear on the data tape. In order to make the algebra of the I operator work out nicely, it is commonly assumed that all computer programs involved are "self-delimiting," i.e. contain a segment specifying their own length.

This relates closely to pattern theory if one restricts attention to the special case where the metric d is defined so that *d(x,y)* is infinite unless $x = y$. In this case, the pattern intensity reduces to:

$$\frac{c(e) - c(p)}{c(e)}$$

If e is a bit string of length N (with basic complexity defined as entity length) and P is a program, this is

$$1 - c(P)/N$$

So, suppose one defines the basic complexity c(P) of a program P as the length of its binary sequence representation. And suppose P_e is a program of minimal length that computes e. Then for $P=P_e$ the intensity $1 - c(P)/N$ will be at its maximum value IN_{max}. The algorithmic information of e is $c(P_e)$, and is hence given by the formula:

$$c(P_e) = N(1 - IN_{max})$$

A straightforward example is the binary sequence:

$$e = 1001001001001001001001001001001001001001 \ldots$$
$$1001001001001001001001001001001001$$

consisting of 1000 repetitions of the string "100". Then we have *c(p) = 200*, while *c(e) = 1000*, and so the intensity of *p* as a pattern in *e* comes out to *[1000 - 200]/1000 = .8.*

It is important to note, though, that while algorithmic information theory assumes the metric d away, this simplification is not adequate for real-world pattern analysis. Most patterns in the real world are inexact. It's true that a process p that computes e approximately is always closely related to another process p' defined roughly as "p plus some random noise embodying the difference between r_p and e." But it is unintuitive and often impractical to consider the random noise as part of the pattern identified in e.

For instance, consider a picture consisting of a black square against a field of quasirandom scribbles:

In pattern theory, we can say that a short program P computing the black square is a pattern in this picture. In straightforward algorithmic information theory we cannot, because this program does not precisely compute the picture in question. Instead, there is a program Q that computes the scribbles in the background; and according to algorithmic information theory, it is the program P' that combines P and Q that is recognized as a pattern in the picture.

Finally, algorithmic information can be considered relative to background knowledge, and this works exactly the same way as in pattern theory. Basically, the algorithmic information $I(x|K)$ is the length of the shortest self-delimiting program that produces x given K as input. If one assumes a three-tape Turing machine model, then it is the length of the shortest self-delimiting program producing x on the data tape and beginning with K on the background knowledge tape.

Structure

A convenient name for the set of all processes p that are patterns in e is the *Structure* of e; it is denoted *St(e)* and is a fuzzy set with degrees of membership determined by the definition of pattern intensity. The more intense a pattern is, the greater its degree of membership in the structure. This is really the point of the definition of pattern intensity. Obviously, reducing something as explicitly multidimensional as a pattern to a single number, the intensity, involves a great amount of information loss. But, when one looks at the structure of an entity, it is good to have a way to prioritize the various patterns in it. The qualitative and quantitative have a sensitive interplay here.

For a more qualitative example, consider the following classic picture:

Suppose one is computing patterns in this picture *relative to a standard human being's knowledge store*, which contains many images of both faces and vases. Then the structure of this picture contains both a pattern representing it as a pair of faces, and a pattern representing it as a vase.

Minimal Structure

One problem with the formal definition of structure given above is that the structure of any nonrandom entity is nearly always going to contain a huge set of patterns. Even a black square is going to have a huge number of patterns in it. Yes, its structure will contain simple programs to compute the whole square. But its structure will also contain programs like the one shown in the figure, that contain subprograms computing *part* of the square, and append the rest of the square (stored as a bitmap) to the output of the subprogram – and all other sorts of things. Depending on how the metric d is scaled relative to the basic complexity measure c, its structure may contain a large collection of pictures that are similar to but slightly smaller than the square — since these are less complex than the square and have a very small distance to it.

To avoid this problem, we may define an alternate set called the *minimal structure*. The minimal structure of e, denoted $St_M(e)$, is the structure $St(e)$ filtered to contain only those elements of $St(e)$ that are *local maxima of pattern intensity*. Since the space of processes is a discrete space, this concept involves a little arbitrariness. We must assume we have a metric d_2 on the space of processes. Then we must posit a radius r, defined so that, intuitively:

$$N_r(p) = \{ q : d_2(p,q) < r \}$$

is the "local neighborhood" of p. One quantitative way to set r is by looking at the radius R of the set of all processes of basic complexity less than the basic complexity of e (these are the processes that are in a sense "potential patterns" in e), and then set $r = h * r$ for some parameter h.

For p to be a local maximum of pattern intensity, we require that p have a pattern intensity *at least equal to* that of any other process q in the neighborhood $N_r(p)$. What this does is to filter out processes that are basically just inferior mutations of other processes that capture "essential patterns" in the entity e.

For example, in the case of the black square, programs that compute part of the square using a clever subprogram, and then append bitmaps to the output of this subcomputation, are unlikely to appear in the minimal structure. In each such case, one will generally be able to find a similar program that computes a slightly larger subset of the square using a similar clever subprogram, thus achieving a higher pattern-intensity. This will depend significantly upon the radius r, of course.

And similarly, under reasonable parameter settings, patterns that consist of regions slightly smaller than the square will never appear in the minimal structure, because there are always going to be similar regions that are closer to the square and hence have a higher pattern intensity.

It is possible that, in this filtering process, one throws out some things that are of interest. In practice, however, there seems little choice but to take this kind of approach (though one may introduce subtler heuristics), because the task of computing even the minimal structure is intractable. Finding the minimal structure can be done by taking an optimization approach, where one searches for patterns that are local maximum of pattern intensity. Under any reasonably general formalization of process space, this is an uncomputable problem, and one can only approach it pragmatically by looking at a limited subset of process space. But requiring the identification of all patterns, not just the local-maxima ones, would seem to make even this limited pragmatic approximative approach untenable.

Complexity

Now we return to the notion of complexity. I have introduced a concept of "basic complexity" above, but this is merely a substrate concept, used to ground the concept of pattern. Here I will use pattern to define the more critical concept of emergent complexity, which we will just call "complexity". The basic idea is that the complexity of an entity is the amount of pattern in that entity.

Of course, "complexity", like any other natural language concept, wraps up a variety of different meanings. The definition given here surely does not capture all of these, but it is a precise and usable concept that captures much of the flavor of the intuitive "complexity" concept. It also has more than one dimension; we will discuss explicitly:

- The "structural complexity" of an *entity;*
- The "static/dynamic" and "purely dynamic" complexity of a *system;*
- The "functional complexity" of a *process.*

Complexity of a Static Entity

What I call the *structural complexity* of an entity may be defined as the total amount of pattern in the entity, i.e. the size of the entity's structure,

$$\text{StructComp(e)} = |\text{St(e)}|$$

However, this is easier to write down and to conceptually describe than to fully formalize. The problem is that the measurement of the size $|\ |$ of a fuzzy collection of patterns is a matter of some difficulty. The structure of any reasonably complex entity is going to contain a *lot* of different patterns, many of which are very similar to each other, and it doesn't seem sensible to simply add up the intensity of many nearly-identical patterns in the same entity e. But nor is there any simple way to "subtract off for overlap" given the lack of a robust algebraic structure on the space of patterns.

If patterns were linear projections, one would deal with overlap by taking the set of patterns St(e) and using Gram-Schmidt orthonormalization to find an orthonormal basis underlying them. The orthonormalization process would produce as set of patterns without overlap. But pattern space is not a normed vector space, so things are much trickier

The best way I have conceived (so far) to deal with this problem is based on *averaging*. In this approach, first proposed in *The Structure of Intelligence,* one essentially averages over all possible ways of subtracting off for overlap.

To see how this goes, consider the famous ambiguous "old woman / young woman" picture:

It is conceptually similar to the faces/vase picture from the previous section. This figure will have more complexity than a comparable picture that showed only a young woman's face, or only an old woman's face, because it has an "old woman's face" pattern and a "young woman's face" pattern, both. These patterns have some overlap, of course. Thus, one can compute both:

$r_{young\_woman\_pattern} = e$

$r_{old\_woman\_pattern} = e$

But to get |St(e)|, one cannot simply sum up:

```
IN(young_woman_pattern, e) + IN(old_woman_pattern,e)
```

because this does not account for overlap. It may be that these two patterns embody some of the same regularities. Instead, one has to look at both:

```
S₁ = IN(young_woman_pattern, e) +
IN(old_woman_pattern,e|young_woman_pattern)
```

and

```
S₂ = IN(young_woman_pattern, e) +
IN(old_woman_pattern,e|old_word_pattern)
```

If these two patterns were the only ones in the structure of the entity, then we would average these two orderings to obtain the structural complexity of the entity:

```
StructComp(e) = .5 ( S₁ + S₂ )
```

Of course, these are by no means the only two patterns in the picture; to make an accurate assessment of |St(e)| it is necessary to consider all the patterns in e and all possible ways of ordering them to subtract off for overlap. This is not a very computationally tractable process, but in a theoretical sense, on top of the incomputable problem of actually recognizing the patterns, this extra step doesn't do much harm! In practice, to estimate this quantity, one looks at a restricted subclass of patterns, and one must estimate the average via random sampling.

In the above example, we assumed there were only two patterns in St(e). This is an oversimplification, but it is less of an oversimplification if instead of St(e) we look at $St_M(e)$, the minimal structure, obtaining the minimal structural complexity:

```
StructComp_M(e)
```

It is *conceivable* that, given proper parameter settings and the right set of background knowledge, St_M of the two-faced picture could come out to contain only the two patterns S_1 and S_2 cited above. Even if not, it is quite plausible that St_M might contain only a few dozen other things. On the other hand it is not conceivable that this might be the case for St(e). In a Novamente or other comparable AI system with robust image-processing capabilities, there would always be a few other prominent other patterns besides S_1 and S_2, as there would be patterns involved in traditional image-compression-style analysis of the image.

Pattern and Complexity in Systems

We have been talking about patterns in particular entities, but very often in practice, one is interested in looking at patterns in systems that change over time. To handle dynamical systems one doesn't need to change the fundamental concepts of pattern theory, but one does need to introduce some additional concepts.

Generally, when studying a system *S* that has states *S(t)* at various points in time t, one is interested in patterns which are recognizable in the system's history, defined as a series:

```
Hist(S) = (S(t),S(t+1),...,S(t+r))
```

Patterns in a system's history might be called "static/dynamic patterns"; they are patterns which incorporate information about both the "static" structure of a system at a given time, and the "dynamic" structure of a system's temporal history. A convenient shorthand notation for the portion of the system's history between times t and s is:

```
S(t,s) = (S(t),S(t+1),...,S(t+s))
```

By Hist(S) generically, we mean S(t,s), where t is the start of the system and s is the end of the system (where "start" and "end" are meaningful at least from the perspective of a given analysis).

Sometimes one may wish to measure exclusively the dynamical patterns in a system – the patterns *by which the system changes*, as apart from the patterns in the system's state at particular points in time. Toward this end, one may define the *dynamic pattern intensity* of a process p in a system S as follows. Suppose that p maps time-series of system states into system states. Then, we may calculate the extent to which p is a pattern in the state of the system S at time t, calculating the degree of pattern relative to the past history of S leading up to but not including S(t). We may then define the dynamic intensity, DynIN(p,S) of a pattern p as the average over all time-points t of the degree to which p is a pattern in S at time t. We can also look at dynamic intensity as restricted to particular time intervals.

Using these ideas, the static/dynamic complexity and purely dynamic complexity of a system may be defined by setting:

```
SDComp(S) = StructComp( Hist(S) )
```

and by defining the membership degree of p in:

```
DynComp(S)
```

to be equal to DynIN(p,S).

To illustrate these concepts, consider the following series of pictures.

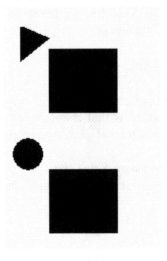

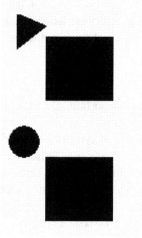

Of course, there is no way to make such a nice pictorial representation of, say, the internal state of a complex cognitive system at a given point in time; but these simple pictures may serve to illustrate the general ideas involved.

In this example, consider a program *draw_square* that produces a square and places it in the appropriate position on the graphical canvas. This is a static/dynamic pattern in the series of states, but it is not a significant dynamical pattern, because it is just as intense in the *set of states* as it is in the *series of states*. A similar statement holds for the programs *draw_circle* and *draw_triangle* that produce circles and triangles and place them in appropriate positions.

Using these programs, one can produce a compressed version of the series of pictures, which basically amounts to

```
S(1) = r draw_square AND draw_triangle
S(2) = r draw_square AND draw_circle
S(3) = r draw_square AND draw_triangle
S(4) = r draw_square AND draw_circle
```

On the other hand, one significant dynamical pattern in the series of pictures is the program of the form

```
A = rdraw_square AND draw_triangle

B = rdraw_square AND draw_circle

St(1) = A

For t=1 to N

If S(t) = A Then S(t+1) = B

Else S(t+1) = A
```

Obviously, this program is much more compact than a list of the compressed version of each system state. If one wants to generate an infinite history then it is extremely small. To generate a finite history of a long length N, the bulkiest part of the program is the number N, which takes about $\log_2(N)$ bits to represent.

Novamente Maps as Dynamic Patterns

In Chapter 15 I reviewed one important example of dynamic patterns in Novamente: maps. A map is precisely a set of Atoms that is involved in a prominent dynamic pattern.

The simplest kind of map is a set of Atoms that tend to all be important together. Then the dynamic pattern is simply the pattern that says "When a few Atoms in the set M is important, so are all the rest." It's easy to see how this allows the representation of the history of Novamente as something simpler. Suppose the history of the system looks like:

```
time  highly important atoms

1     A1, A2, A3, A4, A7, A9, A15
2     A1, A2, A6, A8, A15
3     A4, A5, A7, A11
4     A77, A12, A4, A6
5     A1, A2, A4, A6, A88, A15
6     A15, A6, A7, A9
```

Here $\{A_1, A_2, A_{15}\}$ look to be a very simple map: whenever one is important so are the others. The map-ness isn't absolute, because at time 6, A_{15} occurs without the other Atoms in the map. But it's pretty close to perfect.

So, suppose one names this map M_1. Then, one can compress the history of the system using a program P constructed as follows:

P stores a reduced history Hist'(S), constructed as follows. At each time t at which map M_1 is active, one goes through the parts of the history that refer to the states of the Atoms in map M_1 at time t, and omits the numerical importance value of each such Atom.
P stores the information
- $M_1 = \{A_1, A_2, A_{15}\}$
- M_1 is active at times $T_1 = \{t_1, t_2, t_5, \ldots\}$

P produces the history Hist(S) by running through Hist'(S) and inserting in the records for all Atoms in M_1 at times T_1, a numerical value indicating "high importance."

This program P does not produce the system history Hist(S) exactly, because it doesn't get the importance values of the Atoms in the map M_1 exactly right. Whether P is really a pattern in Hist(S) or not depends on how this error is weighted as compared to the compression obtained by not having to store the specific importance values of the Atoms in the map at the times of map activity.

It should be clear that a similar logic holds for less simple maps. For instance, if one has an oscillatory map such as:

```
time highly important atoms
```

```
1    A1, A2, A3, A4, A7, A9, A15
2    A41, A42, A6, A8, A15
3    A1, A2, A5, A7, A11
4    A41, A42, A4, A6
5    A1, A2, A4, A6, A88, A15
6    A41, A42 15, A6, A7, A9
```

then the temporal pattern "{A1, A2} are highly important at time t, then {A41, 42} are highly important at time t+1, and vice versa" can be used to create a program similar to the program P constructed above, which is a pattern in the history of the system.

Now, these maps are highly localized ones. This means that they will be more prominent patterns in the histories of certain subsystems of the overall system, than in the history of the system as a whole. This means that, in the terminology to be introduced just below, they are *subsystem patterns*.

Most maps spread over more of the system than these simple examples, and not all of them simply indicate that importance levels of their elements are high at certain times. Some maps may involve more fine-tuned patterns in importance or activation. But still the translation of maps into formal dynamic patterns remains relatively straightforward.

In the case of a schema map, for example, it is activations rather than importances that are patterned, and the usual case is not one of repetition. Suppose for example one has a schema computing Sqrt($x^2 + y^2$). Represented as a distributed schema, this involves four SchemaNodes: two for the square, one for the square root, and one for addition. Suppose these are represented in the formal Atom-list componentizing the system as Atoms A_{15}, A_{21}, A_{24} and A_{99}.

Then we will have, for example, patterns such as:

time	*active Atoms*
15	A15, A21,...
16	A99, ...
17	A24, ...

over and over again. The same sequence of activations will be seen every time the distributed schema is executed, because the execution of schema maps is "activation-driven". A compressed version of the system history can be produced by:

Creating a modified history Hist'(S) by deleting the activation values of {A_{15}, A_{21}, A_{24} and A_{99}} at the times when this schema execution pattern is detected

Storing a list T_1 of all times at which the schema execution pattern is initiated, in the system's history

Producing Hist(S) by going through Hist'(S) and, at all times in T_1, inserting a high activation value in the "activation" slow for all the Atoms in the map {A_{15}, A_{21}, A_{24} and A_{99}}

System and Subsystem Patterns

System complexity has many aspects. Above I introduced the basic notion of patterns in a system's history, and a little later I'll formalize the notions of emergence and cooperativity in systems. In this subsection I'll introduce yet another aspect of complexity in systems, which is what I call a *subsystem pattern*. Informally, this refers to a pattern that is a significant pattern in *some parts* of a system at *some times* during the system's history.

First, define the set of *history slices* in the history of the system S, i.e. the set of series of the form {S(t-r),...,S(t)). Define a *slice-set* as a set of history slices of this type. Looking at patterns defined in slice-sets rather than in the whole history, allows one to explicitly focus on at change-patterns that are only relevant to some period in a system's evolution. A subsystem pattern is one that is localized even further: not only to a particular phase of a system's history, but to a particular part of a system.

Subsystem patterns are a special kind of "system pattern," where the latter term is taken to mean any pattern in the history of a system. Any subsystem pattern of reasonable intensity is also going to be a system pattern, because if a pattern P produces subsystem S' $\subset$ S, it can be embedded in a program P' that produces S by:

Storing S-S'
Producing S' using P
Producing S by joining (S-S') and S

If P is a pattern in S', this program P' should be a pattern in S, unless P is such a weak pattern in S (in which case the overhead involved in the non-data-storage segment of very brief program P could push the basic complexity of P' too high)

We can then define a *subsystem pattern* in a system S in two ways: one that is generally applicable, one that applies only to componentized systems.

The general way is as: **A program that is a pattern in some history-slice-set T that is part of S's history**. In this case, since we want to look at patterns that define change in only *part* of the system, it is useful to use a p'th power average instead of an arithmetic average in the distance metric on system state space, with p equal to a fairly large number, say 10. This means that functions agreeing *very closely* with *some parts of S(t)* for *some points of time t* will be rewarded and considered as significant patterns. This is very important because many meaningful systemic patterns will only regard certain parts of the system, rather than the whole system. However, such part-focused patterns will still be systemic patterns according to the definition of pattern, though they will tend to provide less simplification of the relevant history-slice-sets than more broadly supported patterns.

The other way of defining "subsystem pattern" applies only when a decomposition of a system into component parts is known. That is, suppose one has a system that is composed of a number of component parts, so that its state $S(t)$ can be considered as a set of part-states $\{S_i(t), i=1,...,N\}$. Each component part leads to its own tuple $Hist(S_i)$, and hence to its own static/dynamic patterns.

In this case, we can define a subsystem pattern as: **A program that is a pattern in some history-slice-set T that is part of the history of some set of S's components**. In cases where we need to distinguish it from the more general definition, this may be referred to as a *componentwise subsystem pattern*.

Each subsystem pattern (under either definition) will have a maximal slice-set, defined as the slice-set over which its intensity is greatest.   However, this maximal slice-set may be very small, perhaps the one slice in which the pattern is most intense. So it makes sense to define the quality of a slice-set, with respect to a pattern, as a weighted combination of the intensity of the pattern in the slice-set and the size of the slice-set. In this way we can associate a maximal-quality slice-set with each pattern in the system. In measuring the intensity of a subsystem pattern, one can look at its intensity in the system's history as a whole, or in its maximum-quality slice-set.

In the case of the simple maps described in the previous subsection, it's obvious what the maximum-quality slice-set is: the set of times during which the map is active. And it's obvious what subsystems the patterns corresponding to these maps are most pertinent to: the Atoms in the maps themselves. The patterns corresponding to these maps are patterns in the system history of Novamente as a whole, but they are far more prominent patterns in the subsystem consisting only of the Atoms in the map; and either in the whole system or just this optimal subsystem, they are maximally prominent when attention is restricted to the time-points of map activity.

For a more general sort of example, suppose it is the case that every time SchemaNodes and GoalNodes are simultaneously highly active in the system, the system learns a lot (increases the total truth value of its Atoms significantly). This pattern is a system pattern across the whole system, because part of it pertains to all Atoms in the system.

On the other hand, suppose that every time SchemaNodes containing large compound schema are highly activated, this causes PredicateNodes containing large compound predicates to be formed a little later. This is a pattern in the whole system, but it is even more prominently a pattern in the subset of the system consisting of SchemaNodes and PredicateNodes containing large internal compound structures. It will be more easily identified using the componentwise concept of subsystem pattern (which is applicable because Novamente is a system that is cleanly decomposable into components: Atoms and MindAgents).

Derived Hypergraphs

Now I have finally introduced enough formalism to give a rigorous version of the notion of a "derived hypergraph", a formalization of the concept of a Novamente map given in Chapter 15. I will present the idea in the context of Novamente specifically, but the extension to complex systems more generally should be obvious.

Among the Vertices in the derived hypergraph associated with a Novamente system are Vertices corresponding to maps that are patterns in that Novamente system. Each of these MapVertices may be given a weight which denotes the degree to which the map is a pattern in the Novamente system.

The Edges between these MapVertices in the derived hypergraph then denote various relationships between these maps. ExecutionOutput and Execution Edges may be defined for SchemaVertices, according to the logic given in the Introduction. Inheritance Edges between Vertices are introduced following semantics to be given below in the chapters on PLN and attention allocation.

However, MapVertices are not the only ones. One may also introduce Vertices representing mathematical functions that are patterns in maps or sets of maps – and, more abstractly, Vertices representing patterns between these patterns, etc. One may then define an EmergenceEdge, which exists between (A,B,C) if C is an emergent pattern between A and B and has a weight corresponding to the degree of emergence as defined in Section 8 below. The subgraph of the derived hypergraph formed by the EmergenceEdges and corresponding Vertices is what was in (Goertzel, 1993) termed a "network of emergence," and there posited as an essential, evolving structure of long-term memory in humans and other intelligent systems.

In essence, then, the derived hypergraph of a Novamente system is a network of patterns spawned by the Novamente Atomspace as acted on by SchemaNodes and MindAgents. When we formally define "mind" in the following chapter, it will become clear that the derived hypergraph associated with a Novamente system is a significant component of that Novamente system's mind.

Next, although we've been talking about Novamente, it's clear that the same notions apply to cognitive systems more generally. Maps and their interrelationships may be found in any complex evolving system, and I suggest that highly intelligent systems will invariably give rise to rich and complex derived hypergraphs.

Functional Complexity

Next, what about the complexity of a process, a mathematical function F? To define this, first we must specify the notion of pattern as applied to functions, beginning with basic complexity and distance.

Here we will consider only discrete functions, i.e. functions F: X→Y that effectively map a finite space X to a finite space Y. Of course, if one is given a function on spaces X and Y that are countable, one can turn it into a discrete function by placing upper bounds on the complexity of elements in X and Y that one is willing to consider. A discrete function F may be defined by its "graph", i.e. by a set of ordered pairs $\{(x_i, y_i), i=1,...,N\}$ where $x_i \in X$ and $y_i \in Y$. For simplicity, we will assume in this section that X and Y are sets of binary strings, but the ideas given here do not depend on this restriction in any deep way; we will use them in the context of Novamente atoms as well.

We will consider functions to be approximated by computer programs, but we will consider only time-finite programs. To avoid the halting problem, we will simply declare a time T and assume that if a program has not returned a value by time T, it returns a special value DID_NOT_HALT_IN_TIME.

Any discrete function can be computed (exactly or approximately) by a variety of different computer programs P. For simplicity, let us consider programs P that take their inputs on one tape, and give outputs on another tape, and themselves exist as binary strings living on a third tape. To interpret such a program as a mapping from space X to space Y, one then assumes that some binary strings on the input tape correspond to elements of X, and some binary strings on the output tape correspond to elements of Y.

The basic complexity of such a program P may be defined as discussed above, using a combination of runtime and program length and possibly the difficulty of discovery.

There is one subtle point that here, which can be seen by considering a function such as:

$$h(x) = c$$

where c is a very long random bit string ("random" meaning "the shortest program for computing it on the reference computer architecture has length exceeds its own."). In this case, the behavior of F as a function is very simple: it is a constant function. On the other hand, the shortest program P for computing F is very long, because it has to compute the constant output value c. Yet this shortest program P is going to be very fast, and very easy to discover, so its basic complexity should be fairly small according to an integrative complexity measure combining program length, runtime and crypticity. This example does not necessarily indicate a problem with the definition of functional pattern being developed here; it merely indicates the subtlety of the definition of complexity and pattern in this context. Specifically, the issue of weighting the various components of the basic complexity of a process is a thorny one.

In the approach being taken here, to consider P as a *possible* pattern in F (i.e. to feed it into the definition of pattern), we don't need to assume P is a mapping from X into Y. One can feed any program P binary string inputs x_i, and in response to each of these inputs, it will either give an output interpretable as some element of Y, or give some output not interpretable as an element of Y (maybe DID_NOT_HALT_IN_TIME). The problem of sorting out programs P that are actually mappings from X into Y, from the other programs P, is left to the distance measure involved in the definition of pattern, which fortunately is not difficult to specify in a reasonable way.

Suppose F and G are both discrete functions mapping X➔Y. Let N be the number of elements in X, and Mean_p be the p'th-power average (as introduced above). Assume there is a metric d_1 on the range space Y. Then, we can define:

$$D_{(X,Y)}(F,G) = \text{Mean}_p \{d_1(F(x),G(x)), x \in X\}$$

Next, suppose P is a program that maps the bit strings x_i into bit strings living in the set Y. Then P implicitly defines a function $F_P:X➔Y$, and we may define:

$$D_{(X,Y)}(F,P) = D_{(X,Y)}(F, F_P)$$

On the other hand, suppose P does not map all the bit strings x_i given in the graph of F into bit strings living in the set Y. Then we can simply define $D_{(X,Y)}(F,P) = \infty$.

We may then define a program P to be a *functional pattern* in F, using the ordinary definition of pattern, where the basic complexity of P is defined in the standard way as described above, and where the above metric is used as the distance measure.

Definitions of structure, emergence and complexity of functions follow immediately.

For a different sort of example, consider a function F that takes in a picture containing a small shape somewhere in the left half of the screen, and outputs a picture containing the same small shape in a symmetrical position on the right half of the screen. The graph of F contains a series of pairs of images.

Clearly, the program P which is the most prominent pattern in this function F is the one that takes a bit map and flips it around symmetrically, reversing left and right. This program will have a very low basic complexity (it's compact and fast), and its distance to F will be very small.

Another example is a function that takes the input picture and appends a constant random-looking image to it (this is similar to the mathematical function $h(x) = x+c$ discussed above). This does have a prominent pattern, assuming a basic complexity measure involving program size. Because, the program that says "Append the following image to my input" and contains a copy of the appropriate image, is still much smaller than the graph of the function, which contains repeated copies of this random image.

Emergence

Yet another critical concept ensuing from the concept of pattern, is that of *emergence*. This concept has almost as many different common-language meanings as "complexity," and the usual caveat applies here: We're not trying to fully capture the common-language idea of emergence, just to capture a good bit of it in a way that is useful for patternist philosophy and associated practical work.

In this section I will present a notion of emergence that might be called "structural", although it can be applied to systems and functions as well as entities. In essence, it says that a pattern is emergent between two entities to the extent that it is more intense in the *combination* of the two entities than in the two entities considered separately. In the following section, I will introduce a related sort of emergence called *cooperativity* – a concept that applies to any two processes within a dynamical system that generate a lot of interesting phenomena in the system when they're active together, more so than when they're active separately.

A Formalization of the Emergence Concept

To formalize the emergence concept we must add one more primitive notion to our formal vocabulary: joining. Let e+f denote some kind of "join" of the two entities e and f (such as, if e and f are two bit strings, the result of placing e and f end to end). In Novamente the join operator is embodied in a link type called a SetLink, which defines an unordered combination of its targets.

The simplest way to define emergence (as introduced initially in *The Evolving Mind)* is to say that a process *p* is an *emergent pattern* between *e* and *f* to the extent

$$Em(p,e,f) = IN(p,e+f) - [IN(p,e) + IN(p,f)]$$

The shortcoming of this definition is that it weights e and f equally, whereas it may well be that e contributes 10 times more than f to e+f. To remedy this, one may define:

$$Em(p,e,f) = IN(p,e+f) - [w_e IN(p,e) + w_f IN(p,f)]$$

$$w_e= 2 c(e)/[c(e) + c(f)]$$

$$w_f = 2 c(f)/[c(e) + c(f)]$$

A fuzzy set called the *emergence* between e and f:

$$Em(e,f)$$

may then be defined, where the degree of membership of p in Em(e,f) is given by Em(p,e,f). This is the set of all patterns emerging between e and f. We may define the emergent complexity of e and f as:

$$EmComp(e,f) = | Em(e,f)|$$

As with the definition of structure, it may sometimes be interesting to pare down "redundant" programs from the emergence. This can be done in the same way for emergence as for structure, resulting in quantities $Em_M(e,f)$ [minimal emergence] and $EmComp_M(e,f)$ [minimal emergent complexity].

Finally, this formalization of emergence works perfectly well for defining the emergence between two mathematical functions, or two systems. The application to functions requires no modification; one must simply interpret the entities e and f in the above definition of Em(p,e,f) as functions, and interpreting the process p as a functional pattern. The situation with systems is a little more complicated, and will be dealt with a little later.

A Mathematical Example of Emergence

Next I will give a simple mathematical example of emergent pattern. For simplicity, we will suppose in this example that basic complexity of programs is assessed solely in terms of program length, as in algorithmic information theory.

Suppose that:

e_3 is a length-N bit string that is generated by a short program P_3;
e_1 is the string consisting of all the bits in even positions of e_3;
e_2 is the string consisting of all the bits in odd positions of e_3;

- The shortest program for computing e_1 is the program P_1 that computes e_3 and then selects only the bits in even positions;
- A similar minimal program P_2 exists for e_2.

Finally, suppose that the join operator (+) on bit strings is defined as concatenation.

If the hypothetical example seems strange, think for instance of the first N=100000 digits of Pi. There are several short programs for generating Pi. How about the first 50000 even digits of Pi? What's the best way to generate them? Quite possibly, to generate the first 100000 digits of Pi and then extract the even digits. This is not *known* to be a property of Pi, it's just a conjecture —just as the "statistical randomness" of the series of Pi's digits is as yet unproved, though validated by extensive numerical testing..

Pi however is not a *perfect* candidate for our conjectural number e_3, because it's know that there are *many* qualitatively, reasonably short programs to generate Pi. Ideally, one would like a number that could be computed only by *one* reasonably short algorithm. It might well be possible to use algorithmic information theory methods to prove that such a number e_3 exists, but I have not made this effort; hopefully the illustrative example is useful anyway.

Anyway, given the assumptions we have made, it follows there is a program P_{12} that computes e_1+e_2 very compactly, by running P_3 to get e_3 and then doing some simple reorganization of the bits. Suppose:

- P_3 is of length M<<N
- P_{12} is of length M+k where k<<M
- P_1 and P_2 are each of length M+m where m<k<<M

Then the intensity of P_{12} in e_1+e_2 is about

$$IN(P_{12}, \ e_1+e_2) \ = \ 1 \ - \ (M+k)/N$$

whereas the intensity of P_1 and P_2 in the bit strings e_1 and e_2 is given by:

$$IN(P_1, \ e_1) \ = \ IN(P_2, \ e_2) \ = \ 1 \ - \ 2M/N$$

If one assumes the base distance measure D on bit-string space is defined as a normalized Hamming distance[77], then the emergence comes out very simply. The output of P_{12} is $e_1 + e_2$, and $D(e_1+e_2, e_2) = D(e_1+e_2, e_1)$ is going to be very large[78], rendering:

$$IN(P_{12}, \ e_1 \ ) \ = \ IN(P_{12}, \ e_1 \ ) \ \approx 0$$

Thus, one has
$$Em(P_{12}, \ e_1, \ e_2) \ \approx IN(P_{12}, \ e_1+e_2) \ = \ 1 \ - \ (M+k)/N$$

If P_{12} were the only pattern in the minimal emergence $Em_M(e_1 + e_2)$ [not the case if $e_3 = Pi$, as there will be programs qualitatively different from any single program P_3 for generating Pi; but possibly the case for some other examples], we would get a minimal emergent complexity of:

$$EmComp_M \ (e_1 + e_2) \ = \ 1 \ - \ 2M/N$$

Systemic Emergence

What does it mean to talk about the emergence between two dynamical systems? Firstly, if one has two systems S and T, one can speak of:

$$EmIN(S,T) \ =$$

$$|St(Hist(S)+Hist(T))| \ - \ [w_S \ |St(Hist(S))| \ + \ w_T \ |St(Hist(T))|]$$

and

$$Em(p,S,T) \ =$$

$$IN(p,Hist(S)+Hist(T)) \ - \ [\ w_e \ IN(p,Hist(S)) \ + \ w_f \ IN(p,Hist(T)) \ ]$$

But there is also a notion of systemic emergence that goes beyond this, having to do with the emergence of patterns among the parts of a single system.

Suppose one has a system that is composed of a number of component parts, so that its state S(t) can be considered as a set of part-states $\{S_i(t), i=1,...,N\}$. Each component part leads to its own tuple $Hist(S_i)$, and hence to its own static/dynamic patterns. An *emergently complex system* may then be defined as any system in which a great number of emergent patterns exist between the parts. In other words, the emergent complexity of a system should be measured in terms of the size:

$$| \ St(Hist(S)) \ - \ [w_1 \ St(Hist(S_1)) \ + \ ... \ + \ w_N \ St(Hist(S_N))] \ |$$

where the weights are defined appropriately as in the definition of emergence above.

---

[77] The Hamming distance between two bit strings is the number of bits in which they differ
[78] If one assumes a p'th power metric with large p, this won't be true, as partial matches such as the match between e_1+e_2 and e_2 or e_1 will get more credit.

For instance, consider the following very simple dynamical system. Imagine a screen which, at each time-point, has two shapes on it, one toward the right-hand side and one shape toward the left-hand side. The shapes are as follows.

	Right	Left
1	Shape0	Shape1
2	Shape1	Shape2
3	Shape2	Shape3
4	Shape3	Shape4

Here, S_0 and S_1 correspond to the right and left halves of the ShapeWorld canvas, roughly speaking. The time-courses $Hist(S_1)$ and $Hist(S_2)$ are not all that significantly structured, but the time-course $Hist(S)$ has a significant pattern, as reflected by the fact that $S_1(t+1) = S_0(t)$.

Might there be systems that are highly complex but not emergently complex? Indeed there might be, but empirically speaking, this does not seem to be the case in the physical universe, or in the mind. In the known physical and computational universes, complexity seems as a rule to be built up from the interactions of simpler parts.

Patterns Related to System Processes

We have talked about patterns in systems, and emergent patterns in systems, but we have not yet specifically talked about the effects of particular dynamic processes on systems. This is very important for Novamente, as most of the subtlety of the Novamente design comes out of its intended amenability to the free intermixture of various learning dynamics, which is supposed to lead to the emergence of appropriate high-level mind-structures; and I suppose that it is similarly appropriate to any complex intelligent system.

Here I will present formal notions useful for discussing the effects of explicit system processes like MindAgents on Novamente dynamics. We will define the *resultant* of a process as the set of patterns that are caused by it, and the cooperativity of two processes as the patterns that result specifically from the combination of the two processes. I will then introduce the notion of *implicit dynamics*, i.e. patterns that empirically appear to regulate the behavior of the system. Many implicit dynamics come directly out of explicit dynamics like those encoded in MindAgents, but some may be emergent and less directly tied to explicitly encoded rules.

Resultants and Cooperativity

Assume one has a system S (identified by its history), and a set of processes $Pr_1 ... Pr_n$ that act on the system S. Assume the system S is decomposable into parts, and that at each time, each part of the system $S_i(t)$ may be acted on by each process Pr_j to a certain extent, denoted:

$$activity(i,j,t) \in [0,1]$$

If a process Pr_j is the sort of process that "either definitely acts on a component or definitely does not," then $activity(i,j,t)$ is always either 0 or 1. Otherwise it may take an intermediate value, representing an amount of activity.

In Novamente the Pr_j might correspond to explicit system processes (MindAgents), or they might correspond to *implicit processes*, i.e. dynamical subsystem patterns recognized in the system's history. For illustrative purposes we will discuss explicit system processes here, but in later chapters we will discuss cooperativity in the context of implicit dynamics as well.

To take a simple example, we may have:

Pr_1 = clustering

Pr_2 = reasoning

This represents a fairly coarse-grained analysis since in Novamente logical inference is actually carried out by several different MindAgents, but for the purposes of formal analysis we may consider there to be a single inference process. (In general there could be more than one clustering MindAgent in Novamente as well, although we are currently working with a single clustering method.)

Next, the most appropriate way to componentize Novamente is, as discussed above, to break it down to the Atom level. This means that the S_i are Atoms, so that the $S_i(t)$ are effectively (Atom, time) pairs.[79]

Next, for each process Pr_i, we wish to look at the set of (system component, time) pairs that are acted on by that process, defined as:

$T_{i,r}$ = { $S_j(t)$: activity(i,j,t) > r }

This is the subset of the system S – in space and in time – that is directly involved with the process Pr_i. Of course, the activity of Pr_i will have many indirect effects as well, and the notion of cooperativity we are developing here will take these into account. But the first step is to look at the direct effects of a process Pr_j, because these constitute the representation of Pr_j within the history of the system.

How do we get at the indirect effects? Quite simply, we introduce the notion of causality. What we want is to find *subsystem patterns* that are *caused by* $T_{i,r}$. The set of all such subsystem patterns is what we call the *resultant* of process Pr_j, and denote:

Resultant(Pr_j)

It is a fuzzy set, and its size |Resultant$_j$| may be measured with the same complications as with structure and emergence. Also as with structure and emergence, we may define a minimal resultant, Resultant$_{M;j}$, restricting attention to patterns that are local maxima of intensity.

The notion of causality, to which we have referred here, is itself quite subtle. As noted in Chapter 17, and pursued a little further in Appendix A5 below, a full definition of causality is elusive, and even posing an adequate definition is far from a trivial endeavor. A chapter of the PLN book is entirely devoted to this topic. The approach taken in Novamente is to define a fuzzy rather than crisp notion of causality, denoting the statement "the degree of causality between A and B is tv" by:

---

[79]Recall that the subscript i, distinguishing each Atom from the others, should be considered isomorphic to an Atom's Handle (address) in Novamente; Handles are unique over the lifetime of a Novamente system.

```
CausalLink A B <tv>
```

where tv is an uncertain truth value. The details of the definition of a CausalLink are complex. In short, though, the approach involves defining causality as a combination of two factors. A causes B if:

- A probabilistically & temporally implies B, meaning that it's often true that when B occurs, A occurred beforehand;
- There is a known or inferred "simple mechanism" that explains why occurrence of A should (often) lead to occurrence of B.

Note the occurrence of "simplicity" here, in the notion of a "simple mechanism." Essentially this "simple mechanism" is a particular type of pattern. A mechanism for producing A given B is a function, whose simplicity/complexity may be gauged in the manner of any other function, as described above.

I make no claim that this CausalLink formalism is a complete rendition of the human common-language notion of causation, but after careful study I have come to believe it comes reasonably close, and will satisfy the practical needs of AGI development, at least in the Novamente context.

In the present case, we can apply the CausalLink formulation of causality if we assume that the sets $T_{i,r}$ and the relevant subsystem patterns P are represented by Novamente structures (PredicateNodes containing complex internal predicates). The membership of P in the fuzzy set Resultant(Pr_j) is given by:

$$s = \chi_{Resultant\ (Prj)}(P)$$

where s is the strength of the link:

```
CausalLink T_{i,r} P <s>
```

Cooperativity, then, is a phenomenon that occurs when the resultant of the combination of two processes is much greater than the sum of the resultants of the processes considered individually. Conceptually, this is a kind of emergence, but a special kind. Suppose we denote the process defined as the simultaneous execution of Pr_i and Pr_j by:

```
Pr_i & Pr_j
```

Note that this is semantically a little different from the + operator ("joining") used in the definition of emergence above.

Formally, we may define the *cooperative emergence* of two processes Pr_i and Pr_j as the set of patterns that belong to the Resultant(Pr_{ij}) more strongly than either of Resultant(Pr_i) or Resultant(Pr_j). The degree of membership of a pattern p in:

```
CoopEm (Pr_i , Pr_j)
```

is given by:

$$CoopEm(p, \; Pr_i \; , \; Pr_j \;) \; = \; \chi_R$$

where

$$R \; = \; [Resultant(Pri \; \& \; Prj)](p) \; -$$

$$max\{ \chi \; [Resultant(Pri)](p) \; , \chi \; [Resultant(Prj)](p) \; \}$$

For a concrete example of this, let us return to the example of reasoning and clustering, briefly mentioned above. Assume it is true (as indeed seems to be the case) that the two together create better categories (category-embodying ConceptNodes with greater truth value) than one would expect from the results of their separate activity.

Let's think about these categories that are recognized by the system via reasoning/clustering combination, but wouldn't be recognized by the system if reasoning and clustering were used separately. These categories represent patterns recognized in the (internal or external) data available to the system. Now, these category/patterns may be present in the available data even without reasoning and clustering being used together – but recognizing them as ConceptNodes enables them to move from the *implicit* level to the *explicit* level, and hence to be used in the system's explicit processes of reasoning, association-spreading, and so forth. The ensuing cognition done on these ConceptNodes will lead to all sorts of further nodes and links being created, and further strengthening of existing nodes and links where appropriate. The subsystem patterns created by this ensuing activity are part of the set:

$$Resultant(\; Pr_1 \; \& \; Pr_2 \;) \; = \; Resultant \; (clustering \; \& \; reasoning)$$

and, to the extent that they wouldn't occur without the ConceptNodes created by reasoning/clustering collaboration, they are part of:

$$CoopEm(clustering, \; reasoning)$$

In the following Appendix, I will define the mind of a system as the set of patterns in the system that are causally related with the system's intelligence. In this sense, a possibly useful principle of intelligence is that an intelligent system's mind should consist largely of *emergent subsystem patterns* arising via *cooperativity*. Clustering/reasoning synergy is one among many dozens of examples of such synergies that we have discussed in the preceding chapters.

And the high-level mind patterns discussed in Chapter One – the dual network and the self, for example – are envisioned as emergent system patterns arising through cooperativity of *all the system's AI modules*. The collection of AI modules in the system is intended to be a close-to-minimal set capable of leading to cooperative emergence of these high-level emergent structures.

Implicit Dynamics

The processes Pr_i discussed above might be called "explicit dynamics." In Novamente, these correspond to MindAgents, or sometimes aspects of MindAgents or combinations of MindAgents. These are the basic rules by which the Novamente designers/programrs cause a Novamente Atomspace to evolve over time.

It's also interesting, however, to take a more empirical view of Novamente dynamics. What would we think were the dynamics of the system if we were studying the system based only on its recorded history, without knowledge of its codebase? What dynamical rules would we infer from it? This gives rise to the notion of *implicit dynamics*. The implicit dynamics of S may be defined as the set of dynamic patterns in S, as defined above. That is, it is the set of patterns in the way S changes over time, as opposed to patterns in the state of S at any given time.

Any map will correspond to an implicit system dynamic, because, as we have seen, each map corresponds to a dynamic system pattern. But maps are by far the only implicit dynamics in the system. Explicit processes in a system, like inference, clustering and evolutionary learning in Novamente, will correspond with particular dynamic system patterns, particular implicit dynamics. Generally, each MindAgent in Novamente will correspond to many different dynamic system patterns, not just one. Some of these will refer to particular equational rules inside the MindAgents, others to direct consequences of these rules, others to less direct consequences. And then of course there will be some implicit dynamics that do not correspond closely to any particular MindAgents, but are rather more cooperatively emergent in nature.

For instance, importance updating will correspond to system patterns such as the following.

This one says that, when an Atom A is important, and another Atom B is linked to it, B is likely to become more important by a certain amount:

```
"If at time t, Atom A has importance > .3, and Atom A links to
Atom B with a link of weight > .2, then from time t to time t+1,
the importance of atom B will increase by a ratio of at least
1.1."
```

On the other hand, the deduction rule in inference will correspond to system patterns such as the following.

This pattern just says encodes the Atom-level "deduction" pattern of inference:

```
"If at time t, (InheritanceLink A B) and (InheritanceLink B C)
are both in the top 10% Atoms in terms of importance, then there
is a probability p that in the interval (t, t+10),
(InheritanceLink A C) will be acted on by the FirstOrderInference
MindAgent, where p = .4 *
SystemParameter.FirstOrderInferenceImportance"
```

This one states that the deduction pattern of inference is often associated with increases in the weight of evidence of the conclusion of inference:

```
"If at time t, (InheritanceLink A B) and (InheritanceLink B C)
are both in the top 10% Atoms in terms of importance, then there
is a probability p that in the interval (t, t+10),
```

(InheritanceLink A C) will have the weight-of-evidence of its truth value increased from d to at least [2 d-1]/ [2d -1 + kd - k], where k = SystemParameter.InferencePersonalityParameter, and p = .4 * SystemParameter.FirstOrderInferenceImportance"80

These two encode the actual quantitative deductive inference rule:

"If at time t, (InheritanceLink A B <tv>) and (InheritanceLink B C <tv1>) are both in the top 10% Atoms in terms of importance, and (InheritanceLink AC) is not in the system at all, then there is a probability p that in the interval (t, t+10), (InheritanceLink A C <tv2>) will appear in the system, with tv2 as given by the PLN FOI deduction formula, and p = .4 * SystemParameter.FirstOrderInferenceImportance."

"If at time t, (InheritanceLink A B <tv) and (InheritanceLink B C <tv1>) are both in the top 10% Atoms in terms of importance, and (InheritanceLink AC <tv2>) is in the system at time t as well, then there is a probability p that in the interval (t, t+10), (InheritanceLink A C <tv4>) will appear in the system, where tv3 =revision(tv2, tv3) and tv3 = deduction(tv,tv1) are given by the PLN FOI revision and deduction formulas, and p = .4 * SystemParameter.FirstOrderInferenceImportance."

The other reasoning rules will all correspond to similar dynamic system patterns. An implicit dynamic involving two MindAgents would be, for instance, the pattern that important SchemaNodes will tend to be grouped together into distributed schema and then these distributed schema will yield outputs:

"Suppose Atom C SchemaNodes A and B are both in the top 10% of Atoms in terms of importance at time t, and: C matches the input type of A, whereas the input type of B matches the output type of A. Then, with probability p, an Atom X resulting from the schema composition B(A(C)) will appear in the system in the interval (t,t+5). Here p = .3 * (SystemParameter.ApplicationLinkFormationImportance + SystemParameter.SchemaActivationImportance)."

Finally, for an example of an implicit dynamic that is not tied to any one MindAgent or small set of MindAgents, or any particular map, consider:

---

[80] The formula given in this pattern is derived as follows. The standard formula for weight of evidence is $d = n/(n+k)$ where k is the "inference personality parameter." We solve this to get $d(n+k) = n \rightarrow dn + dk = n \rightarrow (d-1)n+dk = 0 \rightarrow n = k \, d/(d-1)$. Now, suppose we add one more piece of evidence, then we have $d' = n+1/(n+1+k) \rightarrow d' = [d/(d-1) + 1] / [d/(d-1) + 1 + k] = [d + d - 1] / d + (d-1)(1+k)] = [2 \, d-1]/ [2d -1 + kd -k]$. The inference step involved in the pattern may add more than this, but it should add at least "one quantum" of evidence.

"Suppose, that during the interval (t-r,t): changes in map A often cause changes in map B, and changes in map A often cause changes in map C, AND maps B and C are strongly associated. Then these same relations are likely to hold during (t,t+s)."

This pattern partially encapsulates the conceptual pattern "dual network structure tends to be stable over time."

Imagine a human scientist with an extremely advanced pattern recognition toolkit, studying a complete record of the history of an evolving Novamente. This person's pattern recognition toolkit would recognize all these patterns and more. It might not be possible for the person to reverse-engineer exactly what the built-in dynamic rules of the system were, but they could understand what the system was doing, by looking at the implicit dynamics. The number of implicit dynamics is huge, but narrowing it down to a "minimal set" by looking for local maxima in pattern intensity space would yield a smaller set.

This is basically how we study the human brain! Fortunately, we do not have to study AI systems this way. In a Novamente, context, however, it is important to remember that the real importance of the explicit dynamics we program into Novamente, lies in the implicit dynamics to which they give rise – individually, in combination with each other, and in combination with each other and the system's environment and evolving knowledge base. These implicit dynamics are present as SchemaVertices in the system's derived hypergraph, even though they may not be explicitly present anywhere in the system's code. The map encapsulation process may recognize them as patterns in the system and then embody them explicitly as SchemaNodes – thus embodying the "cognitive equation" and increasing the system's intelligence-potential via ever-increasing self-awareness.

Probabilistic Pattern Theory

In the preceding sections of this chapter, we have given an overview of pattern theory as presented in previous publications, and applied this theory specifically to the description of various types of static and dynamic patterns in the Novamente system. In this final section we will deviate from this course somewhat, and present an alternate mathematical foundation for pattern theory, which also has some conceptual differences from the approach presented above.

The original motivation for seeking an alternate mathematical foundation for pattern theory was the observation, made briefly in the introduction to this chapter, that the "traditional" formulation of pattern theory as outlined above is guilty of a certain mathematical sterility. The definitions seem to make sound conceptual sense, but it has proved difficult to develop nontrivial algorithms or theorems based on them. This suggests that perhaps they are not the "right" definitions from a mathematical or computational point of view, even though they do serve the purpose of giving a precise and rigorous formulation of the underlying philosophical concepts. With this in mind, in 2004 an alternative approach to the foundations of pattern theory was conceived, in which probability replaces simplicity as the foundational concept.

After the alternate, probabilistic version of pattern theory was developed, a connection was noticed between probabilistic pattern theory and the theory of intensional inheritance in PLN inference. This connection has increased my confidence that the probabilistic formulation of pattern theory is the "right" one.

In this section, I will outline the basic concepts of probabilistic pattern theory, and sketch how some of the Novamente-related pattern-theoretic concepts of the previous sections could be reformulated in probabilistic-pattern-theoretic terms. A more complete reformulation of the previous sections in terms of probabilistic pattern theory will be left to the reader, but there don't seem to be any obvious difficulties lurking.

Probabilistic versus Traditional Pattern Theory

The difference between probabilistic and traditional pattern theory lies right at the foundation. In the traditional pattern theory, as presented above, simplicity is taken as a basic notion, and everything (i.e. pattern, emergence and related concepts) is derived from simplicity plus some basic algebra. In the probabilistic approach, on the other hand, simplicity is derived from probability, and probability is taken as the most basic notion.

The dependence on simplicity in the old pattern theory was a kind of subjectivity (because you had to assume a basic simplicity/complexity measure before making any other definitions); this subjectivity is not removed in the new, probabilistic formulation. Here you have to assume some basic dynamical system with respect to which probabilities are calculated (in a sense that will be explicated below). So at a deep level, the philosophical picture is basically the same, only the mathematical formalization is different.

The primitive notions we need to get started here are:

- multiple possible universes, each one existing through time,
- predicates, which may be true or false of particular universes at particular points in time, or during particular intervals of time,
- the ability to calculate conditional probabilities of events existing in universes (e.g. the probability of event E occurring at time T given that event F occurred at time S, calculated as an average over possible universes).

In a way this is a less simple foundation than the foundations of the traditional pattern theory, but it seems to lead to nicer mathematics, including simple and handy inference rules for making probabilistic estimates regarding the degree to which one entity is a pattern in another

In the rest of this section I will build up the probabilistic definition of pattern, and develop probabilistic definitions of critical associated concepts like emergence and structure.

Association

First we introduce the notion of association. We will say that a predicate F is associated with another predicate G if:

$$P(F \mid G) > P(F \mid \sim G)$$

That is, the presence of F is a positive indicator for the presence of G. The degree of association may be quantified as:

$$ASSOC(F, G) = [P(F \mid G) - P(F \mid \sim G)]^+$$

(where $[x]^+$ denotes the positive part of x, which is x if $x>0$ and 0 otherwise).
The association-structure of G may be defined as the fuzzy set of predicates F that are associated with G, where ASSOC is the fuzzy-set membership function.

Of course, the same definition can be applied to concepts or individuals rather than predicates.

One may also construct the "causal association-structure" of a predicate G, which is defined the same way, except with temporal quantification. One introduces a time-lag T, and then defines:

```
ASSOC(F,G; T) =
[P(F is true at time S+T | G is true at time T) -
P(F is true at time S+T|~G is true at time T)]⁺
```

The temporal version may be interpreted in terms of possible-worlds semantics, i.e:

```
P(F at time S+T|G at time T)) =
the probability that the entity A will be satisfied in a random
universe at time A, given that G is satisfied in that universe at
time T-A
```

```
P(F) =
probability that predicate F will be satisfied in a random
universe, generically
```

If A and B are concepts or individuals rather than predicates, one can define ASSOC(A,B;T) by replacing each of A and B with an appropriately defined predicate. For instance, we can replace A with the predicate F_A defined so that $F_A(x)$ is true iff x is a member of A.

From Association to Pattern

From association to pattern is but a small step. A pattern is a predicate F is something that is associated with F, but is simpler than F. A pattern in an entity A is defined similarly.

We must assume there is some measure c() of complexity. Then we may define the fuzzy-set membership function called the "pattern-intensity", defined as, e.g:

```
IN(F,G) = [c(G) - c(F)]⁺ [P(G | F) - P(G|~F)]⁺
```

There is also a temporal version, of course, defined as:

```
IN(F,G;T) = [c(G;T) - c(F;T)]⁺ * ASSOC(G,F;T)
```

The definition of complexity referenced here can be made in several different ways. Harking back to traditional pattern theory, one can use algorithmic information theory, introducing some reference Turing machine and defining c(X) as the length in bits of X expressed on that reference machine. In this case complexity is independent of time-lag T.

Or one can take a probabilistic approach to defining complexity, via introducing some reference process H, and defining:

```
c(F;T) = ASSOC(F,H;T)
```

In this case we may write:

```
IN[H](F,G;T;T₁) = [ASSOC(F,H;T₁) - ASSOC(G,H;T₁)]⁺ * ASSOC(G,F;T)
```

Notation-wise, typically we will suppress the H and T dependencies and just write IN(F,G) as a shorthand for IN(F,G;T;H); but of course, this doesn't mean the parameters H and T are unimportant. Note the dependence upon the underlying computational model is generally suppressed in algorithmic information theory, as well.

In some cases it is desirable to consider probabilistic pattern intensity as an ordered pair of numbers rather than an individual number. In this case one looks at:

```
ASSOC(F,G;T) =
P(F is true at time S+T | G is true at time T) -
P(F is true at time S+T|~G is true at time T)⁺

SIM[H](F,G;T) = ASSOC(F,H;T) - ASSOC(F,G;T)

IN[H](F,G;T,T1) = SIM(F,G;T1;H)⁺ * ASSOC(G,F;T)⁺
```

The purpose for this formulation is that it turns out to be easier to define "pattern inference" rules based on (SIM, ASSOC) ordered pairs than on composite IN=SIM⁺ ASSOC⁺ values.

It's worth noting that this approach can essentially replicate the algorithmic information theory approach to complexity. Suppose that M is a system that spits out bit strings of length <=N — choosing each bit at random, one by one, and then stopping when it gets to a bit string that represents a self-delimiting program on a given reference Turing machine. Next, define the predicate H as being True at a given point in time only if the system M is present and operational at that point in time. Next, for each bit string B of length <=N consider the predicate F[B](U,T) that returns True if B is present in possible-universe U at time T and False otherwise. Then it is clear that P(F[B]) is effectively zero (in a universe without a program randomly spitting out bit strings, the chance of a random bit string occurring is low). On the other hand, -log(P(F[B] at time S |G at time S-T)) is roughly equal to the length of B. So for this special predicate H and the special predicates F[B], the complexity as defined above is basically equal to the negative logarithm of the program length. Algorithmic information theory emerges as a special case of the probabilistic notion of complexity.

As in traditional pattern theory, we can introduce a notion of relative pattern here: if we assume K as background knowledge, then we can define:

```
ASSOC(F,G|K) = [P(F | G & K) - P(F|~G & K)]⁺
```

and

```
ASSOC(F,G| K ; T) =
[P(F at time S+T | G & K at time T) -
P(F at time S+T|~G & K at time T)]⁺
```

and

$$IN(F,G|K;T) = [c(G|K;T) - c(F|K;T)]^+ * ASSOC(G,F|K;T)$$

$$IN[H](F,G|K;T) =$$
$$[ASSOC(F,H|K;T) - ASSOC(F,G|K;T)]^+ * ASSOC(G,F|K;T)$$

Probabilistic Emergence

In a similar manner, we may define the degree to which a predicate F is an emergent pattern in predicates K and L by defining emergent predictability:

$$EmPred(F;K,L) =$$

$$log(P(K(A) \& L(A) | F(A-T)) - P(K(A)|F(A-T)) * P(L(A) | F(A-T)))$$

Emergent predictability is thus a kind of probabilistic dependency: how dependent are K and L conditional on F?

Emergent pattern then emerges as:

$$Em(F; K, L) = EmPred(F;K,L) * [Pred(F,G) - Pred(K\&L)]^+$$

That is, an emergent pattern is a pattern that is simpler than the combined predicate K & L, and also predicts the occurrence of this combined predicate significantly more than would be predicted from the degree to which it predicts K or L individually.

Probabilistic Structural Complexity

Using these ideas, associated with any concept or predicate in Novamente, we may construct a set called the "structure" of that entity. We may do this in (at least) two ways:

- The associative structure, which consists of the set of all observations and concepts that are associated with the entity;
- The pattern structure, which is the set of all observations and concepts that are patterns in the entity.

Finally, we may define the *structural complexity* of some entity as the sum (subtracting off for overlaps) of the total pattern-intensity of everything in the pattern-structure of that entity. This turns out to be very easy to define probabilistically. To define the structural complexity of G, we can simply look at the expected pattern-intensity in G of a randomly-chosen predicate F. Now, this "random choosing" requires some probability distribution, and we may define this probability distribution conveniently using the same "background" predicate H used in the definition of pattern. That is, we may consider the a priori probability of a predicate F to be IN(F,H;T). Then we may ask: of the predicates F chosen from this distribution, what's the expected pattern-intensity of these predicates as patterns in H?   We then have:

$$St(g) \ = \ \frac{\sum_F \left[IN(F, G \mid H; T) \ * \ IN(F, H; T) \right]}{\sum_F IN(F, H; T)}$$

In this manner one may define all the key concepts of pattern theory in terms of probability theory rather than algorithmic information theory.   The philosophical concepts underlying both formalizations are the same, and there are fairly simple mathematical relationships between the two formalizations, but yet they have a quite different flavor.   Which one is more useful for practical or theoretical work remains to be seen; perhaps each formalization will be useful for different purposes.   At the moment it is the probabilistic formulation that is finding more practical use within the Novamente AI framework, mostly because it combines more smoothly with Novamente's probability-theory-based inference module.

Appendix B
Toward a Mathematical Theory of Mind

In Chapter 3, I gave an informal review of the notions of intelligence and mind that fall naturally out of patternist philosophy. In this Appendix I review these same notions in a more formal and mathematical way. The treatment is still not fully mathematically rigorous, but it is more precise than the purely verbal treatment given earlier.

Potentially, with a fully mathematical treatment of intelligence, one might be able to prove theorems about the relationships between the internal structures and external behaviors of intelligent systems. One can imagine theorems stating something like "If a Novamente system with this many Atoms of such-and-such types, and these MindAgents with these parameter values, is placed in an environment satisfying these properties, then its intelligence measure will be > .32 with probability .83." At present *imagining* such theorems is about all we can do. But perhaps the semi-formal notions presented here will be one small step toward that goal.

After the first version of this chapter was written (which was a few years before completion of this book), I became aware of Marcus Hutter's excellent work, which proceeds along fairly similar (but not identical) lines, but gives rigorous mathematical proofs. His work is referenced a number of times in the main text of this book. Hutter gives a precise mathematical definition of "intelligence" – similar in spirit but different in details from the one given here – and he proves that certain computer programs will be intelligent according to his definition. The catch is that the programs he considers require either infinite (in the case of his AIXI program) or unfathomably large (in the case of his AIXI-tl program) computational resources. The formal ideas introduced here could almost surely be connected with Hutter's more complex ideas via the introduction of appropriate mathematics, but we have not done this work yet. While this would be very interesting from a theoretical perspective, it would not really help us understand Novamente in practice, since Hutter's mathematical apparatus doesn't contain any tools capable of saying anything about the intelligence (under his definition or any other) of systems with resources as profoundly limited as those of, say, Novamente or the human brain.

I believe it would be possible to carry out a Hutter-like development of the precise notions of intelligence pursued here, but if Hutter's work is any guide, this would be a Herculean mathematical undertaking. Such a fully mathematical treatment would occupy a book the length of this one, would take me a couple years to work out, and would be readable only by a handful of specialist researchers. Such work is certainly valuable but it's not something I personally have time to undertake right now.

Defining Intelligence

Recall from above my basic notion of intelligence, which is that:

Intelligence is the ability to achieve complex goals in a complex environment

Roughly speaking, the greater the total complexity of the set of goals that the organism can achieve, and the complexity of the set of environments in which these goals are achieved, the more intelligent it is.

Also critical is the notion of *efficient intelligence*, which is basically "intelligence per unit of space-time resources utilized." In practice, the only kind of intelligence one can hope to build is a highly efficient intelligence. And it is interesting to observe how important this is. If one foregoes the requirement of efficiency, architecting intelligence becomes remarkably easy. I will illustrate this in the following section, by giving a simple design for an AI system that, if implemented, would be highly intelligent according to the definition of intelligence given. The catch is that this system would use incredibly huge amounts of computational resources, more than is feasible given any conceivable hardware substrate.

When one seeks to fully formalize these conceptual definitions of intelligence, one quickly runs into various fairly arbitrary decisions. The best way of mathematically formulating the concept of "the ability to achieve complex goals in a complex environment" is not entirely clear. What is described in this section is one possible approach, which has the merit of relative simplicity.

While our intuitive definition of intelligence refers to the concept of an "environment," in our formal definition we will have more use for the concept of a "world." The difference is that, when talking about a system S, the "environment" implies the part of the world outside of S, whereas the world as a whole includes S as well as its environment. I.e:

world = external environment ∪ internal system state

We will talk about *possible worlds* that a system S may live in, assuming that we have a robust way of identifying *what the system S is* in each possible world. The possible-worlds-based approach in itself is computationally intractable, but afterwards we will discuss some ways to us it to structure practical computations, beginning with the idea of using the system's history as the repository of possible worlds.

To formalize the intuitive notion of a world, we will define a set WS, the space of "world-slices." Basically, a "world-slice" is a time-series of "world-states", i.e:

$$\texttt{W} \in \texttt{WS} \quad \rightarrow \quad (\texttt{W} = \{\texttt{W}(\texttt{t}_\texttt{i}) \,,\dots, \, \texttt{W}(\texttt{t}_\texttt{j})\} \quad)$$

where the t_i are moments in time, and the $W(t_i)$ are drawn from some given space of world-states.[81] A world-slice, unless it is of length 1, will have many component world-slices, which are subseries of it. A world is simply a maximally long world-slice, i.e. one that is not contained in any other world-slice.

In the AGI-SIM simulation world we use with Novamente, for example, the space of external world-states would be the set of states of the AGI-SIM world – including various possible rooms and scenarios. The space of internal world-states is the set of possible Novamente states. A world-slice is a series of Novamente states and simulation-world-states, taken over time.

---

[81] If concreteness is desired, for theoretical purposes the space of world-states may be taken as a set of binary strings.

Now, what are goals and how do they relate to worlds? Each world gives rise to a space of goals that can be achieved in that world. Let:

$$\mathbf{GS(W)} \texttt{ = the space of all goals = } \{f: \mathbf{WSS} \rightarrow [0,1]\}$$

where WSS is a subset of the set of world-slices defined in that world. In other words, we consider a goal as a mapping from a world-slice into a number, where the number indicates the degree of goal satisfaction. Not all goals are pertinent to all world-slices, for instance there may be some goals that are only meaningful over long periods of time, so that they cannot be evaluated at very short world-slices.

In the AGI-SIM simulation world, a simple goal could be to make the human teacher happy. This is a mapping from WS's into numbers, in a very simple and obvious way. Each world-state contains a numerical value indicating the level to which the user has set the happiness slider. So the mapping is: Average the user's happiness level over all world-states in a given world-slice, and that gives you the degree of goal-achievement during the world-slice.

Of course real-world goals are always much more complicated than this. Consider a human with the goals of getting food and water and sleep and sex. This can be represented as a function that assigns a number to each world-slice, representing the amount of food, water, sleep and sex that the human has obtained during the interval to which the world-slice pertains.

Next, to define the degree to which a system achieves complex goals in complex worlds, we will need to define a few preliminary quantities:

- *Achieve(S,G,W)* indicates the expected degree to which system *S* will indeed achieve the goal *G* in the world-slice *W*.
- *Complexity (G,W)* indicates the complexity of the goal G, interpreted as a function evaluated at the world-slices *contained in* world-slice W.
- *Complexity (W)* represents the complexity of the world-slice W.

The complexity of a world-slice is given by the general definition of the complexity of static entities, presented in the previous chapter. The complexity of a goal G is given by the above definition of the complexity of *processes*. A goal is a mathematical function mapping world-slices into numbers, and we discussed above the definition of pattern and complexity for mathematical functions.

We will also introduce a quantity:

$$\texttt{relevance(G,W)}$$

indicating the subjective relevance of each (goal, world-slice) pair under consideration. This term can be set to 1 and one still has a reasonable definition of intelligence, but in practice one may want to measure intelligence specifically in regard to certain goals and environments more than others, and including this term in the equation allows this.

Given all these component quantities, one way of equationally summarizing the intuitive concept of intelligence given above is as follows:

```
Intelligence(S) =
(1/M) Mean_p G∈GS Mean_q W∈WS [ achieve(S,G,W) * complexity(W) *
complexity(G,W) * relevance(G,W) ]
```

The constant M may be chosen so that the maximum intelligence possible is 1, yielding an intelligence measure that always lies in the interval [0,1]. An intelligence level of 1 refers to a system that has a probability 1 of perfectly achieving any goal under consideration over any interval of time in any possible world under consideration. An intelligence level of 0 refers to a system that has a probability 0 of achieving any goal under consideration to any nonzero extent in any possible interval of time in any possible world under consideration. No practical systems will ever achieve intelligences close to 1, assuming a reasonably broad set of world-slices and a reasonably difficult goal.

Note that we have used two different parameter values, p and q, in the p'th power means here. This is because one may wish to use different degrees of "spread" for the average across goals versus the average across worlds. One thing that is clear is that, in this context, one does not want p or q to =1 or anything close to that. One doesn't want to require a system to achieve *every* complex goal well in *every* complex environment to be considered intelligent. But on the other hand, one also doesn't want p or q to =∞, because it's nowhere near good enough for a system to be able to achieve just *one* complex goal in just *one* complex environment. Something intermediate is required to capture the intuitive notion of intelligence in a reasonable way.
Obviously, no matter how one refines the details, this is not going to be a practical definition, since one is never really going to be able to compute an average over all possible goals and worlds! But nevertheless the definition provides some intuitive guidance.

As warned in advance, this definition of intelligence is purely *behavioral*: it doesn't specify any particular experiences or structures or processes as characteristic of intelligent systems. The idea is that intelligence is something systems display (or not); how they achieve it under the hood is another story. It may well be that certain structures and processes and experiences are necessary aspects of any sufficiently intelligent system; in that case, from our perspective, one has interesting theorems about intelligence, but does not necessarily need to modify the definition of intelligence.

General versus Specialized Intelligence

The "relevance" term in the above definition of intelligence provides a way of modulating between completely general and overspecialized modes of intelligence. Highly specialized intelligences are intelligent only in the context of a small set of (goal, world-slice pairs). That is, they are intelligent only if one chooses a "relevance function" with a very small support. More general-purpose intelligences are intelligent for a much larger relevant set.

The degree to which humans are a general-purpose intelligence is subject to debate. The more we understand about the human brain, the clearer it becomes that, in fact, most of our intelligence is fairly narrowly specialized to the problems posed by our physical environment and by the task of socially interacting with other humans. It seems that the human brain consists of a small subsystem devoted to general-purpose pattern recognition, combined with a host of specialized subsystems devoted to particular types of specialized intelligence. Novamente has this same hybrid nature, but is perhaps a bit more firmly focused on general-purpose intelligence, due to its more cognitive and less perceptual/active bias.

Eric Baum, in his excellent book *What is Thought?* (2004), has suggested that replicating human intelligence or anything roughly comparable will be very difficult, because of the many subtle "inductive biases" encoded in human DNA, which make the human brain into a very powerful specialized intelligence, particularly customized for the environment on Earth where humans evolved. In our view, he makes a good point, but overstates it somewhat. I suspect that general architectural and algorithmic specializations (such as for example those embodied in Novamente) provide adequate "inductive bias" to enable pragmatic "reasonably general intelligence" in an Earthly environment. Furthermore, it seems clear that a flexible, self-modifiable architecture such as that of Novamente holds more promise of significantly and progressively expanding its generality-of-intelligence than the less flexible, more environment-overfit human brain.

Efficient Intelligence

Now I return to the notion of "efficient intelligence," informally introduced above. The above semi-formal definition of intelligence does not deal with the problem of limited space and time resources, it's purely behavioral. Given two systems:

- S, which runs on 10000 PC's and has a measured intelligence of .17;
- T, which runs on 2 PC's and has a measured intelligence of .15.

System S is judged more intelligent. Sure, T might become smarter than S if it had a little more hardware, but intelligence as we have defined it is not about "what if," it's about what a system can actually do.

Although this purely behavioral approach to intelligence makes sense, it is also interesting sometimes to look at a related notion of efficient intelligence. Conceptually, this can be defined as the ratio:

$$\frac{Intelligence}{Resource\ Utilization}$$

In practice, the useful definition of resource utilization is context-dependent. In the case of AI systems running on contemporary computer hardware, one tends to think in terms of processor power and RAM required, so one can define resource utilization pragmatically as a weighted average of the processing power and RAM used. In a physical system, on the other hand, one often thinks of energy as the primal resource – and if one considers not only the software aspect of an AI program, but also the hardware it runs on, then one can use energy expenditure as a measure of resource utilization here as well. A program that requires more processors and more machines will generally use more energy. The equation:

$$\frac{Intelligence}{Energy}$$

has a certain philosophical zing to it, although I don't see any direct applications for it at present; in practice it seems more useful to think in terms of resource utilization in terms of processing power and RAM.

A Near-Minimal AI Design

I've tried to very carefully distinguish intelligence from efficient intelligence. In this section I will make this point even more vividly, by sketching a design for a very powerful AGI that is enormously inefficient and commensurately simple.

Hopefully, this will drive home the point that space and time efficiency are not mere "technical considerations" – they're what AGI is all about. If you ignore them, the problem of creating intelligence (according to our definition) becomes trivial. But the luxury of ignoring space and time constraints was not permitted to evolution in creating the human brain, and is not permitted to us as AI designers; and nor will it be permitted to future, highly-intelligent AI's as they set about redesigning themselves.

Whether this kind of "purely theoretical AI design" is of any pragmatic interest for AI design is, at this stage, largely a matter of opinion. There are those who believe that the best approach to AI is to begin with a simple approach that works under the assumption of near-infinite computational resources, and incrementally make it more efficient. There are others who believe that the assumption of near-infinite computational resources brings one into a fantasyland of no relevance to practical system design. One problem, of course, is that a posited AI design, if it is too resource-wasteful, may not even be *physically possible*. Our view lies somewhere between these extremes.

The essence of "infinite resources AI" was worked out by Ray Solomonoff in the 1960's, and goes by the name of Solomonoff Induction. The mathematical theory of Solomonoff induction is complex, but the basic idea can be simply expressed in an informal manner. What we'll describe here is just one way of embodying the Solomonoff induction idea in a concrete AI design; there are many others.

Suppose you have a system with: some goals, receptors and actuators allowing it to perceive and act in an environment, a large amount of memory and processing power, and flexibility to rewrite most of its memory however it likes. The goals and the environment can be made as complex as one likes. Then the system can achieve arbitrarily great intelligence, defined in terms of satisfying its goals in its environment, as follows.

Suppose that the system's memory is divided into two parts: the metaprogram, and the current program. Since we are positing extremely ample computational resources, there is no problem allowing the current program and the metaprogram to operate simultaneously at all times.

The metaprogram works as follows. It creates and maintains a huge table whose rows are of the form:

Time t	$H(t)$ = Percepts observed up to time t	$A(t)$ = Actions taken at time t	$G(t)$ = Goal satisfaction up to time t

At each point in time, it searches the space of all programs P of size less than N (where N is very, very large), and for each program P it assesses two quantities:

- $Size(P)$ = The size of P;

- *Utility(P)* = The estimated degree to which the system would have achieved its goals if *P* had been its "current program" during its entire past history.

The metaprogram then chooses a program *P* that maximizes a combination of *Size* and *Utility*, such as:

$$(c_1 \; Size(P))^{q} \; * \; (c_2 \; Utility(P))^{2-q}$$

In other words, it asks itself: *"What is a simple program P so that, if I had obeyed P in the past, I would have had a high probability of achieving my goals?"*

For each *P* it is estimating an expectation value, using probability theory; this is a lot of work, but that's OK, because it has infinite computing power at its disposal. The *Size* factor incorporates a variant of Solomonoff's big insight, which is that minimizing program length avoids overfitting. Without the *Size* factor, the metaprogram would just choose a program *P* that was a lookup table mapping particular inputs into particular outputs, which then would generalize very poorly into the future.

The most desirable program *P* is then selected as the current program, and allowed to govern the system for the immediate future – until the metaprogram finds something better.

Obviously, the metaprogram is acting here as a general pattern recognition engine. In finding the optimal program *P*, it is searching for patterns in the past history of the system, patterns of historical perceptions and current actions that have led to goal-achievement.

Now, the problem with this approach to AI is fairly obvious: the memory and processing requirements are absurdly severe! However, it is useful as an illustration of the importance of efficient intelligence. What it shows is that the difficulty achieving real AI is basically a consequence of memory and processing efficiency issues. If these weren't an issue, AI would be easy. It would require just a few hundred lines of code in a high-level programming language.

The discussion in this section has been informal rather than fully mathematically rigorous. However, related mathematical ideas have been pursued by other theorists, with full formal rigor. Two examples are the AIXItl algorithm by Marcus Hutter (2004) and Juergen Schmidhuber's Goedel Machine (2005). These authors have proposed algorithms fairly similar to the one described above, and Hutter in particular has done a careful mathematical job of proving that (to informally paraphrase his results) his system can perform as intelligently as any other program, though possibly using vastly more memory and processing time.

In essence, what Hutter's work (in the spirit of the ideas in this section) shows is that creating computational intelligence is easy in principle. However, creating *efficient computational intelligence* is not easy at all. The two problems turn out to be very different. One can create computational intelligence, ignoring efficiency completely, just by writing a program that searches program space to find the smartest programs, and then uses them! But this kind of approach is largely useless given realistic bounds on space and time resources. It's not entirely useless, in the sense that search through program space is one part of real intelligence. But, in a pragmatic AI approach, search through program space has got to proceed differently: a system can only search for small programs carrying out particular specialized functions, and must then piece these programs together according to a methodology other than program space search.

As a footnote to the above discussion, one approach to making practical AI would be to start with a totally impractical approach like that of the previous section, and try to reduce the memory and processing requirements. Perhaps the most straightforward way to do this is to use evolutionary programming – genetic programming or its various variants like MOSES, discussed above. In evolutionary programming, one begins with a goal, and searches to find a program that fulfills this goal. The search algorithm is not a brute force search, but a cleverly self-focusing search that relies on successive refinement of a population of potential solutions using evaluation, plus probabilistic inference or mutation and crossover operations.

Unlike simple Solomonoff induction, evolutionary programming is a tractable approach to AI in some cases. When the goal is simple and the environment is not too complex, it works fine. But it doesn't scale well enough to be used as the basis for a large-scale, real-world AI system. Evolution times, on modern networked clusters of tens of machines, are hours to days for moderately complex problems; for the problem of driving a real system toward complex goals, millennia might be an optimistic estimate.

Of course, evolutionary programming is not the only approach one could take to juicing up Solomonoff induction. One could also, for instance, use recurrent backpropagation neural networks given an appropriate mapping into the input and output neurons of the nets. However, this would be even *less* efficient.

So, then, is the Solomonoff induction approach entirely worthless from a practical perspective? The answer, in my view, is: almost, but not quite. Solomonoff induction does embody a valid insight into pragmatic AI engineering, but an insight that is only practically useful after a huge amount of preliminary design and engineering work has been done, along quite different lines.

If one configures a Novamente system to be self-modifying, the resulting system has basically the same structure as the Solomonoff AI described above. There is a "current program" controlling the system, and a "metaprogram" that changes the current program based on meta-level learning. I believe this sort of design is in fact critical for achieving maximally intelligent AI – AI much more intelligent than its creators. But, in order for this to work in practice, the metaprogram needs to be doing something better than exhaustive search of all programs, and the current program needs to be something other than a randomly selected program, or else it will not generate even remotely interesting data for the metaprogram to look at.

Defining Mind

Having dealt somewhat thoroughly with intelligence, it is now time to turn to its sister concept: mind.

In Chapter 6 I presented a high-level philosophical analysis of mind in terms of four levels: primal, reactive, relational and synergetic. Here I will focus only on the relational and synergetic levels. Suppose we have an intelligent system S. Then, as a first stab, we may define:

> *The internal aspect of S's mind is the set of patterns in S.*
> *The external aspect of S's mind is the set of patterns emergent between S and its environment.*

These definitions do not touch the experiential level; and the physical level is only implicit, because it is assumed there is some physical system S that is operating and hence leading to these patterns that are part of S's mind.

Even if one accepts the conceptual approach underlying these definitions, there is something missing from them: an appropriate way of quantifying fuzzy membership in the mind. Not all patterns in an intelligent system are equally "mind-ish", even if they all do contribute to the system's intelligence to some extent. For each pattern P considered in S's mind's internal or external aspects, one needs to quantify how much P is actually correlated with S's intelligent behavior.

In a purely mathematical sense, there are many ways to do this. One may for instance posit a system S/P which is the system most similar to S but lacking the pattern P, and gauge the intelligence of S/P. The difference:

```
intel(P|S) = intelligence(S) - intelligence(S/P)
```

may then be posited as a measure of the contribution of P to the intelligence of S. This may be scaled into a measure of the degree to which P belongs to the internal or external aspect of the mind of S. In the following section I will describe a more complicated but also more practical (and perhaps more philosophically satisfactory) alternative to this measurement.

One issue is how to define the "relevant set" required by the definition of intelligence, when evaluating intelligence within this definition of mind. Perhaps the most sensible approach is to consider the actual environment in which a system has existed, as its relevant set.

Like the above definition of intelligence but even more so, this is not a practical definition. One cannot use it to construct a "mindscope" that observes an entity and draws a picture of its mind. Furthermore, it may well need substantial revisions (tweaks to the math that don't affect the philosophy, analogous to the differences between traditional and probabilistic pattern theory) before it can be used to derive nontrivial conclusions about intelligent systems. However, it gives us a reasonably solid conceptual framework within which to think about such entities as "the mind of Novamente", "the mind of Ben Goertzel", "the mind of the Goertzel family", "the mind of the emerging global brain mindplex", etc.; and that is worth something.

The SMEPH Approach to Modeling Minds

These ideas are helpful in articulating more specific approaches to mind modeling. For example, in my own thinking about concrete AGI implementation, I have found it helpful to have a general conceptual/mathematical framework that is more rigorous and precise than the psynet model (given in Chapter 15 above) yet less specific than any particular software design. With this in mind, I have created a general approach to AI that I refer to via the somewhat unaesthetic acronym SMEPH, which stands for Self-Modifying Evolving Probabilistic Hypergraphs. SMEPH is a fairly specific formalism for describing intelligent systems, which is consistent with patternist philosophy and the psynet model but provides more guidance regarding the analysis and construction of particular intelligent systems.

The basic ideas underlying SMEPH are threefold, as the acronym would suggest:

- To use a specific mathematical structure called a "generalized hypergraph" to model intelligent systems;
- To study the way hypergraphs change over time (i.e. they way they evolve" — the word "evolution" is used here in a general sense, rather than specifically in the sense of evolution by natural selection, although that is an aspect of SMEPH as well when one delves into the details);
- To use probability theory to study the relationships between the parts of the hypergraph.

A hypergraph is an abstract mathematical structure (Bollobas, 1998), which consists of objects called Vertices and objects called Edges, which connect the Vertices. In computer science, a "graph" traditionally means a bunch of dots connected with lines (i.e. "vertices" connected by "edges", or "nodes" connected by "links"). A hypergraph, on the other hand, can have Edges that connect more than two Vertices; and SMEPH's hypergraphs extend ordinary hypergraphs to contain additional features such as Edges that point to Edges instead of Vertices; or Vertices that, when you zoom in on them, contain little hypergraphs. Properly, SMEPH's hypergraphs should always be referred to as "generalized hypergraphs," but this is cumbersome, so we will persist in calling them "hypergraphs" instead. In a hypergraph of this sort, edges and vertices are not as distinct as they are within an ordinary mathematical graph (for instance, they can both have edges connecting them), and so it is useful to have a generic term encompassing both Edges and Vertices; for this purpose, in SMEPH and Novamente, we use the term "Atom."

A "weighted, labeled hypergraph" is a hypergraph whose Edges and Vertices come along with labels, and with one or more numbers that are generically called "weights." The label associated with an Edge or Vertex may sometimes be interpreted as telling you what "type" of entity it is. On the other hand, an example of a weight that may be attached to an Edge or Vertex is a number representing a probability, or a number representing how important the Vertex or Edge is to the system.

Hypergraphs may come along with various dynamics. For instance, one may think about:

- Dynamics that modify the properties of Vertices or Edges in a hypergraph (such as the weights attached to them);
- Dynamics that add new Vertices or Edges to a hypergraph, or remove existing ones.

The SMEPH approach to intelligence is centered on a particular collection of Vertex and Edge types. The key Vertex types are ConceptVertex and SchemaVertex, the former representing an idea or a set of percepts, and the latter representing a procedure for doing something (perhaps something in the physical world, or perhaps an abstract mental action). The key Edge types are ExtensionalInheritanceEdge (which, linking one Vertex or Edge to another, indicates that the former is a special case of the latter), ExtensionalSimilarityEdge (which indicates that one Vertex or Edge is similar to another), and ExecutionEdge (a ternary edge, which joins {S,B,C} when S is a SchemaVertex and the result from applying S to B is C). So, in SMEPH, one is often looking at hypergraphs whose Vertices represent ideas or procedures, and whose Edges represent relationships of specialization, similarity or transformation among ideas and/or procedures.

ExtensionalInheritance and ExtensionalSimilarity Edges come with probabilistic weights indicating the extent of the relationship they denote (e.g. the ExtensionalSimilarityEdge joining the cat ConceptVertex to the dog ConceptVertex gets a higher probability weight than the one joining the cat ConceptVertex to the washing-machine ConceptVertex). The mathematics involving these probabilistic weights becomes quite involved — particularly when one introduces SchemaVertices corresponding to abstract mathematical operations, a step that enables SMEPH hypergraphs to have the complete mathematical power of standard logical formalisms like predicate calculus, but with the added advantage of a natural representation of uncertainty in terms of probabilities, as well as a natural representation of networks and webs of complex knowledge.

SMEPH hypergraphs may be used to model and describe intelligent systems (such as human mind/brains, for example). One can (in principle) draw a SMEPH hypergraph corresponding to an individual intelligent system, with Vertices and Edges for the concepts and processes in that system's mind. This leads to what is called the "derived hypergraph" of that system, a notion defined in terms of pattern theory. Morst simply, one may say that a ConceptVertex in the derived hypergraph of a system corresponds to a structural pattern that persists over time in that system; whereas a SchemaVertex corresponds to a multi-time-point dynamical pattern that recurs in that system's dynamics. Drawing the derived hypergraph of an intelligent system is one way of depicting the mind of that system – this follows from the definition of a mind as the set of patterns in an intelligent system, and the fact (which follows from mathematical pattern theory) that the patterns in the system can be read off from the derived hypergraph.

Extending the notion a little further, one may posit that any intelligent system has *both internal and external derived hypergraphs*, incorporating respectively:

- Vertices defined as patterns in the system
- Vertices defined as patterns in the system, or emergent between the system and its environment

Both the Vertices and the Edges in these derived hypergraphs will be part of the system's mind. In fact, these derived hypergraphs are essentially synonymous with the mind of the system, since the mind is defined as the set of patterns in the system, and these patterns are the Vertices of the derived hypergraph. The derived hypergraph is a particular way of *organizing* the mind of an intelligent system.

As well as being used to conceptually model intelligent systems, SMEPH hypergraphs may also be used as the foundation of an AGI design. In this case, a SMEPH hypergraph is used explicitly as the medium for the (long and short term) memory of an intelligent system, and its thought processes are explicitly described and implemented as dynamics modifying this hypergraph. Such a SMEPH-based intelligence will also have a derived hypergraph, which will not be identical to the hypergraph it uses for explicit knowledge representation. However, an interesting feedback arises here, in that the intelligence's self-study will generally lead it to recognize large portions of its derived hypergraph as patterns in itself, and then embody these patterns within its concretely implemented knowledge hypergraph. The Novamente AI system is the second in a series of AGI-oriented AI systems specifically based on the SMEPH framework; the first was the Webmind AI Engine (Goertzel, 2001; Goertzel et al, 2001).

The Mind of a Novamente Instance

As an example, it's interesting to ask: What does the above perspective say about the mind of an AI software system? According to this framework, the mind of an AI system is not the source code, it's not the state of the system in RAM at a given time; it is:

- The patterns existent in the system's RAM state, at a given time and over time, which are related to the system's intelligent behavior;
- The patterns emergent between the system's RAM state and the outside world, at a given time and over time, which are related to the system's intelligent behavior.

In the case of Novamente, these patterns include the emergent patterns we call "maps" – sets of Atoms that tend to be active during the same interval of time – but also other types of patterns as well. The degree to which these patterns belong to the mind was defined above – though this definition is admittedly difficult to apply in practice!

This definition of mind does not capture all aspects of an AI system's "mind," in common-language terms. For instance, it doesn't capture "what it feels like to be an AI," the experiential aspect. But it tells us something. It tells us that the role of all the data structures and dynamics that we create, as AI designers and engineers, is to give rise to RAM patterns that are associated with intelligence.

Roughly speaking, we can estimate the mind of a Novamente system by:

- Listing all the Atoms in the system, and ranking them according to long-term-importance (the LTI being a rough indicator of how useful an Atom is to the system's intelligence);
- Recognizing (static/dynamic) patterns in the system by running the existent "map encapsulation" algorithms on the system, to a greater extent than is necessary for the system's intelligent functioning, and estimating the importance of each of these maps;
- Looking at a recent-past archive of the system's environment and its own internal history, and recognizing emergent patterns there, to a greater extent than is necessary for intelligent experiential interaction.

This estimates the mind of a Novamente as:

- The Atoms in the system's actual Atomspace, plus
- Some other Atoms representing emergent patterns among the Atoms in the Atomspace, that would be in the Atomspace themselves if the system judged it had the resources to store them there.

This method will miss patterns in the system that the system itself is not smart enough to recognize. This leads to another possible approach, which is to use one Novamente system to look for patterns in another. If the analyzer Novamente is smarter than the analyzee Novamente, then one can get a better estimate of the analyzee's mind this way.

Measuring Self-Modification

Finally, having defined intelligence and mind in a semi-formal and hopefully conceptually satisfactory way, I now turn to a related question: What does it mean to say that a system modifies itself in accordance with the goal of increasing its intelligence? This is a key question for Novamente, since one of the long-term goals of the system is to achieve deep self-modifying AI – in which a major focus of the system's intelligence is on systematically improving its own intelligence.

The first step towards clarifying the notion of self-modification is to create a time series indicating the amount of intelligence increase the system experiences at each time. There are many ways to do it; I will describe here one simple approach. Let $I(t)$ denote the intelligence of the system in the recent past of time t (a weighted average of the intelligence over various intervals (r,t)). Then, let:

$$II(t,u) = (I(u) - I(t))^+$$

denote the intelligence increase over the interval (t,u).

The critical question, then, is: what are the patterns in the system that are causally related with *intelligence increase*. I.e., what are patterns P so that IN(P,S,r,s) is causally related to II(t,u) for t>s?

A system may be said to display self-modifying intelligence to the extent that there are purely patterns in S that are strongly causally related to intelligence increase. It is not a given that powerful such patterns will exist, just because S's intelligence has been increasing. It's possible, though not likely, that intelligence increase can occur at random. And, more plausibly, there is another way for a system's intelligence to increase: by external agents continually improving the system's mind, either by teaching it or by directly manipulating its store of declarative and procedural knowledge. But in the latter case, the really powerful causal relations will involve not patterns in S but in S∪ E, where E is the system's environment.

One can certainly look for patterns P so that IN(P, S∪ E, r,s) is causally related with II(t,u) for t>s. These patterns, however, are not patterns by which S self-modifies in the direction of increasing intelligence. Rather, they are patterns by which S and the environment work together to increase S's intelligence. In practical terms, these patterns may be of just as much interest as purely self-modificatory ones.

Varieties of Self-Modification

It may sometimes be interesting to distinguish two types of intelligence-oriented self-modification, which I call *noetic* and *dynamic*. Noetic[82] self-modification is the generic kind, as defined just above. Dynamic self-modification is a little subtler: it tries to assess the degree to which changes in S's dynamics are causally related with increases in system intelligence. It is a special case of noetic self-modification.

The tricky part of defining dynamic self-modification is defining dynamics. One want to look at the dynamic rules the system's mind observably follows, not the dynamics of the underlying "brain" level, which may be only loosely and indirectly related to mind dynamics. Once one has defined what the dynamics of a system are at a certain point in time, in a suitably general way, then the dynamic self-modification can easily be defined.

Toward this end, recall the notion of *implicit dynamics*. The implicit dynamics of S were defined above as the set of dynamic patterns in S – i.e. the set of patterns in the way S changes over time, as opposed to patterns in the state of S at any given time.

At any given time t, we may identify a set DP(t) of dynamic patterns that have been prominent over the recent past relative to t (by weighted-averaging over intervals (s,t)). We may then define the dynamic pattern change:

$$DPC(t,u) = (DP(u) - DP(t))^+$$

The degree to which S is dynamically self-modifying may be defined as the amount of causal relationship between DPC(r,s) and II(t,u), where t>s. How much do changes in the system's implicit dynamics cause increases in the system's intelligence.

---

[82] This word means "having to do with knowledge"

In any complex dynamical system, both kinds of self-modification will be going on all the time. However, some self-modifying systems are more self-modifying than others, and our specific conjecture is that:

- A system that merely carries out the self-modification necessarily involved in learning will have a lot of noetic self-modification, and not all that much of the other two kinds;
- A system that modifies its learning dynamics substantially based on experience, will have significant dynamical self-modification as well as intelligent noetic self-modification;
- A system that reconfigures its basic "source code" will have extremely large intelligent noetic and dynamical self-modification.

It would be very nice to prove mathematical theorems corresponding to these intuitive conjectures. But, this seems formidably difficult. And of course, it's quite possible that when this task is seriously approached, it will lead to technical modifications of the given definitions.

In Novamente terms, the above conjectures lead to the following more specific conjectures and observations:

- Noetic self-modification is implicit in all Novamente activity. What Novamente does is change its state based on what it learns and infers.
- Dynamical self-modification is implicit in noetic self-modification, but, in Novamente, it becomes much more significant when schema learning becomes a major part of the system's dynamics. Schema learning is the mechanism by which Novamente explicitly modifies procedures by which it does things. When the system begins modifying its own basic cognitive dynamics, then dynamical self-modification becomes truly prevalent.
- Dynamical self-modification becomes truly overpowering when the system becomes able to rewrite its own sourcecode and transform itself into something entirely different than it once was – in this case it is modifying its own substructure in a direct and obvious way.

Finally, while developing the concepts of noetic and dynamical self-modification, I also attempted to define a related concept of *substructural self-modification*. This was intended to measure the degree by which the system modifies its own "basic components." This attempt was not entirely successful but I think the underlying idea is interesting enough to merit briefly reporting here.

The intuitive concept is relatively simple. For instance, if a human learns new thought processes, this modifies his brain's dynamics to a certain extent. On the other hand, if a human augments his brain with a neurochip that enables him to do arithmetic and algebra at high speeds, this modifies his knowledge base and dynamics to a very large extent.

One would like to say that the neurochip expands the dynamical repertoire of the system in some fundamental sense. Once the chip is there, the system can do a lot of things it could not have done otherwise: for instance, compute $1757775555*466464444$ in a tenth of a second.

On the other hand, it is not clear to me at this point how this is really qualitatively different from noetic and dynamical self-modification. After all, learning a new cognitive skill allows one to carry out activities that could not have been carried out otherwise – just as does the installation of a neurochip.

The difference seems to be mainly one of extent. A neurochip suddenly allows a huge amount of new "implicit dynamics." It thus allows the formation of all sorts of system states that would have been impossible otherwise. The upshot seems to be that the substructural modification implies dramatic self-modification.

In the physical world, without modifying its basic components, there is going to be a ceiling to the intelligence that any finite system can assume. "Substructural self-modification" in the intuitive sense is thus necessary for dynamical and noetic self-modification to continue beyond a certain point.

To formalize the notion of substructural self-modification, it seems to be necessary to take a slightly different point of view. For instance, one can consider a physical system in terms of an hierarchy of abstract state machines $M_1, M_2,...,M_k$, where each state of M_i is defined as a set of states of M_{i-1}, and the transitions between states of M_i are consistent with the transitions permitted between states of M_{i-1}. In the case of a computer program we may have, roughly speaking:

M_1 = computer hardware
M_2 = operating system
M_3 = programming language
M_4 = AI program
M_5 = Easily mutable parts of AI program

Generally speaking, intelligent modifications to lower levels of the hierarchy will tend to cause higher degrees of noetic and dynamical self-modification. "Substructural" self-modification as in the neurochip example occurs at a lower level than modification of learning algorithms via experiential adaptation. In Novamente, modifications to the underlying source code occur at a lower level than modifications to schema operating with a fixed-source-code system.

Appendix C
Notes on the Formalization
of Causal Inference

This brief Appendix follows up on the ideas of Chapter 17, describing in a little more depth how one may formalize an approach to causal inference based on the combination of predictive implication and pattern-recognitive inference of plausible causal mechanisms.

Recall from Chapter 17 that we may say that if A causes B, there is likely to be a function ("schema") S so that:

```
ImplicationLink
      AND
            InheritanceLink X A
            InheritanceLink Y B
      ExecutionLink S X Y
```

and so that S is *as simple as possible*. Of course, a schema S that is just a lookup table of ordered pairs *(X,Y)* satisfying *(Inheritance X A) AND (Inheritance Y B)*, isn't a meaningful "causal mechanism" mapping A into B. On the other hand, a small and compact schema that effectively summarizes a large set of ordered pairs *(X,Y)*, gives a genuine reason to believe that *A* and *B* may be "really connected."

Yet more specifically, following the ideas in Appendix A1 we may define:

```
PatternIntensity(S, A, B)
```

as the intensity of *S* as a pattern in this set of ordered pairs *(X,Y)*, given the store of knowledge in the inferring mind at the given point in time as background information. What we are looking for, in a candidate causal mechanism, is a schema S satisfying the above implication *and* with a large pattern-intensity.

Note that the approximation metric d(,) in the definition of pattern plays a significant role here via inference, and is implicitly defined via SimilarityLinks. This is because if we have:

```
                  ExecutionLink S W Z'

      (InheritanceLink W A) AND (InheritanceLink Z B)

            SimilarityLink Z Z' <t>
```

then this increases the strength of:

```
ImplicationLink
      AND
            InheritanceLink X A
            InheritanceLinkY B
      ExecutionLink S X Y
```

via a couple inference steps.

The existence of such a mechanism, in itself is of course not evidence for a causal relationship. However, many cases of human causal ascription involve the combination of the identification of such a mechanism, with the construction of a predictive implication relationship.

I suggest that a decent fraction of the human notion of causation can be captured by creating a CausalLink relationship type, defined by the following relationship:

```
EquivalenceLink
      Node: A, B
      SchemaNode: S
      AND
            EvaluationLink PatternIntensity (S,A,B) <t₁>
            PredictiveImplicationLink A B <t>
      CausalLink A B <t OR t₁>
```

I have described this here in the context of event causality, but the same approach works for process causality, as will be described in the following section. The general definition of a CausalLink is the same in all cases; the difference is in the details of how the relevant schema-patterns are defined.

Formalizing Inference of Process Causality

What do we mean when we say that one process X is causally related to another process Y? What we mean, basically, is that if we know what X has done in the past, this helps us tell what Y is going to do in the near future. Rather than proposing a formalization of causality that breaks down into "predictive implication + causal mechanism," in this case, it seems more intuitive to have a single causal inference technique that incorporates both aspects. This is the "generalized functional causality" (GFC) approach, defined as follows.

Suppose we have two processes X and Y, represented by the numerical time series:

```
X = x(1),…, x(n)
Y = y(1),…, y(n)
```

In the GFC approach, we seek a function f, so that:

```
y(n) = f( x(n-1),…,x(n-k); y(n-1),…, y(n-k))
```

This formula is easily exploited in the case where x(i) and y(i) are numerical time series. However, it has meaning in a non-numerical context as well. In general, if X and Y are ConceptNodes, we can rewrite the above relation as a relation between time-stamped nodes, so that we have

$$Y_t = f(\{(X_t)\}, s \in [t, t-k])$$

Here Y_t is defined as the member of category Y with time-stamp t, i.e. by by the relations:

```
MemberLink Y_t r Y
```

```
ExecutionLink atTime Y_t t
```

Here, f is not necessarily a numerical function but is rather any schema that maps a set of time-stamped nodes into a single time-stamped node.

The quality of the schema f is determined by:

- Its simplicity, and
- The added value that using X's past provides in predicting the state of Y at time t (beyond the predictive power contained in the past of Y itself).

To assess the added predictive power, what we want is:

```
ImplicationLink <t>
    f({Y_s }∪ {X_s }, s ∈[t, t-k] )
    Y_t
```

```
ImplicationLink <t₁>
    f({Y_s }, s ∈[t, t-k] )
    Y_t
```

$$D = t - t_1 > 0$$

where the "accuracy in predicting" is given by the value D. On the other hand, the simplicity of the schema f can be computed in terms of the size of the schema expression inside the SchemaNode. Schema learning may be used to find schema satisfying the appropriate conditions.

What is very nice about this approach is that the two key aspects of causality — predictive implication and mechanism inference — are wrapped up in one. We are looking for a function f that will allow us to predict one process based on the other one, and that represents a simple "candidate causal mechanism" by which this prediction may take place. The strength of the CausalLink between X and Y must then be defined as an appropriate weighted combination of the simplicity and accuracy of the optimal f.

References

- Adams, Bryan, Cynthia Breazeal, Rodney Brooks, and Brian Scassellati. Humanoid Robots: A New Kind of Tool, *IEEE Intelligent Systems*, Vol. 15, No. 4, July/August, 2000, pp. 25-31.
- Alexander, R. McNeill (1996). *Optima for Animals.* Princeton University Press.
- Amit, Daniel (1999). *Modeling Brain Function*, Cambridge University Press
- Anderson, J. R. (1976). *Language, memory, and thought*. Hillsdale, NJ: Erlbaum.
- Anderson, J. R. (2005) Human symbol manipulation within an integrated cognitive architecture. *Cognitive Science, 29*(3), 313-341.
- Anderson, J. R. *Learning and Memory.* 2nd ed. N.Y.: Wiley, 2000. Chap. 2 and some of 3
- Anderson, J. R., Matessa, M., & Lebiere, C. (1997). ACT-R: A theory of higher level cognition and its relation to visual attention. *Human Computer Interaction*, 12(4), 439-462.
- Ashby, Ross (1960). *Design for a Brain.* New York: Wiley
- Augros and Stanciu (1987). *The New Biology.* Shambhala
- Aurobindo, Sri Ghose and Robert McDermott (2001). *The Essential Aurobindo.* Lindisfarne
- Baddeley, Alan (1999). *Essentials of Human Memory.* Taylor and Francis Group.
- Baddeley, Alan D. (1999). *Essentials of Human Memory.* Taylor and Francis.
- Baez, John (2002). The Octonions. Bull. Amer. Math. Soc. **39 (2002), 145-205**.
- Bakhtin, Mikhail (1984). *Problems of Dostoevsky's Poetics.* University of Minnesota Press.
- Barbour, Julian (1999). *The End of Time.* Oxford University Press.
- Barwise, Jon and John Etchemendy (1988). *The Liar: An Essay on Truth and Circularity*, Oxford: Oxford University Press
- Bateson, Gregory (1979). *Mind and Nature: A Necessary Unity.* New York: Ballantine Books
- Batten, David (2000). *Discovering Artificial Economies.* Westview Press.
- Baum, Eric (2004). *What Is Thought?* Cambridge MA: MIT Press Beckett, Samuel (1996), *Company.* New York: Riverrun Press
- Beckett, Samuel (1964), *How It Is.* San Francisco: Grove Press
- Beckett, Samuel (1970), *Endgame and Act Without Words.* San Francisco: Grove Press
- Beckett, Samuel (1983). *Worstward Ho.* San Francisco: Grove Press
- Beckett, Samuel (1997). *Waiting for Godot.* San Francisco: Grove Press
- Benioff, Paul (1998). Foundational Aspects of Quantum Computers and Quantum Robots. *Superlattices and Microstructures* 23, 407-417
- Bennett, Charles H. (1995). *Quantum information and computation*, Physics Today, October 1995, 24-30
- Bennett, Charles H. (1990). How to define complexity in physics and why. In Wojciech Zurek, editor, *Complexity, Entropy and the Physics of Information*, Addison-Wesley, Reading, Mass.

- Berge, Claude (1999). *Hypergraphs*, Elsevier Science
- Bertallanfy, Ludwig von (1993). *General System Theory*, George Braziller Inc., New York
- Bloom, Alfred (1981). *The Linguistic Shaping of Thought*. Hillsdale NJ: Erlbaum
- Boden, Margeret (2001). *The Creative Mind*. New York: Routledge
- Bohm, David (1994). *Thought as a System*. New York: Routledge
- Bollobas (1998). *Modern Graph Theory*. New York: Springer
- Borgelt, Christian (2005). Keeping Things Simple: Finding Frequent Item Sets by Recursive Elimination. Workshop Open Source Data Mining Software (OSDM'05, Chicago, IL), 66-70. ACM Press, New York, NY, USA 2005
- Borges, Jorge (1999). The Garden of Forking Paths, in *Fictions*. New York: Penguin
- Boroditsky, L. & Ramscar, M. (2002). The Roles of Body and Mind in Abstract Thought. *Psychological Science*, 13(2), 185-188
- Breton, Andre' and Paul Eluard (1992). *The Automatic Message: The Magnetic Fields and the Immaculate Conception*. New York: Serpent's Tail Press
- Brooks, Rodney (1999). *Cambrian Intelligence*. Cambridge MA: MIT Press
- Buber, Martin (1971). *I and Thou*. New York: Free Press
- Burnet, C. MacFarlane (1962). *The Integrity of the Body*. Cambridge MA: Harvard University Press
- Cabeza, Roberto and Alan Kingstone (2001). *Handbook of Functional Neuroimaging of Cognition*, The MIT Press
- Calvin, William and Derek Bickerton (2000). *Lingua ex Machina*. MIT Press.
- Campbell, Donald T. (1974), "Evolutionary Epistemology." In *The philosophy of Karl R. Popper*, edited by P. A. Schilpp, 412-463. LaSalle, IL: Open Court.
- Card, Charles R. (1995). The Emergence of Archetypes in Present-Day Science and Its Significance for a Contemporary Philosophy of Nature. *Dynamical Psychology*, http://www.goertzel.org/dynapsyc/
- Cassimatis, N. Grammatical Processing Using the Mechanisms of Physical Inferences. In *Proceedings of the Twentieth-Sixth Annual Conference of the Cognitive Science Society*, 2004.
- Chaitin, G. (1986). *Algorithmic Information Theory*. New York: Addison-Wesley
- Chalmers, David (1995). Facing Up to the Problem of Consciousness. *Journal of Consciousness Studies* 2(3):200-19, 1995
- Chomsky, Noam (1961). *Aspects of the Theory of Syntax*. Cambridge MA: MIT Press
- Cox, R.T. (1961). *The Algebra of Probable Inference*, Johns Hopkins University Press, Baltimore, MD
- Crutchfield, James P. and Karl Young (1990), ``Computation at the Onset of Chaos'', In *Complexity, Entropy, and the Physics of Information, SFI Studies in the Sciences of Complexity*, Vol VIII, Ed. W.H. Zurek, Addison-Wesley
- Darwin, Charles (1997). *The Origin of Species*. New York: Gramercy Press
- Davis, Martin (1982). *Computability and Unsolvability*. Dover Books.
- Dawkins, Richard (1990). *The Selfish Gene*. Oxford University Press.
- de Garis, Hugo, Ravichandra Sriram, and Zijun Zhang (2003). "Quantum Generation of Neural Networks", *Int. Joint Conf. on Neural Networks*, Portland, Oregon, USA, July 20-24, 2003.
- de Garis, Hugo, Ravichandra Sriram, and Zijun Zhang (2003a). "Quantum Computation vs. Evolutionary Computation : Could Evolutionary Computation Become Obsolete?", *Congress*

on Evolutionary Computation, Canberra, Australia, Dec. 8-12, 2003

- de Garis, Hugo, and Michael Korkin (2002). THE CAM-BRAIN MACHINE (CBM): An FPGA Based Hardware Tool which Evolves a 1000 Neuron Net Circuit Module in Seconds and Updates a 75 Million Neuron Artificial Brain for Real Time Robot Control, *Neurocomputing*, Elsevier, Vol. 42, Issue 1-4,
- Dennett, Daniel (1991). *Brainstorms.* Cambridge MA: MIT Press
- Dennett, Daniel (1992). *Consciousness Explained.* Viking.
- Dennett, Daniel (1996). *Darwin's Dangerous Idea.* Simon and Schuster.
- Dennett, Daniel (1998). Two Contrasts, in *Brainchildren, Essays on Designing Minds,* MIT Press and Penguin
- Dennett, Daniel (2003). *Freedom Evolves.* Viking.
- Deutsch, David (1985). Quantum theory, the Church-Turing principle and the universal quantum computer. Proc. R. Soc. Lond.
- Deutsch, David (1997). *The Fabric of Reality.* Allen Lane
- Deutsch, David (2000). Machines, Logic and Quantum Physics. *Bulletin of Symbolic Logic* 3, September 2000.
- Devaney (1989). *An Introduction to Chaotic Dynamical Systems.* New York: Perseus
- DeWitt, Bryce S. (1972). *The Many-Universes Interpretation of Quantum Mechanics in Proceedings of the International School of Physics "Enrico Fermi" Course IL: Foundations of Quantum Mechanic*s, Academic Press
- Dick, Philip K. (1991). *In Pursuit of Valis: Selections from the Exegesis.* Underwood.
- Dick, Philip K. (1991). *The Three Stigmata of Palmer Eldritch,* Vintage
- Dick, Philip K. (1991). *Ubik,* Vintage
- Dick, Philip K. (1991). *Valis,* Vintage
- Dick, Philip K. (1992). *The Man in the High Castle,* Vintage
- Dick, Philip K. (1996). *Do Androids Dream of Electric Sheep?,* Del Rey
- Dick, Philip K. (1990). We Can Remember It for You Wholesale (The Collected Short Stories of Philip K. Dick, Vol. 2). Carol Publishing Company.
- Dixon, Geoffrey (1994). *Division Algebras: Octonions Quaternions Complex Numbers and the Algebraic Design of Physics.* Springer-Verlag.
- Douthwaite, Julia V. (1997). *The Wild Girl, Natural Man and the Monster: Dangerous Experiments in the Age of Enlightenment,* University of Chicago Press
- Dreyfus, Hubert (1979). *What Computers Can't Do,* MIT Press
- Dupré, John (Ed.). (1987). *The latest on the best : Essays on evolution and optimality.* Cambridge, MA: The MIT Press
- Dupre', John and Nancy Cartwright (1988). Probability and Causality: Why Hume and Indeterminism Don't Mix. Nous 22: 521-536
- Dupre', John (2001). *Human Nature and the Limits of Science.* Oxford University Press.
- Edelman, Gerald (1987). *Neural Darwinism.* New York: Basic Books
- Edelman, Gerald (1990). *The Remembered Present.* New York: Basic Books
- Egan, Greg (1995). *Quarantine.* Eos
- Epstein, Seymour (1980). The Self-Concept: A Review and the Proposal of an Integrated Theory of Personality, *Personality: Basic Aspects and Current Research* 82, 91 (Ervin Staub ed. 1980)

- Everett, H. (1957). Relative State Formulation of Quantum Mechanics, *Reviews of Modern Physics* vol 29, (1957) pp 454-462.
- Feldbaum, Christiane (Editor) (1998). *WordNet: An Electronic Lexical Database.* Cambridge MA: MIT Press
- Ferber, Jacques (1999). *Multi-Agent Systems.* Addison-Wesley
- Feyerabend, Paul (1975). *Against Method.* London: Verso
- Feynman, Richard (1997). *Surely You're Joking, Mr. Feynman.* New York: W.W. Norton and Co.
- Feynman, Richard (2000). Feynman Lectures on Computation. New York: Perseus
- Freeman, Walter (2001). *How Brains Make Up Their Minds.* Columbia University Press.
- French, Peter A. T.E. Uehling and H.K. Wettstein, eds. (1984). Causation and Causal Theories. Minneapolis: University of Minnesota.
- Gallop, J.C. (1991). SQUIDs, the Josephson Effect and Superconducting Electronics. Adam Hilger.
- Gao, Ying (1994). Investigations on the mechanism of the Belousov-Zhabotinsky oscillating reaction. Cuvillier Verlag.
- Gardner, Howard (1993). *Multiple Intelligences.* Basic Books.
- Garey, Michael R. and David S. Johnson (1979). *Computers and Intractability: A Guide to the Theory of NP-Completeness*, W.H. Freeman. ISBN 0716710455. A9.1: LO1–LO7, pp.259–260.
- Gazzaniga, Michael (1989). "Organization of the Human Brain," *Science*, Sept., pp. 947-956
- Geach, P.T. (1957). *Mental Acts.* New York: Routledge
- Gibson, William (2004). *Pattern Recognition*, Berkley Publishing
- Gleason, Jean (2004). *The Development of Language.* Allyn and Bacon.
- Goertzel and Cassio Pennachin (2005). The Novamente AI Engine. In *Artificial General Intelligence*, Ed. by B. Goertzel and C. Pennachin, New York: Springer Verlag
- Goertzel, Ben (1993). *The Structure of Intelligence.* New York: Springer-Verlag
- Goertzel, Ben (1993a) *The Evolving Mind.* New York: Gordon and Breach
- Goertzel, Ben (1994). *Chaotic Logic.* New York: Plenum
- Goertzel, Ben (1997). *Evolutionary quantum computing.* Dynamical Psychology, www.goertzel.org/dynapsyc/1997/Qc.html
- Goertzel, Ben (1997). *From Complexity to Creativity.* New York: Plenum
- Goertzel, Ben (2001). *Creating Internet Intelligence.* New York: Plenum
- Goertzel, Ben (2003). Hebbian Logic Networks, *Dynamical Psychology*, www.**goertzel**.org/dynapsyc/2003/HebbianLogic03.htm
- Goertzel, Ben (2003a). Mindplexes. *Dynamical Psychology.* http://www.goertzel.org/dynapsyc/2003/mindplex.htm
- Goertzel, Ben (2004a). Patterns of Awareness. *Dynamical Psychology.* http://www.goertzel.org/dynapsyc/2004/HardProblem.htm
- Goertzel, Ben (2004b). The Virtual Multiverse Theory of Free Will.. *Dynamical Psychology.* http://www.goertzel.org/dynapsyc/2004/FreeWill.htm
- Goertzel, C. Pennachin, A. Senna, T. Maia, G. Lamacie (2004). Novamente: An Integrative Architecture for Artificial General Intelligence, *Proceedings of IJCAI 2003 Workshop on Cognitive Modeling of Agents.* Acapulco, Mexico

- Goertzel, S. Bugaj, C. Hartley, M. Ross and K. Silverman (2000). The Baby Webmind Project. *Proceedings of AISB 2000*, London, England.
- Goertzel, Ben and Stephan Bugaj (2005). Stages of Cognitive Development in Uncertain AI Systems. Article submitted for publication.
- Goertzel, Ben and Stephan Bugaj (2006). *The Path to Posthumanity.* Academica Press.
- Goethe, Johann (1962). *Faust.* Anchor Books
- Goldberg, David (1988). *Genetic Algorithms for Search, Optimization and Machine Learning.* New York: Addison-Wesley.
- Goldstone, R. L., Feng, Y., & Rogosky, B. (2005). In D. Pecher & R. Zwaan (Eds.) *Grounding cognition: The role of perception and action in memory, language, and thinking.* Cambridge: Cambridge University Press.
- Gould, Stephen Jay (2002). *The Structure of Evolutionary Theory.* Belknap Press.
- Graham-Rowe, Duncan (2001). Look Who's Talking. *New Scientist,* vol 171 issue 2303, 11/08/2001, page 34Grof, Stanislaw (2001). *LSD Psychotherapy.* Amazon.com
- Grof, Stanislaw (2001). *LSD Psychotherapy.*
- Grossberg, Stephen (1988). *The Adaptive Brain.* Elsevier Science.
- Grossberg, Stephen, Leif Finkel and David Field (2004). *Vision and Brain.* Elsevier Science.
- Gruska., Jozef (1999). *Quantum Computing,* McGraw-Hill International (UK) Limited 1999.
- Haikkonen, Pentti (2003). *The Cognitive Approach to Conscious Machines.* Imprint Academics
- Hameroff, Stuart (1987). *Ultimate Computing.* New York: North-Holland.
- Hanh, Thich Nhat (1999). *The Miracle of Mindfulness.* Beacon Press.
- Harnad, S. (1990). The Symbol Grounding Problem. *Physica D*, 42, pages 335-346.
- Harnad, S. (2002). Symbol Grounding and the Origin of Language. In: M. Scheutz (ed.) *Computationalism: New Directions.* Cambridge MA: MIT Press, pages 143-158.
- Hartmann, Ernest (1991). *Boundaries in the Mind.* Diane Publishing.
- Haskell and Robert Feys (1958). *Combinatory Logic*, North-Holland
- Hawking, Stephen and Roger Penrose (2000). *The Nature of Space and Time.* Princeton University Press.
- Hawkins, Jeff (2004). *On Intelligence.* Times Books.
- Hebb, Donald (1948). *The Organization of Behavior.* North-Holland.
- Hecht-Nielsen, R. Mechanization of Cognition. In *Biomimetics* (ed. Bar-Cohen, Y.) CRC Press, Boca Raton, FL, 2005
- Hegel (1975). *Logic.* Oxford University Press.
- Helstrop, Tore and Robert Logie (1999). *Imagery in Working Memory and Mental Discovery,* Taylor and Francis Group.
- Hillis, Daniel (1986). *The Connection Machine.* Cambridge MA: MIT Press.
- Hofstadter, Douglas (1979). *Godel, Escher Bach.* New York: Basic Books
- Holland, John (1992). *Adaptation in Natural and Artificial Systems*, MIT Press
- Hume, David (1986). *A Treatise of Human Understanding.* New York: Penguin
- Hunt, R. R., & Ellis, H. C. (2003). *Fundamentals of cognitive psychology* - 6th Edition. McGraw-Hill.
- Hutter, Marcus (2004). *Universal Artificial Intelligence.* New York: Springer
- Huxley, Aldous (1990). *The Perennial Philosophy.* Perennial.
- J. Schmidhuber. Optimal Ordered Problem Solver. *Machine Learning,* 54, 211-254, 2004

- Jaynes, Julian (2000). *The Origin of Consciousness in the Breakdown of the Bicameral Mind.* Mariner Books.
- Jensen, Arthur R. (1999). The G Factor: the Science of Mental Ability, *Psycoloquy*: 10,#23
- Jibu, M. & Yasue, K. (1995). *Quantum brain dynamics and consciousness.* Amsterdam and Philadelphia: John Benjamins. Jolicoeur, Gluck and Kosslyn, 1984)
- Jones, R. Tambe, M., Laird, J., Rosenbloom, P. (1993). Intelligent automated agents for flight training simulators. In *Proceedings of the Third Conference on Computer Generated Forces and Behavioral Representation.* University of Central Florida. IST-TR-93-07.
- Joos, E., H.D. Zeh, C. Kiefer D. Giulini, J. Kupsch, and I.-O. Stamatescu (2003). *Decoherence and the Appearance of a Classical World in Quantum Theory.* New York: Springer.
- Joyce, James (1999). *Finnegan's Wake.* New York: Penguin.
- Jung, Carl (1991). *Archetypes and the Collective Unconscious.* Bollingen.
- Kampis, George (1991). *Self-Modifying Systems in Biology and Cognitive Science.* New York: Plenum
- Kant, Immanuel (1990). *Critique of Pure Reason.* Prometheus Books.
- Kauffmann, Louis (1996). Time and Paradox, online at http://www.math.uic.edu/~kauffman/Papers.html
- Kauffmann, Stuart (1993). *The Origins of Order: Self-Organization and Selection in Evolution.* New York: Oxford University Press, 1993
- Kim, Y.-H., R. Yu, SP Kulik, Y. Shih, MO Scully (2000). Delayed Choice Quantum Eraser.
- Koppel, Moshe (1987). "Complexity, Depth, and Sophistication," *Complex Systems* **1** , 1087
- Koza, John (1991). *Genetic Programming.* MIT Press.
- Kurzweil, Ray (1999). *The Age of Spiritual Machines.* New York: Penguin
- Kurzweil, Ray (2005). *The Singularity is Near.* New York: Penguin
- Laird, J.E., A. Newell, and P. S. Rosenbloom (1987). Soar: An architecture for general intelligence. Artificial Intelligence, 33(1):1—6. Luck, S. J., E. K. Vogel, and K. L. Shapiro, "Word Meanings Can be Accessed but not Reported During the Attentional Blink." *Nature*, 383 (October 17, 1996): 616-618.
- Lakatos, Imre and Paul Feyerabend and Matteo Motterlini (1999). For and Against Method. Chicago: University of Chicago Press
- Lakoff, George (1990). *Women, Fire and Dangerous Things.* University of Chicago Press.
- Langton, Christopher (1990). Computation at the Edge of Chaos: phase transitions and emergent computation. 1990. In *Emergent Computation*, Ed. by Stephanie Forrest, Cambridge MA: MIT Press
- Langton, Christopher (1997). *Artificial Life: An Overview.* Cambridge MA: MIT Press.
- Leibniz, Gottfried Wilhelm (1985). *Best of all possible worlds.* Open Court Publishing
- Leibniz, Gottfried Wilhelm (1991). *Monadology.* University of Pittsburgh Press.
- Lenat, Doug and R.V. Guha (1990). *Building Large Knowledge-Based Systems: Representation and Inference in the Cyc Projec*t, Addison-Wesley
- Leslie, A. M. and S. Keeble. "Do six-month-old infants perceive causality", Cognition, 25, 1987.
- Libet, B., A. Freeman and K. Sutherland (2000). *The Volitional Brain: Towards a Neuroscience of Free Will.* Imprint Academic.
- Lima de Faria, A. (1990). *Evolution without Selection.* Elsevier Science.

- Looks, Moshe (2005). *Learning Computer Programs with the Bayesian Optimization Algorithm*, MS Thesis, Washington University in St. Louis, 2005
- Looks, Moshe, Ben Goertzel and Cassio Pennachin (2004). Novamente: An Integrative Architecture for Artificial General Intelligence, *Proceedings of AAAI 2004 Symposium on Achieving Human-Level AI via Integrated Systems and Research,*, Washington DC
- Looks, Moshe, Ben Goertzel, and Cassio Pennachin (2005). "Learning Computer Programs with the Bayesian Optimization Algorithm," *Proceedings of Genetic and Evolutionary Computation Conference (GECCO)*, Washington DC, 2005
- Luck, S.J., Vogel, E. K. and Shapiro, K. L. (1996). *Word meanings can be accessed but not reported during the attentional blink.* Nature 383 616-618)
- Lucy, John (1992). *Grammatical Categories and Cognition.* Cambridge University Press.
- Lynch, G. (1986). *Synapses, Circuits, and the Beginnings of Memory*, MIT Press, Cambridge, Massachusetts.
- MacDorman, K.F., Ishiguro, H. & Kuniyoshi, Y. (2001) Cognitive developmental robotics as a new paradigm for the design of humanoid robots. Robotics and Automation, 37, 185-193
- Mandelbrot, Benoit (1982). The Fractal Geometry of Nature. San Francisco: W. H. Freeman.
- Mandler, G. (1985). *Cognitive Psychology: An Essay in Cognitive Science*, Erlbaum Press, Hilldale NJ
- Mandler, George (1975). *The Psychology of Emotion.* Wiley.
- Manning, Christopher and Heinrich Schutze (1999). *Foundations of Statistical Language Processing.* MIT Press.
- Many-Worlds Interpretation of Quantum Theory (2006). *Stanford Encyclopedia of Philosophy*, http://plato.stanford.edu/entries/qm-manyworlds/
- Maturana, Humberto and Francisco Varela (1992). *The Tree of Knowledge.* Shambhala.
- McCulloch, Warren S. (1965). *Embodiments of Mind.* Cambridge University Press.
- McLuhan, Marshall (2005). *The Medium is the Massage.* Gingko Press.
- McTaggart, J. M. E. 1908. "The Unreality of Time," *Mind*, New Series **68**: 457-484.
- Merkle, Ralph (1998). "How Many Bytes in Human Memory?" *Foresight Uptate,* No. 4, October 1998
- Metzinger, Thomas (2004). *Being No One.* Cambridge MA, MIT Press.
- Michotte, A (1962). The perception of causality. Methuen, Andover, MA
- Miller, George A. (1957). The Magical Number Seven, Plus or Minus Two: Some Limits on Our Capacity for Processing Information. *The Psychological Review*, vol. 63, pp. 81-97
- Miller, P. (2001). *Theories of Developmental Psychology.* New York: Worth Publishers
- Mindpixel (2005). Entry in Wikipedia. http://en.wikipedia.org/wiki/Mindpixel
- Minsky, Marvin (1988). *The Society of Mind.* New York: Simon and Schuster.
- Mitchell, Tom (1997). *Machine Learning.* McGraw-Hill, New York.
- Mountcastle, Vernon (1998). *Perceptual Neuroscience: The Cerebral Cortex*, Harvard University Press
- Nason, S. and Laird, J. E. (2005). Soar-RL, Integrating Reinforcement Learning with Soar, *Cognitive Systems Research*, 6 (1), pp. 51-59
- Newell, A. and Simon, HA (1976). Computer science as empirical enquiry. *Communications of the ACM,* 19:113-126
- Nietzsche, Friedrich (1968). *The Will to Power.* New York: Vintage
- Nietzsche, Friedrich (1997). *The Twilight of the Idols.* Hackett Publishing.

- Nietzsche, Friedrich (2001). *The Gay Science.* Cambridge University Press.
- Niles, Ian and Adam Pease. Towards a Standard Upper Ontology. In *Proceedings of the 2nd International Conference on Formal Ontology in Information Systems* (FOIS-2001), Ogunquit, Maine, October 2001
- Opper, S. and H. Ginsburg (1987). *Piaget's Theory of Development.* New York Prentice-Hall 1987
- Paz, Octavio (1991). Collected Poems. New Directions Press.
- Pearl, Judea (1988). Probabilistic Reasoning in Intelligent Systems. New York: Morgan Kauffman
- Pearl, Judea (2000). Causality: Models, Reasoning and Inference. Cambridge University Press
- Peirce, C. S. (1892) "The Law of Mind." Reprinted in Hartshorne, Charles, Paul Weiss and Arthur W. Burks eds, *The Collected Papers of Charles Sanders Peirce*, Cambridge: Harvard University Press, 1980, 6.102-6.163.
- Peirce, Charles S. (1935). *Collected Works, Vol. 8: Scientific Metaphysics.* Cambridge MA: Harvard Press
- Pelikan, Martin (2002). *Bayesian Optimization Algorithm: From Single-Level to Hierarchy.* PhD Thesis, Department of Computer Science, University of Illinois at Urbana-Champaign
- Penrose, Roger (1996). *Shadows of the Mind.* Oxford University Press
- Penrose, Roger (2002). *The Emperor's New Mind.* Oxford University Press
- Perelson, Alan (2002). *The Warriors Within.* New York: Perseus Press.
- Perus, Mitja, Horst Bischof, Tarik Hadzibeganovic (2005). A Natural Quantum Neural-Like Network. *Neuroquantology*, September 2005
- Piaget, Jean, Malcom Piercy and D.E. Berlin (2001). *The Psychology of Intelligence.* Routledge.
- Pick, Arnold (1973). *Aphasia.* New York: Thomas.
- Pietelli-Palmarini, Massamo (1996). *Inevitable Illusions.* New York: Wiley.
- Pinker, Steven (2000). *The Language Instinct.* New York: Bantam
- Plotkin, H.C. (1982). Learning, development, and culture. In *Essays in evolutionary epistemology.* John Wiley and Sons.
- Popper, Karl (2002). *Conjectures and Refutations.* New York: Routledge
- Prigogine, Ilya (1984). *Order out of Chaos.* New York: Bantam.
- Proust, Marcel (1982). *Remembrance of Things Past.* Vintage Books.
- Quantum Eraser, *Phys. Rev. Lett.* 84, 1 Koza, John (1993). *Genetic Programming.* New York: MIT Press.
- Reigler, A. (2005). Constructive Memory. *Kybernetes* vol.34, nos. 1/2, 2005, pp. 89-104.
- Reisberg, D. (2001). *Cognition: Exploring the Science of the Mind.* 2nd ed. New York: Norton.
- Rhodes, Ross (2001). *A Cybernetic Interpretation of Quantum Mechanics*, **at** http://www.bottomlayer.com/
- Ridley, Matt (1998). *The Origins of Virtue.* New York: Penguin
- Rimbaud, Arthur (1967). Complete Works, Selected Letters. University of Chicago Press.
- Rizzolatti and Gallese (1988). "Mechanisms and Theories of Spatial Neglect," in Handbook of Neuropsychology v.1, Elsevier, NY

- Robinson, Abraham and Andrei Voronkov (2001). *Handbook of Automated Reasoning.* MIT Press.
- Rosen, Robert (2002). Life Itself. New York: Columbia University Press
- Rosenfield, Israel (1989). *The Invention of Memory.* New York: Basic Books
- Rowan, John (1990). *Subpersonalities: The People Inside Us,.* Routledge
- Roy, Deb and Niloy Mukherjee (2005). Towards Situated Speech Understanding: Visual Context Priming of Language Models. *Computer Speech and Language*, 19(2), pages 227-248
- Russell, S. and P. Norvig. (1995). *Artificial Intelligence: A modern approach.* Prentice Hall, Upper Saddle River, New Jersey
- Santore, John F. and Stuart C. Shapiro (2003). Crystal Cassie: Use of a 3-D Gaming Environment for a Cognitive Agent. In R. Sun, Ed., *Papers of the IJCAI 2003 Workshop on Cognitive Modeling of Agents and Multi-Agent Interactions*, Acapulco, Mexico, August 9, 2003, 84-91
- Schmidhuber, Juergen (1997). *A Computer Scientist's View of Life, the Universe, and Everything.* LNCS 201-288, Springer, 1997
- J. Schmidhuber (2005). Completely Self-Referential Optimal Reinforcement Learners. In W. Duch et al. (Eds.): *Proc. Intl. Conf. on Artificial Neural Networks ICANN'05*, LNCS 3697, pp. 223-233, Springer-Verlag Berlin Heidelberg
- Schopenhauer, Arthur (2005). *The World as Will and Representation.* Longman.
- Searle, J. (1983). *Intentionality: An Essay in the Philosophy of Mind.* New York, Cambridge University Press.
- Shapiro, Stuart (2000). An Introduction to SNePS 3. In Bernhard Ganter & Guy W. Mineau, Eds. *Conceptual Structures: Logical, Linguistic, and Computational Issues.* Lecture Notes in Artificial Intelligence 1867. Springer-Verlag, Berlin, 2000, 510—524.
- Singer, W. (2001). Consciousness and the Binding Problem. *Ann N Y Acad Sci.* 2001 Apr., 929:123-46.
- Sleator, Daniel and Davy Temperley (1993). Parsing English with a Link Grammar. *Third International Workshop on Parsing Technologies*, Tilburg, The Netherlands
- Smigrodzki, Rafal, Ben Goertzel, Cassio Pennachin, Lucio Coelho, Francisco Prosdocimi, W. Davis Parker Jr. (2005). "Genetic algorithm for analysis of mutations in Parkinson's disease." Artif Intell Med. 2005 Nov;35(3):227-41.
- Smith, Tony (2006). *Quaternions, Octonions and Physics.* http://www.valdostamuseum.org/hamsmith/QOphys.html
- Solomonoff, Ray (1964). "A Formal Theory of Inductive Inference, Part I", *Information and Control*, Vol 7, No. 1, pp. 1-22, March 1964.
- Solomonoff, Ray (1964a). "A Formal Theory of Inductive Inference, Part II", *Information and Control*, Vol. 7, No. 2, pp. 224-254, June 1964a.
- Sommers, Frederic, George Englebretsen and Harry Wolfson (2000). *An Invitation to Formal Reasoning.* Ashgate Publishing.
- Sowa, John F. (1984). *Conceptual Structures: Information Processing in Mind and Machine*, Addison-Wesley, Reading, MA.
- Sowa, John F. (2000) *Knowledge Representation: Logical, Philosophical, and Computational Foundations*, Brooks/Cole Publishing Co., Pacific Grove, CA
- Spencer-Brown, G. (1994). *Laws of Form.* The Cognizer Company.

- Srikant, R. and R. Agrawal. Mining Generalized Association Rules. In *Proc. of the 21st Int'l Conference on Very Large Databases*, Zurich, Switzerland, September 1995.
- Stcherbatsky, Th. (1958). Buddhist Logic, The Hague, Netherlands: Mouton & Co.
- Sternberg Robert (1988). *The Triarchic Mind.* New York: Viking
- Stockmann, H.J. (1999). *Quantum Chaos: An Introduction.* Cambridge University Press.
- Sutton, R. and A. Barto (1998). *Reinforcement Learning.* Cambridge MA: MIT Press
- Thelen, E. and E. Bates. Connectionism and dynamic systems: Are they really different? Introduction to J. Spencer & E. Thelen (Eds.), *Connectionism and dynamic systems. Special Section, Developmental Science*, 6(4), 378–391., 2003
- Turing, A.M. (1950). Computing Machinery and Intelligence. *Mind* 49: 433-460
- Umilta, C. (1988). "Orienting of Attention," in Handbook of Neuropsychology v.1, Elsevier, NY
- Underwood, G, ed. (1993). *The Psychology of Attention.* Vol 1. Aldershot: Elgar.
- Varela, Francisco (1978). Principles of Biological Autonomy. New York: North-Holland
- Varela, Francisco J.; Thompson, Evan; Rosch, Eleanor (1991). *The embodied mind: Cognitivescience and human experience.* Cambridge MA: MIT Press
- Voss, Peter (2005). The Essentials of General Intelligence. In Goertzel and Pennachin (Ed.), *Artificial General Intelligence*, Springer-Verlag
- Walley, Peter (1991). *Statistical Reasoning with Imprecise Probabilities.* New York: Chapman and Hall
- Wang, Pei (1995). *Non-Axiomatic Reasoning System.* PhD Thesis, Indiana University
- Wang, Pei (2005). Experience-grounded semantics: a theory for intelligent systems. *Cognitive Systems Research*, Vol. 6, No. 4, Pages 282-302
- Wang, Pei (2006). *The Logic of Intelligence.* Unpublished manuscript, online at http://www.cogsci.indiana.edu/farg/peiwang/papers.html
- Wason, P.C. (1966). Reasoning. In B. M. Foss (Ed.) *New Horizons in Psychology*, Penguin
- Weng, J. Developmental Robotics: Theory and Experiments". *International Journal of Humanoid Robotics*, vol. 1, no. 2, 2004.
- Wheeler, John (1989). *Frontiers of Time.* New York: North-Holland.
- Whitehead, Alfred North (1997). *Process and Reality.* New York: Free Press
- Whorf, Benjamin Lee (1964). *Language, Thought and Reality.* Cambridge MA: MIT Press
- Wiener, Norbert (1965). *Cybernetics.* Cambridge MA: MIT Press
- Wigner, E. P. (1962). In The Scientist Speculates (Edited by I. J. Good.)
- Wilson, E. O. (2000). Sociobiology: The New Synthesis.
- Wixted, John (2004). The Psychology and Neuroscience of Forgetting, *Annual Review of Psychology*, Vol. 55: 235-269
- Wolf, Fred Alan (1981). *Taking the Quantum Leap.* New York: Harpercollins
- Wolfram, Stephen (2002). *A New Kind of Science.* Wolfram Media.
- Woodward, James (2003) Making Things Happen: A Theory of Causal Explanation, Oxford University Press
- Youssef, Saul (1995). *Quantum Mechanics as an Exotic Probability Theory*, proceedings of the Fifteenth International Workshop on Maximum Entropy and Bayesian Methods, ed. K.M.Hanson and R.N.Silver, Santa Fe, August, 1995.
- Zak, M. (1990). "Creative Dynamics Approach to Neural Intelligence," *Biological Cybernetics*, Vol. 64, No. 1, pp. 15-23.

- Zak, M. (1991). "Terminal Chaos for Information Processing in Neurodynamics," *Biological Cybernetics*, Vol. 64, pp. 343-351.
- Zeilinger, Anton (1999). *Experiment and the foundations of quantum physics.* Rev.Mod.Phys., p.S-288
- Zukav, Gary (2001). *The Dancing Wu-Li Masters*. New York: Harper
- Zurek, Wojciech (1991). Zurek Decoherence and the Transition from the Quantum to the Classical, *Physics Today*, 36-44